# ENVIRONMENTAL ENGINEERING FOR EXPLORATION AND PRODUCTION ACTIVITIES

**Mary Jane Wilson**
*Editor*
President and Chief Executive Officer, WZI Inc.

and

**Jesse D. Frederick**
*Editor*
Vice President, WZI Inc.

**First Printing**
**Henry L. Doherty Memorial Fund of AIME**
**Society of Petroleum Engineers Inc.**

**Richardson, Texas**
**1999**

# SPE Monograph Series

The Monograph Series of the Society of Petroleum Engineers was established in 1965 by action of the SPE Board of Directors. The Series is intended to provide authoritative, up-to-date treatment of the fundamental principles and state of the art in selected fields of technology. The Series is directed by the Society's Monograph Committee. A committee member designated as Monograph Editor provides technical evaluation with the aid of the Review Committee. Below is a listing of those who have been most closely involved with the preparation of this monograph.

# Monograph Committee (1999)

Mary Jane Wilson, WZI, Chairman
Jesse Frederick, WZI
Medhat Kamal, Arco E&P Technology
Mark Miller, U. of Texas
Ken Newman, CTES L.S.
Dan O'Meara Jr., U. of Oklahoma
Sammi Raza, Saudi Arabian Oil Co.
David Underdown, Chevron Production Technology Co.



**ISBN 1-55563-085-5**

# Preface

This monograph represents the collective effort of dedicated oil and gas environmental professionals. It was put together in an effort to bring to the field a single point of reference that could be used to solve problems or to direct further study.

Many individuals in our industry have extensive experience that reflects entire careers of solving environmental problems and working on emerging technology, laws, and regulations. In this monograph, we brought together a representative sample of the finest practicing professionals, academics, and regulators. The contributing authors, facilitators, and editors additionally sought the input of their peers. The effort and contribution of these outstanding engineers and scientists, as well as the organizations that allowed their participation, can be appreciated only in understanding the demand for their precious time. For this, we thank all who contributed, including the SPE staff.

These years of experience are thus made available to the willing reader.

Change is inevitable and we must be prepared to expand on that which we have learned. As editors, we challenge our peers to enjoy, use, review, critique, and revise this monograph as need dictates.

**Mary Jane Wilson**
**Jesse D. Frederick**

# Table of Contents

# Chapter 1
# Introduction

**R. Lyn Arscott,** E&P Forum

## 1.1 Petroleum Engineering and Environmentalism

It is doubtful whether such industry pioneers as Drake, Lucas, and Muscat could have imagined the range of technical disciplines a modern petroleum engineer must understand. For many years, petroleum engineers appreciated the basic principles of geology, geophysics, and construction engineering because they are integral parts of the recovery process. Today, petroleum engineers must also be familiar with such esoteric creatures as *Mysidopsis bahia* for bioassays of discharged produced water and with *Pseudomonas* bacteria for remediation of reserve pits. Among various other scientists (such as ergonomists, economists, hydrologists, ecologists, oceanographers, and microbiologists), petroleum engineers interface with biologists for habitat management and environmental impact statements; with atmospheric physicists for air-dispersion issues; and with industrial hygienists, toxicologists, physicians, and epidemiologists concerning the effects of toxic chemicals and various diseases on worker health and safety. In addition to these hard "technical disciplines," the softer disciplines (such as psychology, community relations, and politics) cannot be ignored. All these subjects belong in the arsenal of the modern petroleum engineer.

## 1.2 The Environmental Revolution

Environmentalism early in this century was characterized by the conservationism of John Muir and Theodore Roosevelt. In 1962, Rachel Carson[1] drew attention to the toxic effects of synthetic chemicals in our environment with her book *Silent Spring*, which pioneered wide public support for environmental concerns. The first U.S. Earth Day was in 1970, which was also the year that one of the largest environmental bureaucracies in the world was born—the U.S. Environmental Protection Agency (EPA). The 1970's saw the birth of most major U.S. environmental legislation for air, water, waste, remediation, and endangered species. Since then, this legislative framework (voted by politicians) has been transformed into rules and regulations written by the regulatory agencies. By the early 1990's, the U.S. had more than 10,000 pages of federal regulations affecting its industry, with an associated cost of compliance for all U.S. industries of about U.S. $120 billion. A 1993 American Petroleum Inst. (API) study estimated that the U.S. E&P industry probably will increase compliance spending from U.S. $8 billion/yr in 1990 to between U.S. $26 billion and $33 billion/yr in Year 2000. This translates to the oil and gas industry's share of total U.S. industry environmental expenditures leaping from 7% in 1990 to between 14 and 17% in 2000. The environmental revolution experienced in the U.S. and other developed countries is rapidly spreading worldwide. Major incidents, such as Bhopal in 1984, the *Piper Alpha* disaster in 1988, and the *Exxon Valdez* spill in 1989, have fueled public support for tough controls on industry and for punitive accountability.

The petroleum industry really started to deal with the environmental revolution in the mid– to late 1980's. The first Society of Petroleum Engineers (SPE) forum on the environment was held in August 1989. The first SPE International Conference on Health, Safety and Environment in Oil and Gas Exploration and Production was held in The Hague, The Netherlands, in November 1991. The first joint SPE/EPA conference was held in San Antonio, Texas, in March 1993. The first world environmental summit was held in Rio de Janeiro in 1990. A positive outcome of this summit was the clear message to the world economies that wealth must be created to fund the protection of the health of the population and the environment. The less-industrialized Southern Hemisphere gave the message to the more-industrialized Northern Hemisphere that if the North is interested in saving such environmental treasures as the rain forests, they should help subsidize the economies of the South. The relationship between environmental protection and economic growth will be debated for many years. The Rio meeting brought forth important treaties covering global environmental issues, such as global climate change, ozone depletion, biodiversity, and ocean pollution.

## 1.3 Management in the Environmental Era

Much of the current legislation and regulation was formed in an era of confrontation between industry and government. Command and control by government, or prescriptive regulation, was the *modus operandi*. On the other hand, industry has learned by experience that goal setting, management systems, and incident-free operations are the keys to improved performance. Numerous industry associations have proposed comprehensive management practices for designing and implementing environmental programs. The Chemical Manufacturers Assn. (CMA) was an early leader with a program called "Responsible Care." The API has a program called "Strategies for Today's Environmental Partnership." The E&P Forum has issued comprehensive guidelines, and the Intl. Standards Organization has issued a series of standards of management practice with its 14000 series. Many of the practices included in these programs are justified by economic incentives, such as the need to reduce the cost of spills or releases of toxic materials or to reduce fines or lawsuits. Simple reduction of waste can save significant amounts of valuable materials. Improved government/industry partnership can result in more cost-effective regulations. In addition to economic considerations, implementation of health, safety, and environmental (HSE) management practices is designed to prevent criminal penalties (incarceration of managers and other employees). Improved relations with communities can result in the timely approval of reasonable permit conditions. Many companies have integrated their HSE practices with a total-quality-management process. All these programs consider the special concerns of communities and public-interest groups.

E&P permits often require detailed studies of the environmental impact of the operations and related transportation, including the effect on flora and fauna, air emissions, water discharge, waste disposal, groundwater contamination, native populations, and archeological remains. Operations must meet permit requirements and all other applicable government regulations. The petroleum industry operates by permission of the public—no permission, no permits. All operators, both large and small, must behave in a manner acceptable to the public. What happens to the least of us affects all of us. The petroleum industry appears to be a particular target for criticism by some special-interest groups. This attitude, sometimes called "petrophobia," means that the petroleum industry must make extra efforts to operate in a safe and environmentally sound manner. Improvements in technology and operating practices are necessary to meet our environmental goals.

## 1.4 Monograph Overview

**1.4.1 Introduction.** A modern petroleum engineer cannot survive without understanding and managing HSE risks. This monograph covers the technical considerations of protecting the environment during oil and gas exploration and development. The following major categories are necessary to understand the issues.

1. Pollution prevention.
2. Habitat preservation.
3. Environmental-impact analysis.
4. Terrestrial and ocean environments.
5. Pollution evaluation.
6. Risk assessment.
7. Oilfield-pit management, closure, and remediation.
8. Legal and regulatory structures for environmental protection.

**1.4.2 Pollution Prevention.** The hierarchy of pollution prevention is source reduction, recycling, treatment, and disposal. Oilfield wastes, discharges, and emissions should be reduced at the source as much as possible. For example, closed mud systems reduce water use and eliminate the need to close and remediate reserve pits, thus reducing the risk of groundwater contamination. Bulk delivery of chemicals eliminates the need to dispose of packaging or drums. If source reduction does not eliminate a waste stream, the waste material should be recycled (for example, used lubricating oil should be sent back to a refinery). If recycling is not practical, treatment should be applied to render the waste nonhazardous before safe disposal, which is preferred to disposal in an approved hazardous-waste landfill. Once an engineer understands the environmental and financial benefits of pollution prevention, innovation can be a powerful force in thinking of ways to reduce the waste stream and releases. Chaps. 2 and 6 discuss typical waste streams and options for pollution prevention and evaluation.

**1.4.3 Habitat Preservation.** Compatible use of land by industry, by farmers, or for recreation is possible if industry cooperates with other stakeholders and conducts its operations in an environmentally sound manner. Impact on plants and animals can be mitigated by such techniques as saving topsoil, land banking, or reserving acreage for wildlife. Directional drilling from single pads reduces clearing of land or filling in a wetland. Special vegetation management can enhance the food chain and thus support wildlife. Transporting material, such as oil or sulfur, by pipeline is less intrusive than transporting by truck or rail. Chap. 3 details these and other ideas.

**1.4.4 Environmental-Impact Analysis.** All projects have some impact on the environment or community. Defining the preproject baseline requires data on air, water, soil, biological, archeological, and societal issues. Project impact and mitigation options can then be calculated and analyzed before permits are obtained. Chap. 4 covers this subject.

**1.4.5 Terrestrial and Ocean Environments.** Water is ubiquitous in our environment. The sources of contamination to surface and subsurface water must be understood and controlled. The fate and effect of contamination is complex. Remediation can take many forms, from removal of floating or suspended materials to in-situ bacterial processes. These issues are the subject of Chap. 5.

**1.4.6 Pollution Evaluation.** E&P operations vary worldwide, and those that are cost-effective and protect the environment are indicative of proper management. Chap. 6 discusses the practice of pollution evaluation, which recognizes potential sources of releases and potential pollutants related to E&P operations. E&P operations typically include exploration (e.g., seismic-data acquisition), drilling and completion, production operations, and support activities. Understanding the sources and pollutants associated with these operations can aid in preventing adverse environmental impacts and increase overall cost-efficiency.

**1.4.7 Risk Assessment.** There is no such thing as zero risk resulting from human activity. There is, however, an acceptable risk. Chap. 7 discusses the debate as to what is "acceptable" and why the "perception" of acceptability is different from the "measurement" of risk. The oil industry has a very mixed record in dealing with public outrage.

**1.4.8 Oilfield-Pit Management, Closure, and Remediation.** The oil and gas exploration business is conducted in both pristine and urban areas. These operations have a finite life before the property must be returned to an acceptable state as defined by the landowner, government, or local community. This may include plugging and abandoning wells, closing pits, dismantling offshore platforms or onshore facilities, and appropriate revegetation. Restoring the location to usable habitat or to agricultural, commercial, or residential use is a necessity in our industry. Chap. 8 treats this subject.

Remediation comes into play after pollution prevention has taken place and impact and risk analysis has identified undesirable impacts to sensitive receptors or habitat. Chap. 8 focuses on a specific area of concern—oilfield pits—and shows the reader where to seek guidance on remediation efforts other than pit closures.

**1.4.9 Legal and Regulatory Structures for Environmental Protection.** A sound knowledge of the laws and regulations that apply to operations is essential to minimize noncompliance incidents, to keep the cost of fines down, and to reduce the risk of legal actions against managers and employees. Chap. 9 describes regulatory structures in various parts of the world.

## Reference

1. Carson, R.: *Silent Spring*, Houghton Mifflin, New York City (1962).

## General References

"Environmental Investments: The Cost of a Clean Environment," report to Congress, U.S. EPA, Washington, DC (November 1990).

"Costs to the Petroleum Industry of Major New and Future Federal Environmental Requirements," Discussion Paper No. 070R, API, New York City (October 1993).

# Chapter 2
# Pollution Prevention

**Warren J. Hahn,** Vastar Resources Inc.
**J. Alan Abel,** Chevron Research and Technology Co.
**Robert A. Fisher,** Atlantic Richfield Corp.
**M. Scott Mansholt,** Texaco
**Rosemary A. Martinez,** Amoco Corp.
**Robert J. Redweik,** Shell Western E&P Co.

## 2.1 Overview

This chapter provides information on the principles, goals, benefits, and components of a successful pollution-prevention program. A comprehensive list of potential sources and types of pollutants generated in E&P operations and specific examples for eliminating or reducing them also are included.

The definitions of pollutants, pollution, and pollution prevention used in this chapter are similar to those used by the American Soc. for Testing and Materials (ASTM).[1] A pollutant is defined as any substance that creates an adverse human-health or environmental effect when introduced into the environment. Pollution is the introduction or release of a pollutant into the environment (air, land, or water). Pollution prevention is the methods and techniques for reducing or eliminating pollution, including source reduction, reuse, recycling, reclamation, or modification of operating practices to reduce the quantity and/or toxicity of air emissions, solid-waste generation, and waste-water discharges. This is similar to the definition of pollution prevention developed by the American Petroleum Inst. (API), which states that "Pollution prevention is a multimedia concept that reduces or eliminates discharges to air, water, or land. Pollution prevention includes the development of more environmentally acceptable products, changes in processes and practices, source reduction, beneficial use, and environmentally sound recycling."[2] Although more restrictive definitions of pollution prevention (restricted to specific methods of reducing pollution generation) are often used, this chapter uses the term pollution prevention in the most comprehensive sense, similar to API and ASTM definitions.

**2.1.1 Elements of a Pollution-Prevention Program.** A pollution-prevention program is a management expectation that facilities will strive to reduce environmental impact and includes mechanisms to promote and support development and implementation of pollution prevention throughout the facility. The pollution-prevention process is the actual implementation of the steps necessary to carry out pollution prevention and generally includes steps similar to those for a quality-improvement process. Although the contents of individual pollution-prevention programs vary, several key elements typically are present in all programs. These generally include top-management support, characterization of pollution generation and management costs, project evaluation and implementation, and periodic pollution-prevention assessments.

***Top-Management Support.*** A successful pollution-prevention program requires that senior management provide support for the program and that the support be adequately documented and communicated with such materials as formal executive directives. Making pollution prevention a cornerstone of a company's environmental policy reinforces management commitment and helps ensure every employee's involvement in the company's pollution-prevention efforts. Setting of goals by management can help focus company efforts and provide guidance to personnel on achieving the desired results in all areas of operations, including pollution prevention.

Employee recognition is also an important part of management support of a pollution-prevention program because it encourages employee efforts and helps reinforce management commitment. Employees who make outstanding contributions toward reducing the generation of pollutants can be recognized with individual awards, and their accomplishments can be described in articles in company/facility newsletters and publications.

Management support of pollution prevention can be reinforced further by establishing a policy that includes pollution prevention in routine employee training. While training efforts can be incorporated into existing training or added as a separate training, they should inform employees about pollution-prevention methods.

***Characterization of Pollutant Generation and Management Costs.*** Each operation must develop inventories of pollutants to characterize pollutant quantities and rates of release. These data can be used to identify pollution-prevention opportunities, to track progress toward pollution-reduction goals, to assess overall program effectiveness, and to report to management.

Because pollution-prevention projects must compete for resources with all other projects, full identification of potential economic benefits of pollution prevention is important. To determine these benefits, all costs associated with the generation of pollutants, including those for management, treatment, disposal, or any other mitigation costs, should be identified and allocated for in the inventory.[3] This usually can be done with existing cost-accounting systems. The assigned costs of environmental impact are divided by the quantities of pollutants generated to produce a "unit cost" for each pollutant. With a unit cost, the savings of a proposed project can then be estimated easily by multiplying the projected pollutant reduction by the unit cost.

***Project Evaluation and Implementation.*** Implementation of pollution-prevention projects that are technically and economically feasible is the ultimate goal of a pollution-prevention program. Generally, prevention opportunities are identified during initial pollution-prevention assessments, periodic follow-up assessments, or routine optimization studies. Once identified and evaluated, prevention opportunities should be prioritized to ensure that resources are used effectively to provide the maximum benefit for an operation. Solutions then need to be developed to address the root cause for each pollution-prevention opportunity. Subsequently, each solution must be evaluated for economic and process feasibility. This evaluation may include detailed analysis of technical components, worker safety, maintenance requirements, compatibility with existing equipment, labor, and serviceability along with a comprehensive economic evaluation.

***Periodic Pollution-Prevention Assessments.*** Regular performance evaluations are used to ensure that operations continue to remain profitable and competitive. These evaluations should include regular progress reviews of pollution-prevention efforts and associated cost savings. The pollutant inventories described earlier can be used as the starting point for subsequent evaluations. These evaluations also can serve to collect feedback and identify potential areas for improvement. Both quantitative and qualitative measures can be used to assess program effectiveness.[4] The following are examples of the measures used to evaluate the success of a pollution-prevention program.

1. Quantitative reductions in pollutant generation and associated environmental impact management costs.
2. Number of pollution-prevention ideas submitted.
3. Number of pollution-prevention ideas implemented.
4. The creativity of the pollution-prevention ideas.

**2.1.2 Benefits of a Pollution-Prevention Program.** Pollution-prevention programs have been shown to provide substantial benefits to companies that have carried them out. The underlying concept in pollution prevention is the use of good engineering practice to provide maximum utility at minimum cost. The scope of pollution prevention has evolved as knowledge of potential impacts of pollution and costs associated with pollution have increased. Basically, release of pollutants (once considered "free") now has social, ecological, and monetary costs, which should be considered part of engineering design and analysis. Pollution prevention provides a more effective mechanism for reducing costs than traditional "end-of-the-pipe" or "top-of-the-smokestack" treatment technologies. Eliminating pollution through source control or recycling is often more efficient than dealing with its consequences. Some consequences of pollutant generation that potentially can be avoided are described next.

***Economic Impacts.*** Potential economic benefit is a significant driving force for pollution prevention. Environmental impact management costs have increased significantly in recent years, changing the balance between the cost of prevention and that for treatment. As a consequence, many older facilities, designed when costs associated with pollutant generation were lower, can now reduce costs by identifying opportunities to incorporate pollution prevention retroactively. With new facilities, pollution prevention can be included in the initial design to provide the most cost-effective operations.

Identification of costs and benefits is a routine part of the engineering analysis of any project. However, identification of cost savings that result from implementing pollution-prevention projects is especially important because these projects must compete with other projects for available capital and engineering resources.

To assess potential financial benefits of a pollution-prevention project accurately and secure allocation of the necessary resources, the total costs of environmental impact management must first be identified correctly. Most E&P operations currently have accounting systems that capture most of the costs resulting from pollutant generation. However, these systems do not always allocate the costs to a single, established cost center or account to allow simple assignment to the sources of the pollutants. Also, because of the complexity of E&P operations, other related costs may be overlooked. The following examples describe how costs can be hidden.

1. Total costs for environmental impact management are combined and assigned to an "environmental" cost center (e.g., maintenance, overhead, or environmental operations) and are not associated with the specific operation generating the pollutant.
2. Total costs of environmental administration are combined and assigned to an overhead cost center.
3. The value of products lost as environmental releases are ignored and never quantified.
4. Corporate and other external resources required as the result of pollutant generation are not billed to the generating operation.

A comprehensive analysis of the costs of generating and subsequently handling a pollutant will correctly identify the total costs associated with environmental impact management. Various types of costs should be considered in this analysis.

1. Collection costs (maintenance, labor, and equipment).
2. Storage costs (facilities, bin/tank demurrage, gauging, and monitoring).
3. Treatment costs (dewatering, volume-reduction equipment, incineration, and biological treatment).
4. Disposal costs (tipping fees, transportation, insurance/reserves, auditing, sewer charges, and labor).
5. Regulatory fees (waste taxes and emission fees).
6. The value of lost product/raw materials.
7. Administrative costs (external/internal, clerical, operations and maintenance, area supervision, environmental staff, purchasing, and labor).
8. Compliance costs (recordkeeping, preparing reports, and audits).

Once the total environmental impact management costs have been quantified accurately, elimination of all or a percentage of these costs resulting from implementation of a pollution-prevention project will be one of the project's most important benefits.

***Regulatory Impacts.*** Increased knowledge of the impacts of pollution on human health and the environment has resulted in increased regulation of air emissions, waste generation, and waste-water discharges. One potential benefit of pollution prevention is that whenever a pollutant release is eliminated, the need to consider the associated regulations governing that release also is eliminated. Some of the regulatory benefits of reducing the quantity of pollutants generated are reduction/elimination of resources needed to ensure compliance, of exposure to noncompliance and associated fines, of reporting and recordkeeping, and of regulatory taxes and permitting fees.

***Other Impacts.*** Other benefits of pollution-prevention efforts exist that are not associated with either easily quantifiable costs or regulatory issues. Their impact on decision making can be significant but varies with each individual situation. Some of these benefits are reduction of long-term liability for environmental remediation, of employee exposure, and of ecological impacts of facility operations and an increase in community goodwill.

## 2.2 Source Analysis

**2.2.1 Inventories.** An essential first step in any pollution-prevention program is development of an inventory of pollutants that quantifies air emissions, solid-waste generation, and waste-water discharges. This list is necessary to identify pollution-prevention opportunities and serves as a reference point for all subsequent pollution-prevention activities.

***Rationale and General Procedure for Conducting Inventories.*** The primary reasons for conducting an inventory are to provide data necessary to identify pollution-prevention opportunities, to set priorities, to monitor change, and to gather data for reporting or communication. The inventory should provide valuable information without use of excessive resources. Establishment of clear guidelines for personnel responsible for carrying out an inventory ensures that the necessary resources are used. For a simple inventory, the range of pollutants and parameters to be evaluated and any other requisite information to be collected on processes, procedures, or conditions should be specified. An inventory must provide data to identify high-priority releases, such as the highest-quantity releases, the source of the most toxic kinds of releases, and the pollutants with the highest environmental impact management costs. Conducting a pollutant inventory is not a one-time event if the purpose of the inventory is to track or demonstrate pollution-prevention progress. Inventory data over time are necessary to show trends in quantities, costs, and toxicity or relationships between metrics. Finally, inventory data can be used for both internal communication or required agency reports.

The task of information collection and data management can be substantial, but it is important that the level of effort at least be sufficient to facilitate identification and evaluation of pollution-prevention opportunities. Otherwise, an inventory has limited value.

***Information Required.*** Different types of data can be useful in providing guidance for conducting pollution inventories.

*Type of Release.* A description of the type of pollutant, including its source and characteristics, provides insight into the scope of environmental impact management and the range of materials to be handled. Specific characterization of pollutants, whether by source, quantitative analysis, or qualitative description, can help classify or group similar kinds of pollutants to allow for a common evaluation. As an example, oily sludges from production separators and storage tanks often have similar characteristics and are generated from related sources that may have common pollution-prevention options.

*Quantities.* The quantity of pollutants generated is important because the cost and/or environmental impact of a pollutant source generally is proportional to its quantity. It also provides a simple measure of progress. Two examples include determination of volatile organic compounds (VOC's) released from a dehydration unit in pounds and barrels of waste water generated during a tank cleaning.

*Sources and Locations.* Identifying the location or sources of pollutants provides insight into possible methods for prevention or reduction. It is important to link costs and environmental impacts to each specific pollutant source to provide the information necessary to evaluate the economic incentives for reducing or eliminating the source and, ultimately, to ensure that resources are applied where they provide the most benefit. An example is determination of specific wells that produce naturally occurring radioactive material (NORM) scale and subsequent application of corrective measures to only those wells. Identifying sources may highlight different management practices used for the same or similar types of pollutants in different locations and result in opportunities to share less-well-known prevention techniques.

*Time or Frequency.* The frequency of generation of each source of pollutants should be determined to evaluate variability in generation rates or variations in characteristics. This information is important because it can help identify the root cause, such as process upsets, maintenance activities, or routine operations. Decisions concerning the most effective prevention technique for a particular source may depend on whether that variability was recognized and whether it can be linked to facility operations. For example, when a facility that cleans its tanks for inspection every 2 years observes an increase in oily sludges generated during the cleaning year, it is a reasonable assumption that the root cause of that increased generation is the tank-cleaning activities. Another example is an increase in the oil content of produced water when chemical addition to the production stream is interrupted, suggesting a chemical-injection-pump malfunction.

*Destination.* Data describing the current ultimate destination or disposition of pollutants can be an important indicator of management costs and/or potential environmental impacts. For example, waste-handling records or manifests identify whether tank bottoms were managed in an on-site recycling facility or taken to an off-site treatment facility, each of which has its own unique costs and long-term liabilities.

*Baseline, Reference, and Normalized Data.* Additional information may be necessary to show the context in which data are collected. Baseline data are generally the data from the first comprehensive inventory and provide a starting point for comparison with future inventories to provide a measure of progress. Normalizing of data is useful because it allows comparisons of data under different conditions. Normalization is accomplished by dividing the quantity in inventory by some other measure, such as facility production. An example would be calculation of bbl water produced/1,000 bbl oil produced, or normalizing produced water to net oil production. Using this approach allows data at two oil production rates to be compared on an equivalent basis. That is, if produced-water rates double and the amount of oil produced also doubles, the operation of the field has not changed significantly. Normalized data are used routinely throughout E&P operations. For example, usual practice is to normalize drilling costs to hole depth. Another example is evaluation of waste-disposal costs from a drilling operation on a per-foot-of-hole basis. Evaluating these costs before and after implementing pollution prevention allows comparison of drilling operations at different depths or in different numbers of wells to determine whether changing drilling-mud additives impacts drilling costs.

***Data Sources.*** To reduce costs, data for an inventory should be gathered from existing data-collection systems when possible.

*Production Records.* These data are core records for operations and can provide significant information. Volumetric production records often can provide data on material quantities, while written operator logs often can provide anecdotal information on the cause of releases. These may be overlooked during audits because they are generally outside routine environmental responsibilities.

*Produced-Water/Waste-Water Records.* Operating records for produced- and waste-water processing systems generally provide volumes and some compositional data for water discharges. While these are maintained primarily to demonstrate satisfactory operation of a processing system, they often can provide insight into variations in water discharges and any associated generation of solid waste.

*Tank Gauging.* Total tank throughputs are often key inputs for estimating air emissions.

*Operating Data.* Operating data, such as pressures or temperatures, may indicate unusual conditions that affect pollutant generation, such as nonroutine operations and process upsets.

*Environmental Reports.* Generally, environmental reports provide the only existing pollutant inventories. These reports are submitted to various media-specific agencies that regulate each facility and include air-emissions and waste-water-discharge reports. Reports, such as groundwater-monitoring reports, can identify historical or existing releases that are not identified in other reports.

*Waste-Disposal Invoices.* Solid-waste-disposal invoices provide such information as quantity, cost, and type of waste management.

*Estimates.* Some pollutants are not easily quantifiable and must be estimated with available estimation techniques, such as AP-42.[5] [Vol. 1 of Ref. 5 contains emission factors for stationary point and area sources. Point sources are sources with stacks or vents, and area sources are sources with nonspecific emission points (e.g., region-wide pesticide use). Vol. 2 addresses mobile sources (e.g., automobiles and trucks).] Care should be taken to ensure that these estimates are as accurate as possible. They should be replaced with actual measurements whenever possible.

*Laboratory Results.* Laboratory analyses are a traditional source of general environmental data and also may be a source for toxicity data.

Other data may need to be gathered specifically for pollution-prevention inventories. These data can be obtained primarily through interviews with operations personnel.

***Interpretation.*** A pollution-prevention program must be based on valid data. For example, data that overestimate costs and underestimate benefits result in missed pollution-prevention opportunities and data that underestimate costs and overestimate benefits can result in misallocation of pollution-prevention resources. Therefore, once data have been collected, they should be subjected to quality assurance. After the data have been determined to be accurate, they should be used to identify the impacts of each pollutant. At the least, data interpretation should include prioritization of pollutants according to quantity, risks, costs, and other applicable factors; analysis of current handling practices for each pollutant; and analysis of current costs and impacts assigned to each pollutant.[6] The results of data interpretation are the reference point necessary to start the pollution-prevention process (i.e., actual implementation of projects or practices resulting in pollution prevention).

**2.2.2 Pollution Sources in the E&P Industry**. Many E&P activities have the potential to generate pollutants. Pollution prevention is most effective when it is incorporated into all phases of operations, from initial planning through plugging and abandonment.

**TABLE 2.1—COMMON E&P INDUSTRY AIR EMISSIONS**

| Emissions | Type of Generating Activity |
|---|---|
| VOC | Fugitive emissions, product storage, process venting |
| $H_2S$ | Well-casing gas venting, flaring |
| $SO_x$ | Sour-gas combustion |
| $NO_x$ | Flaring, internal-combustion engines |
| CO | Flaring, internal-combustion engines |
| $PM_{10}$ | Internal-combustion engines, construction, road travel |
| Halon | Fire-prevention-system maintenance |
| CFC | Building-/auto-cooling-system service |

$H_2S$ = hydrogen sulfide, $SO_x$ = sulfur oxides, $NO_x$ = nitrogen oxides, CO = carbon monoxide, $PM_{10}$ = particulate matter with $< 10\ \mu m$ diameter, and CFC = chlorofluorocarbon.

Each activity that can generate pollutants also represents a prevention opportunity. Activities associated with exploration that can be pollutant sources are seismic surveying, drillsite construction, drilling and support operations, and completion or abandonment operations. Activities associated with production operations that can be pollutant sources include facility construction, routine production operations, well workovers, facility maintenance, and facility abandonment. Some pollutants and the activities that produce them are discussed next by media.

***Air Emissions.*** Reducing air pollution is an important part of any pollution-prevention program. The three main sources for air pollutants are (1) release of hydrocarbons and related materials (e.g., $H_2S$) during processing, (2) combustion byproducts, and (3) release of materials used in support of operations. The use of pollution-prevention techniques to achieve reductions of air emissions often can be cost-effective and practical. For example, in the case of volatile hydrocarbon emissions, pollution prevention can provide a net economic benefit when the value of the hydrocarbons captured is greater than the cost of preventing their loss to the air. Also, in some regulatory environments, reductions of air emissions below the levels required for compliance can be "banked" or credited to the company and used later or sold when "credits" are needed for permitting a new facility. **Table 2.1** shows typical air-pollution sources for the E&P industry.

***Water Discharges.*** The largest single water discharge generated in E&P is produced water. Pollution prevention through beneficial reuse of produced water in waterflood operations is widely practiced in the industry. Other types of waste waters generated in E&P (**Table 2.2**) also are capable of causing environmental impact if not properly managed.

***Solid Wastes.*** Solid wastes are the most familiar type of pollutants, and many past pollution-prevention efforts have been focused on this medium. **Table 2.3** lists solid wastes commonly generated from E&P.

## 2.3 Source-Reduction and -Elimination Techniques

**2.3.1 Planning/Materials Management.** ***Purchasing Procedures.*** Many pollutants generated in E&P can be reduced by modifying purchasing procedures. Purchasing materials that provide equivalent performance but have less toxicity eliminates the hazardous nature of any solid wastes generated and reduces potential environmental impacts. Closely matching purchase volumes to operational requirements or writing purchase contracts to permit the return of unused materials to the original supplier also eliminates excess materials as a source of solid waste. Requiring that purchased materials come in minimal or recyclable packaging significantly reduces this as a source of solid waste. Another pollution-prevention purchasing technique is requiring that chemicals be supplied in bulk quantities to eliminate the need to manage and dispose of multiple containers.[7] Efficient purchasing procedures reduce pollutant generation and also reduce purchasing costs and lower operating expenses. Table 2.4 shows examples of source reduction through purchasing practices.

**TABLE 2.2—COMMON E&P INDUSTRY WATER DISCHARGES**

| Discharge | Type of Generating Activity |
|---|---|
| Produced water | Production operations |
| Scrubber blowdown | Scrubber operations |
| Completion fluids | Well completions |
| Well-treatment fluids | Well workovers |
| Cooling water | Production-facility operation |
| Rig wash | Drilling operations |
| Hydrotest water | Pipeline, flowline testing |
| Storm water | Runoff from drilling, production sites |
| Sewage | All operations |

***Storage and Distribution Procedures.*** Proper storage procedures for material prevent generation of such solid wastes as off-specification, contaminated, or unidentifiable materials created by unnecessary weathering of containers. To reduce the risk of damage to containers and subsequent spills, proper transportation and handling procedures should be used when moving materials. Common causes of spills are use of underrated lifting equipment or attempting to lift too much material in a load. Not only can such practices cause spills, but more importantly, they can cause injuries to personnel. Containers should also be properly secured during transportation to prevent spills. **Table 2.5** provides examples of storage and distribution procedures for source reduction.

***Substitution.*** Substituting nonhazardous or less hazardous products for hazardous products may not affect the amount of residual material generated, but it does minimize the potential impact to the environment caused by residuals. Additionally, substitution with minimal hazard products and materials generally increases the number of handling options for residuals, decreases the costs and liabilities of handling residuals, and reduces exposure to employees handling residuals. For example, substituting less toxic solvents for chlorinated solvents reduces disposal or recycling costs and lowers the administrative burden for handling solvent residuals. **Table 2.6** lists examples of source reduction through product/material substitution.

**2.3.2 Process Changes.** ***Equipment/Process Modifications.*** Modification of equipment and/or processes is another way to reduce generation of pollutants. Process modifications can include changing such variables as temperature, pressure, volume, rate, or residence time in a process so that smaller volumes of pollutants are generated. These modifications may or may not require capital expenditures and may take place before or after construction. Modifications and/or enhancements to process control systems also prevent pollution, especially at remote E&P facilities. Control and/or alarm systems alert operations personnel to process conditions that can cause the release of pollutants, allowing them to take appropriate measures to prevent releases. High-level alarms at many remote, unmanned tank batteries alert off-site personnel in time for them to respond and prevent tanks from overflowing. In addition to reducing pollution generation, process changes in older facilities often improve efficiency and lower operating costs. **Table 2.7** gives examples of source reduction through equipment/process modifications.

***Pollution-Control Systems.*** Generation of pollutants from many of the processes in E&P operations can be prevented or minimized through use of pollution-control systems. Examples of such control systems are vapor-recovery units and leak-detection systems. Implementation of control systems usually requires capital expenditures for the control equipment. **Table 2.8** gives examples of source reduction with control systems.

**2.3.3 Maintenance.** ***Routine Maintenance and Repair.*** Routine maintenance and repair are prudent operating practices that ensure that operations run smoothly. Effective maintenance results in equipment that is less likely to develop leaks or cause a facility upset, thereby reducing the occurrence of leaks and spills. While pol-

| TABLE 2.3—COMMON E&P INDUSTRY SOLID WASTES | |
|---|---|
| Solid Waste | Type of Generating Activity |
| Oily soil | Leaks and spills |
| Tank bottoms | Crude oil storage |
| Solvents | Cleaning, production analysis |
| Lube oil | Vehicle, engine maintenance |
| Oil filters | Vehicle, engine maintenance |
| Water filters | Production maintenance |
| Glycol filters | Maintenance of gas production and processing facilities |
| Acid | Well treatment |
| Caustic | Scrubber operations |
| Excess cement slurry | Drilling and completion operations, construction |
| Paint waste | Drilling and production maintenance |
| Empty drums | Drilling and production operations |
| Sandblast media | Drilling and production maintenance |
| Production chemicals | Production operations |
| Scale | Production maintenance |
| Refractory brick | Facility abandonment |
| Fly ash | Production maintenance |
| Glycol | Gas processing |
| Sweetening catalyst | Gas processing |
| Mercury | Production operations |
| Methanol | Production operations |
| Amine | Gas processing |
| Drilling fluids | Drilling operations |
| Drilling cuttings | Drilling operations |
| PCB's | Production-facility maintenance |
| Batteries | Vehicle, equipment maintenance |
| Asbestos | Facility abandonment |
| Grease | Well workovers, equipment maintenance |
| Trash | All operations |
| Oily rags | All operations |
| Absorbent material | All operations |
| NORM | Production maintenance, well workovers |
| Pesticides/herbicides | Production operations |
| $H_2S$ scrubber waste | Scrubber operations |
| Hydraulic fluid | Production-equipment maintenance |
| Laboratory wastes | Production analysis |
| Radioactive tracers | Drilling operations |
| Activated carbon | Production operations |
| Paraffin/pigging waste | Pipeline/flowline maintenance |
| Concrete | Production-facility abandonment |
| Pipe dope | Drilling operations |
| Thread protectors | Drilling operations |
| Antifreeze | Vehicle, equipment maintenance |
| Aerosol cans | All operations |
| Molecular sieve | Production-facility operations |

PCB = polychlorinated biphenyl.

lutants can be generated during maintenance procedures, this can be minimized with proper management practices. Reducing the frequency of oil changes as much as possible minimizes the quantity of used oil and used oil filters generated as well as the time and money spent managing them. All maintenance procedures should be evaluated to determine whether their frequency and/or intensity can be reduced. Other examples appear in **Table 2.9.**

***Cleaning and Housekeeping.*** The terms cleaning and housekeeping often are used interchangeably; however, in this chapter, cleaning denotes removal of foreign material from contaminated equipment and housekeeping denotes worksite materials and equipment management so that contamination does not occur in the first place. Installation and maintenance of drip pans to collect oil from leaking equipment is an example of housekeeping. Cleaning equipment allows pollutants to be removed and properly managed before release off site. Like routine maintenance and repair activities, cleaning can be a source of pollutants if not conducted properly. Efforts should be made to contain and manage properly residuals from cleaning, such as wash waters.[8] **Table 2.10** provides examples of cleaning/housekeeping source-reduction methods.

**2.3.4 Operational Changes.** ***Containment.*** Spills can occur periodically; therefore, it is important to have a backup system to prevent or minimize the generation of leak or spill-cleanup residuals. Containment minimizes the amount of contaminated soil from a spill and mitigates the threat of spilled material entering a waterway. **Table 2.11** shows examples of containment measures for source reduction.

***Segregation.*** Segregation reduces the amount of solid waste or waste water that requires handling and increases the amount that can be reused or recycled. For example, oily solid wastes should be segregated from regular trash so that the entire waste stream will not require handling as oily waste. Paper, glass, and plastic usually can be recycled if segregated, but not if mixed with oily wastes. Segregation is also very effective for storm-water runoff. Diverting storm-water runoff so that it does not contact areas with hydrocarbons can prevent the runoff from becoming a source of pollution. **Table 2.12** gives other examples.

***Sampling Procedures.*** Pollutants generated during routine sampling procedures often are overlooked. Several types of source-reduction/pollution-prevention techniques previously discussed can be incorporated into sampling procedures to ensure that generation of pollutants is reduced or eliminated. Containment should always be provided under sampling outlets, and, when possible, sampled material should be returned to the process after being analyzed. Purge volumes should be minimized by techniques such as closed-loop sampling. When sample material cannot be returned to the process, only the volume needed for the analysis should be collected. **Table 2.13** shows two examples of pollutants that can be minimized through the use of proper sampling procedures.

## 2.4 Reuse Techniques

After all measures have been taken to reduce the volume and/or hazard of a pollutant, reuse alternatives should be evaluated. A pollutant is "reused" if it can be returned to its original process (purpose) by cleaning, regeneration, or some other method. Reuse techniques often have the dual benefit of reducing environmental impact management costs and reducing the need to purchase new materials.[9]

**2.4.1 Regeneration**. Several types of potential pollution sources from E&P operations can be reused after regeneration. Regeneration may take place on site or, more commonly, at an off-site commercial regeneration facility. Before using any off-site facility for regeneration, companies should ensure that it is in compliance with applicable regulations. A management system can be put into place for periodic review of these facilities to minimize future liability. Spent activated carbon is an example of a material commonly sent off site for regeneration, then returned and reused for the same purpose. The costs for regenerating spent activated carbon usually are significantly less than those for purchasing new and disposing of spent activated carbon. **Table 2.14** lists pollutants that can be reused after regeneration.

**2.4.2 Cleaning/Decontamination**. Some materials used in E&P processes for filtering can be backwashed and cleaned to remove contaminants and, subsequently, reused. Materials used in cleaning up spills and leaks can also be cleaned and reused if reusable cleanup

**TABLE 2.4—SOURCE REDUCTION THROUGH PURCHASING PRACTICES**

| Pollutant | Purchasing Practice |
|---|---|
| Solvents | Purchase low-toxicity/-hazard solvents and purchase only amount needed; return unused portion. |
| Acid | Purchase only amount needed; return unused portion. |
| Caustic | Purchase only amount needed; return unused portion. |
| Empty drums | Purchase supplies in bulk containers; include requirement in contract that containers must be returned to vendor. |
| Production chemicals | Purchase only amount needed; return unused portion. |
| Methanol | Purchase only amount needed; return unused portion. |
| Grease | Avoid greases containing chlorinated solvents and purchase only amount needed; return unused portion. |
| Pesticides/herbicides | Plan to prevent excess: purchase only amount needed; return unused portion. |
| Cement slurry | Purchase only amount needed; return unused portion. |
| Packaging | Require minimal or recyclable packaging. |
| Thread protectors | Add requirement in contract that protectors must be returned to vendor. |
| Paint | Purchase only amount needed; return unused portion. |
| Well-treatment fluids | Purchase only amount needed; return unused portion. |

**TABLE 2.5—SOURCE REDUCTION THROUGH STORAGE/DISTRIBUTION PROCEDURES**

| Pollutant | Storage/Distribution Procedure |
|---|---|
| Contaminated soil | Protect containers from weathering, store only with secondary containment, and properly secure containers during transportation. |
| Production chemicals | Keep container labels from weathering to ensure contents can be identified. Use safe lifting procedures when handling to prevent container ruptures. |
| Storm water | Cover storage areas to prevent runoff. |

**TABLE 2.6—SOURCE REDUCTION THROUGH PRODUCT/MATERIAL SUBSTITUTION**

| Pollutant | Product/Material Substitution Procedure |
|---|---|
| Solvents | Use nonhazardous solvents. |
| Paint | Use minimal-hazard paint products and thinners. |
| Pipe dope | Use heavy-metal-free pipe dope. |
| Antifreeze | Use minimal-hazard antifreeze (propylene glycol). |
| $SO_x$ emissions | Use low-sulfur fuels. |
| Refrigerants (CFC's) | Use non-CFC refrigerants. |
| Oil-based drilling fluids | Use synthetic drilling fluids. |
| Radioactive tracers | Use nonradioactive alternatives. |
| Laboratory wastes | Use minimal-hazard laboratory materials. |
| Aerosol cans | Purchase materials in nonaerosol form. |

$SO_x$ = sulfur oxides and CFC = chlorofluorocarbon.

materials are used. Solvents, including painting and laboratory solvents, may be filtered and recovered on site with small distillation units. Used oils generated in the process of maintaining and lubricating production, drilling, workover, and gas-plant equipment are another example of a potential pollutant that can be reused. Although most facilities collect used oils and recycle them into the production, continuous on-site regeneration may be preferable for such applications as large stationary internal-combustion engines.[10] **Table 2.15** lists pollutant sources that can be reused after cleaning or decontamination.

**2.4.3 Closed-Loop Systems.** Because closed-loop systems are an integral part of a process and the materials being reused are not exposed to the environment, they usually are not subject to any waste or recycling restrictions. Preplanning for closed-loop systems is important because retrofitting may not be economically feasible compared with other types of management options. Properly designed closed-loop systems for drilling-fluids-waste minimization have been shown to reduce solid-waste volumes significantly.[11] Drilling fluids should be designed to minimize degradation of drilled solids that are difficult to remove and increase fluid waste. The combination of fluid-inhibition and/or flocculation technology or other solids-separation technology with drilling-fluids systems has also been shown to be beneficial.[12] Another example of a closed-loop system is the return of empty containers to vendors for reuse. This may require changes in purchasing procedures to select suppliers who reuse their own containers. Another option is to send containers to a vendor who cleans, tests, and recertifies them for reuse. These options require coordination among purchasing, operations, and environmental personnel. Containers usually must be classified as empty and closed (bungs in place) before vendors will accept them. **Table 2.16** presents other examples of closed-loop systems.

## 2.5 Recycling Techniques

In certain instances, source reduction or reuse may not be technically or economically feasible. While recycling is not as effective as reducing the generation of pollutants at the source, it still prevents pollution. Thus, recycling is the next step in the pollution-prevention process, and recycling opportunities should be evaluated for any pollutants that are unavoidably generated.

**TABLE 2.7—SOURCE REDUCTION THROUGH EQUIPMENT/PROCESS MODIFICATION**

| Pollutant | Equipment/Process Modification |
|---|---|
| Tank bottoms | Design completions for lower solids production; keep solids suspended in tanks with mechanical agitation; maximize heater/treater retention time; and recycle. |
| Solvents | Use high-pressure wash system instead of solvents for cleaning. |
| NORM | Segregate processing of noncompatible produced waters and minimize pressure drops to minimize precipitation. |
| Molecular sieve | Install carbon filter upstream to extend life. |
| Produced water | Design completions for lower water production. |
| Cooling water | Change to air cooling. |
| $NO_x$, CO emissions | Change to low-$NO_x$ burner; use electric instead of internal-combustion engines; use natural gas instead of diesel engines. |
| VOC emissions | Route hydrocarbon emissions to combustion or compression. |
| Completion fluids | Route to produced-water treatment system. |

$NO_x$ = nitrogen oxides and CO = carbon monoxide.

**TABLE 2.8—SOURCE REDUCTION WITH CONTROL SYSTEMS**

| Pollutant | Control System |
|---|---|
| Tank bottoms | Control with chemical treatment in process. |
| Scale | Control with chemical treatment. |
| Asbestos | Encapsulate in place. |
| VOC emissions | Install vapor-recovery system; implement routine inspection program for connections/seals/hatches. |
| $SO_x$ scrubber waste | Automate control system to increase scrubber efficiency. |
| NORM | Control with chemical treatment. |
| $H_2S$ scrubber waste | Automate control system to increase scrubber efficiency. |
| Paraffins/pigging waste | Control with chemical/solvent treatment. |
| $NO_x$, CO emissions | Install electronic ignition and air/fuel ratio controllers; install catalytic converters. |
| $H_2S$, $SO_x$ emissions | Install scrubber or sulfur-recovery unit. |
| $PM_{10}$ | Control emissions by paving lease roads. |

$SO_x$ = sulfur oxides, $NO_x$ = nitrogen oxides, CO = carbon monoxide, and $PM_{10}$ = particulate matter with $< 10\ \mu m$ diameter.

**TABLE 2.9—SOURCE REDUCTION THROUGH MAINTENANCE AND REPAIR**

| Pollutant | Maintenance and Repair Activity |
|---|---|
| VOC emissions | Maintain connections/seals/hatches to minimize fugitives. |
| Contaminated groundwater | Conduct regular mechanical-integrity tests on well tubulars. |
| Contaminated soil | Perform routine maintenance and repair leaks. |

**TABLE 2.10—SOURCE REDUCTION THROUGH CLEANING/HOUSEKEEPING**

| Pollutant | Cleaning/Housekeeping Procedure |
|---|---|
| Rig wash | Sweep up solids before washing; use a high-pressure, low-volume wash unit. |
| Storm water | Keep equipment and process areas clean; cover process and storage areas. |

**TABLE 2.11—SOURCE REDUCTION THROUGH CONTAINMENT**

| Pollutant | Containment Measure |
|---|---|
| Contaminated soil | Provide containment (e.g., drip pans) around equipment with the potential to leak (e.g., valves and flanges); provide containment around process areas, storage areas, tanks, wellheads, and loading racks. |
| Storm water | Contain spills to prevent contamination of storm water; provide containment for storm water. |

Recycling is defined in this chapter as reclamation of a pollutant so that its residual value can be obtained by its use for a different purpose. Recycling may also involve use or reuse of a pollutant as a substitute for a commercial product (e.g., burning for energy recovery) or as a feedstock in an industrial process. When considering recycling options, the following hierarchy should be used: (1) recycling on site, (2) recycling off site at another owned/operated facility, and (3) recycling off site at a third-party facility.

On-site recycling is the preferred option when both feasible and practical. With proper planning and design of facilities, recycling of pollutants can occur completely within, and as an integral part of, a process. Maintaining complete control over the recycling process allows better control of cost as well as liability.

When on-site recycling is neither feasible nor practical, off-site recycling at other facilities owned or operated by the company/organization should be considered. Although the generator of the pollutant can still take advantage of many of the on-site benefits, the cost and liability exposures are usually higher with the off-site option. The increased exposure results primarily from the handling, packaging, and transportation requirements of the pollutant to be re-

| TABLE 2.12—SOURCE REDUCTION THROUGH SEGREGATION | |
|---|---|
| Pollutant | Segregation Procedure |
| Oily rags | Separate from trash so they can be cleaned and reused. |
| Storm water | Divert from contaminated areas. |
| Filters | Segregate filters that can be recycled. |
| Trash, refuse | Segregate recyclable materials (aluminum, paper, glass). |

| TABLE 2.13—SOURCE REDUCTION THROUGH SAMPLING PROCEDURES | |
|---|---|
| Pollutant | Sampling Procedure |
| Sample wastes | Use in-line analyzers instead of performing shake-out tests. |
| Laboratory wastes | Collect only amount needed for analysis. |

| TABLE 2.14—REUSE THROUGH REGENERATION | |
|---|---|
| Pollutant | Regeneration Practice |
| Solvents | Send to solvent reclaimer. |
| Pumps, valves | Send to reconditioner. |
| Activated carbon | Send to regenerator. |
| Molecular sieve | Send to regenerator. |
| Sweetening catalyst | Send to regenerator. |
| Batteries | Use rechargeable batteries. |

| TABLE 2.15—REUSE THROUGH CLEANING/DECONTAMINATION | |
|---|---|
| Pollutant | Cleaning/Decontamination Practice |
| Oil filters | Use reusable oil filters. |
| Water-injection filters | Use reusable filters. |
| Sandblast media | Collect media, filter, and reuse. |
| Oily rags | Send to oily rags service for cleaning. |
| Solvents | Filter/distill on site and reuse. |
| Lubrication oils | Filter on site and reuse. |
| Absorbent material | Use reusable absorbent material. |

cycled. In addition, off-site management of the pollutant may increase regulatory requirements. In general, on-site recycling is less regulated than off-site recycling options. If unable to recycle on site or at off-site owned/operated facilities, the final option is to use a third-party recycler. Compared with the other options, potential cost and liability exposures increase with the use of a third-party recycler; however, these exposures typically are less than those associated with treatment and disposal. All types of recycling allow beneficial use of potential pollutants, thereby conserving resources.

**2.5.1 Reclamation.** Reclamation is a recycling technique where a pollutant is processed and treated to remove or use particular or select components. Unlike regeneration, the pollutant being recycled by this technique is not used or reused directly as originally intended. An example of this is oily produced sand that can be reclaimed for use as road amendment in some areas.[13] **Table 2.17** lists pollutants that can eliminated by recycling through reclamation. Some types of reclamation involve processing beyond what is normally available at an E&P facility and therefore are done off site.

**2.5.2 Exchange/Substitution.** Some pollutants generated in E&P operations can be substituted in another process with little or no processing required. Like reclamation, the pollutant being recycled by this technique is not used or reused directly as originally intended. Unlike reclamation, however, little or no processing is required. Opportunities for exchange or substitution are rare because a commercially available virgin material meeting certain specifications is usually what must be replaced. However, when these opportunities are feasible, they can result in substantial savings because they reduce both environmental impact management and materials purchasing costs. Some modification of a process that uses a substituted material may be required. **Table 2.18** lists some pollutants that can be recycled through exchange or substitution.

Recycling opportunities are available for many materials used by the oil and gas industry. Many regulatory agencies have begun to sponsor and support exchange networks where generators can be matched with companies that need certain feedstocks and substitutes for raw materials. Generators should also consider that a market for recycled material is necessary for recycling to be economically feasible. Therefore, where practical, recycled materials should be purchased in lieu of virgin materials and products made solely from virgin materials.

## 2.6 Pollution-Prevention Training

**2.6.1 General Environmental Training.** General environmental training usually includes such topics as an organization's environmental protection philosophy, regulatory compliance requirements, compliance resources and tools, and operation of control equipment. If possible, training sessions should be tailored to the participants' roles and responsibilities in the organization's environmental compliance program. This will depend on the scope of an organization's training program. Ideally, managers, technical staff, field operations management, and general field operations personnel should have specialized training focusing on issues relevant to their roles and responsibilities. However, group training has the benefits of emphasizing teamwork and demonstrating management commitment. Whatever the scope of the program, training sessions should include material on waste management and pollution prevention.

**2.6.2 Waste Management Plan Training.** Waste management plans lay out standard practices for specific wastes and serve as operating guides and regulatory/policy references for waste handling.[14] Specific training on waste management plans may be an organizational or regulatory requirement. In other instances, waste management plan training may simply be part of general environmental training sessions. However, because waste management is an integral part of pollution prevention and because pollution prevention is likely to be discussed throughout the waste management plan, training sessions on these plans provide a good opportunity to cover and stress pollution prevention in general.

**2.6.3 Pollution-Prevention Training.** Whether a part of a general environmental training session, a stand-alone training session, or simply the subject of a weekly "tailgate" meeting at a facility, pollution-prevention training should encourage (cheerlead) the pollution-prevention philosophy; stress continued improvement ("a pollution preventer's work is never done"); and reiterate the pollution-prevention hierarchy of reduction, reuse, recycling, treatment, and disposal. As with general environmental training, pollution-prevention training should be designed to fit the participants' roles and responsibilities in pollution prevention.[15] These training sessions are also good opportunities to point out successes in an organization's pollution-prevention efforts. Information on volume reductions and savings from reduced handling costs and regulatory fees provides justification for a program and enhances employee motivation. If an organization has a pollution-prevention incentive program, it should be discussed.

## 2.7 Audit and Prevention Programs

Many programs outside of routine operations and related training can be used to enhance pollution-prevention efforts within an organization. Environmental compliance audits, employee involvement/incentive programs, environmental liability studies, and materials-exchange programs all can involve pollution prevention.

**TABLE 2.16—REUSE WITH CLOSED-LOOP SYSTEMS**

| Pollutant | Containment Measure |
|---|---|
| Empty drums/containers | Return drums/containers to vendor for reuse. |
| Pallets | Return to vendor for reuse. |
| Produced water | Reinject for reuse in enhanced recovery. |
| Drilling fluids | Use closed-loop mud system; return muds to vendor for reuse. |
| Hydrotest water | Reuse water for other hydrotests. |

**TABLE 2.17—RECYCLE THROUGH RECLAMATION**

| Pollutant | Reclamation Practice |
|---|---|
| Oily soil | Use as asphalt road amendment or berm material. |
| Tank bottoms | Treat in high-temperature dehydration facilities; use as asphalt road amendment or berm material; process in refinery coker; send to commercial recycler for oil recovery. |
| Lube oil | Put in production (if petroleum based); send to commercial oil recycler or refinery. |
| Paraffins | Send to fuel-blending facility. |
| Oil filters | Send to commercial oil-filter recycler. |
| Empty drums/scrap metal | Send to scrap-metal recycler. |
| $H_2S$ scrubber waste | Send to reclaimer for use in agricultural products. |
| Concrete | Crush, use as aggregate or fill material. |
| Glycol | Send to commercial recycler. |
| Mercury wastes | Send to commercial mercury recycler. |
| Amine | Send to commercial recycler. |
| Batteries | Send to battery recycler. |
| Grease | Send to fuel-blending facility. |
| Trash | Start program to segregate and recycle paper, plastic, and other materials. |
| Molecular sieve | Send to cement kiln. |
| Hydraulic fluid | Put into production (if petroleum based); send to recycler. |

**TABLE 2.18—RECYCLE THROUGH EXCHANGE/SUBSTITUTION**

| Pollutant | Exchange/Substitution Practice |
|---|---|
| Caustic from NGL scrubbing | Reuse as sulfur dioxide scrubber reagent; use for neutralization. |
| Solvents | Use for paraffin treatment. |
| Acids | Use for neutralization. |
| Drill cuttings | Use as fill material. |
| Fly ash | Use in concrete. |
| Produced water | Use for agricultural/other beneficial use if water quality acceptable. |
| Cooling water | Use for agricultural/other beneficial use if water quality acceptable. |
| $H_2S$ emissions | Reinject in reservoir for pressure maintenance. |
| Sandblast media | Use for aggregate in road amendment. |
| Drilling fluids | Use as low-permeability liner (bentonite); use as spud mud on next well. |
| Scrubber waste | Use as oxygen scavenger. |

NGL = natural gas liquids and $H_2S$ = hydrogen sulfide.

**2.7.1 Environmental Compliance Audits.** Audits are frequently conducted by organizations to assess compliance with applicable environmental regulations. Depending on the scope, an audit can be a good measure of the effectiveness of a pollution-prevention program and may help identify its achievements and/or shortcomings. For example, an audit with several findings in the area of solid-waste management should signal a need to improve pollution-prevention efforts.

There are at least two levels of environmental compliance audits: management practice ("big picture") and local or personal inspection. Management-practice reviews may require a knowledgeable outside observer of the process, whereas local or personal inspections of an operation may be conducted during routine daily activities. More structured local reviews may be performed by peer reviewers from another location, professional inspectors, or government agency personnel.

The general steps in an audit are to determine what is expected, to evaluate how expectations are carried out, and to review the documentation. In an overview of programs, auditors may draw on general guidelines and professional experience. In a local performance inspection, auditors may get information from either the implementing personnel or another knowledgeable person. Auditors often use a checklist to guide the inspection and document the review.

Several levels of evaluation are part of a compliance review.

1. Program Expectations. Identification of management expectations and the requirements of the environmental program.

2. Personal Implementation. Determination of specific responsibilities of individual employees in managing environmental compliance and whether they are meeting those responsibilities.

3. Documentation. Determination of whether documentation requirements of the environmental program are being met.

4. Adequacy. Evaluation of performance in terms of the relevant degrees of adequacy of environmental programs, which can range from accepted good practices, to local policy, to government regulation or legal obligations.

An audit alone only identifies weaknesses or errors in environmental programs and practices and does not ensure that the necessary corrections are made. A follow-up plan is necessary to ensure that all findings are mitigated and to demonstrate management commitment to improvement. The lack of a follow-up plan can result in an audit becoming evidence for legal action against a facility.

When corrections are made, they relieve the symptoms of error but may not address the error itself. Fundamental change results only from recognizing root causes of program shortcomings or from identifying common causes of performance errors. This is the desired result of program reviews, especially from the pollution-prevention perspective.

**2.7.2 Employee Involvement/Incentive Programs**. A key component of any successful pollution-prevention program is employee involvement at all levels of the organization. Management commitment, as previously discussed, is a very important component of successful programs. Participation of operations personnel is also essential. Experience has shown that an employee who works at the point of generation of a pollutant generally has the greatest knowledge about the generating process and often is best suited to identify initially that a pollutant is being generated and to devise a method of preventing it.

Employee involvement in pollution-prevention programs has two main components: knowledge/awareness and commitment. Successful pollution-prevention programs need to provide employees with both components. Employees cannot be active participants in a pollution-prevention program without having knowledge about the impacts of pollution and management's commitment to preventing it. Publicizing a pollution-prevention program is an essential first step in enlisting employee involvement. This must be supplemented by training to provide information about the significance of types of pollution and about pollution-prevention methods before effective employee action can occur. Once employees become aware that a particular practice has potential environmental consequences, they generally are very creative in identifying methods to modify the practice to prevent or reduce those consequences. Employee training can be supplemented with other means of communication, such as facility/company newsletters highlighting pollution-prevention activities.

Providing employees with knowledge about pollution leads to development of employee commitment. This commitment can be reinforced and developed more rapidly by providing employee incentives and/or recognition. Employee recognition/award programs often are used to encourage employee efforts toward achieving other company goals and profitability. Pollution prevention can be added to existing incentive programs. If no employee incentive program exists, one can easily be developed to reward employee pollution-prevention efforts. An award program can provide gifts or other special recognition that employees will find valuable (such as preferred parking). Awards with visibility to other employees (such as jackets) can also help publicize pollution prevention and provide continuing recognition for the wearer. Other forms of recognition also provide employee incentives and program publicity. Company newsletter articles recognizing contributions of individual-employee pollution-prevention efforts are a powerful tool that shows management support, informs employees about pollution prevention, and provides an incentive to other employees to develop or expand their own pollution-prevention efforts.

**2.7.3 Environmental Liability Studies**. Operating companies concerned with planning for the future abandonment of producing facilities and properties often conduct studies to quantify expected costs. Although these studies focus on future liabilities, like environmental compliance audits, they also can identify immediate pollution-prevention opportunities. While the method for these studies may vary, they usually include the following components.[16]

1. Preliminary meetings with facility personnel to review operations.
2. Initial data gathering (facility diagrams, operating data, maps).
3. Regulatory review.
4. On-site data gathering and facility inspections.
5. Off-site data compilation and summary-report preparation.

Because the objective of a liability study is primarily to identify future problems, many currently "hidden" or overlooked pollution sources may be discovered. An example of such a source is a small leak that, if not repaired, may result in generation of large volumes of cleanup wastes when the facility is eventually abandoned. Operations personnel may not recognize the significance of such pollution sources, but, as is the case with environmental compliance audits, review of facilities by external personnel may yield a new perspective on environmental liabilities.

During interviews with operations personnel, reviewers check to see whether practices for minimizing future liability are being used. Many of these practices are also beneficial to pollution-prevention efforts; examples are improvement of housekeeping, efforts to reduce soil contamination (install containment, repair leaks, remove idle equipment), establishment of a soil-remediation program, implementation of a risk-based well-abandonment program, and other risk-minimization practices (e.g., use of contained areas for tank-bottoms handling and workover pits).

Summary reports from liability studies should also include a section outlining actions that can be taken to minimize future cleanup activities and, consequently, generation of cleanup wastes.

**2.7.4 Materials-Exchange Programs.** ***Internal Programs.*** As Secs. 2.4 and 2.5 discuss, reuse or recycling of residuals rather than disposal can be used to prevent pollution. A program to promote materials exchanges between two or more facilities or operating units within an organization provides more opportunities for reuse and recycling. Examples of materials-exchange projects in E&P are transfer of produced water between fields, transfer of used lubricating oil from production facilities to refineries, transfer of hydrotest water between business units, and transfer of spent caustic between business units. The following basic steps are included in developing a materials-transfer program.[17]

1. Identify operations or business units to be involved in the program.
2. Conduct an inventory of residual-material streams.
3. Identify processing capabilities in each operation or business unit.
4. Match residual-material streams with the most cost-effective processing capabilities.
5. Resolve business and regulatory issues relevant to the proposed residual-material transfers.
6. Implement the transfers.

Development of a materials-transfer program can enhance communication between organizations in a company, help organize the significant cooperative efforts necessary to complete all the steps in the program, and implement projects between business units that have had no prior history of cooperation. Some of the "assets" in the various industry segments that potentially can be used in processing residuals are land, steam, emulsion-treating facilities, injection wells, and waste heat from E&P units; tankage, low-cost transportation to other facilities, and land from pipeline units; hydrocarbon processing and waste-water treatment from refining units; and transportation, handling, and waste-water treatment from terminal units.

Screening materials for exchange is similar to what normally is required when evaluating waste management options with the additional requirement that both business units must be satisfied with the outcome. Examples of some of the analytical screening parameters for residual materials are total petroleum hydrocarbons, specific-organic-constituent analyses (e.g., volatile and semivolatile organic compounds), flash point, and oil/water/solids content plus any applicable test methods required by federal, state, or local regulations.

Compensation is a key issue in transfer of residuals. Ideally, the compensation between business units should consider all costs for conducting operations, including any increase in maintenance costs and any costs incurred for handling residuals, along with any value resulting from beneficial reuse, and then should distribute the net benefits evenly between the two units.

Regulatory issues are also an important consideration with materials-exchange projects. Any project to be implemented must be in compliance with any applicable regulations. However, because product yield often is increased, materials-exchange projects can be classified as "process-efficiency-enhancement" projects rather than "residuals-treatment" projects. Thus, these projects serve to replace environmental impact management with operational enhancement, which can have fewer regulatory or permitting requirements.

***External Programs.*** Several regulatory agencies sponsor materials-exchange programs. Through these programs, periodic listings of residuals available and materials needed are published. Most of the issues with external exchange of materials are identical to those with internal transfers except for liability, which is a major issue with external transfers. However, these programs can also be an effective way to recycle or reuse materials and prevent pollution.

## References


1. *Standard E 1609-94, Standard Guide for Development and Implementation of a Pollution Prevention Program*, ASTM, West Conshohocken, Pennsylvania (1994).
2. "Industry Experience with Pollution Prevention Programs," Publication No. 849-31700, API, Washington, DC (1993).
3. "Total Cost Assessment, Accelerating Industrial Pollution Prevention through Innovative Project Financial Analysis," U.S. Environmental Protection Agency (EPA), Washington, DC (May 1992).
4. "Environmental Performance Indicators, Methods for Measuring Pollution Prevention," Publication No. 331, API, Washington, DC (September 1994).
5. *AP-42, Compilation of Air Pollution Emission Factors*, U.S. EPA, Washington, DC (January 1995).
6. "Waste Minimization Opportunity Assessment," Report EPA/625/7-88/003, U.S. EPA, Washington, DC (1988).
7. Dorfman, M.H., Muir, W.R., and Miller, C.: "1992 Environmental Dividends, Cutting More Chemical Waste," Inform Inc., New York City (1992).
8. "Waste Minimization in the Oil Field," Railroad Commission of Texas, Austin, Texas (August 1994).
9. "Waste Minimization in the Petroleum Industry, A Compendium of Practices," Publication No. 849-00020, API, Washington, DC (November 1991).
10. "Environmental Guidance Document, Onshore Solid Waste Management in Exploration and Production Operations" Publication No. 811-10850, API, Washington, DC (January 1989).
11. Lal, M.: "Economic and Performance Analysis Model for Solids Control," paper SPE 18037 presented at the 1988 SPE Annual Technical Conference and Exhibition, Houston, 2–5 October.
12. Thurber, N.E.: "Waste Minimization for Land-Based Drilling Operations," *JPT* (May 1992) 542.
13. Cornwell, J.R.: "Road Mixing Sand Produced From Steamdrive Operations," paper SPE 25930 presented at the 1993 SPE/EPA Exploration & Production Environmental Conference, San Antonio, Texas, 7–10 March.
14. "Developing Area-Specific Waste Management Plans for Exploration and Production Operations," first edition, API, Washington, DC (1991).
15. "Keepin' It All Clean in the Oil Patch," Natl. Environmental Training Assn., Houston (1995).
16. Johnstone, J.E., Neuser, J.M., and Fritts, J.M.: "Estimating and Reducing Companywide Property Abandonment Costs," *SPE Advanced Technology Series* (March 1995) 180.
17. Fisher, R.A. *et al.*: "Integrated Management of Petroleum Industry Wastes," paper presented at the 1995 U.S. Dept. of Energy/Tulsa U. Annual Intl. Petroleum Environmental Conference, New Orleans, 25–27 September.


## SI Metric Conversion Factor

$$\text{bbl} \times 1.589\,873 \quad \text{E}-01 = \text{m}^3$$

# Chapter 3
# Habitat Preservation

**D.W. Chamberlain,** Atlantic Richfield Corp.
**James Johnstone,** Arco E&P Technology
**June Lindstedt-Siva,** Arco
**Marcia H. Wolfe,** M.H. Wolfe and Assocs.
**Ted St. John,** Trees of Life Nursery

### 3.1 Introduction

The discovery of oil in the hills of Pennsylvania in the early 1860's had a tremendous impact on the countryside surrounding Oil Creek. Early photographs show hills cleared of timber and brush to make way for the derricks, tanks, and buildings. The impact of petroleum production on the habitat was not a consideration. Workers' primary concerns were drilling producing wells and making their fortunes before the oil stopped flowing. The first regulations relating to environmental concerns of oil and gas activities were written in the 1920's. These dealt with authority to limit oil and gas fields that would cause damage or loss to nearby property and placed a moratorium on building derricks in beach recreational areas.

Oil has been found in virtually every type of habitat in the world. Before the start of the 20th Century, oil was found in Baku on the shores of the Caspian Sea and in the Central Valley of California and the jungles of Dutch East Indies. Since the turn of the century, oil has been found and developed in Middle Eastern deserts, South American and African rain forests, the Arctic tundra of North America and Russia, and under the oceans in the Gulf of Mexico and the North Sea.

Early habitat conservation in oil and gas fields was coincidental to business pressures. Dikes were placed around tanks to enable recovery of oil. Leaks were repaired to conserve product. Surface-rights landowners requested payment for damage and/or losses to crops, which necessitated optimization of roads and drilling locations.

Development of extensive environmental regulations in the 1960's and 1970's highlighted the importance of habitat conservation. The discovery of oil on the North Slope of Alaska during this period proved to be a turning point in how the industry took proactive steps to study and mitigate impacts on the habitat. Operators brought in biologists to study wildlife populations, and, together, they developed plans to minimize the impacts of development on migrating caribou herds and nesting birds. Methods to revegetate tundra were developed along with new construction techniques to minimize impacts of roads and facilities. This process, one of the first examples of voluntarily conducted environmental planning, became an integral part of the final project design.

Oil and gas regulatory agencies played the first role in mitigating oil-development impacts on habitat. Agencies adopted rules to encourage proper waste disposal, to require cleanup of oil spills, and to cover sumps and well cellars. These measures provided basic protection for wildlife.

In the U.S., impacts on habitats were reduced further as various environmental acts were passed. The Natl. Environmental Policy Act of 1969 required the preparation of detailed environmental impact reports.[1] The Endangered Species Act of 1973 was designed to protect and conserve threatened and endangered species and the ecosystems on which they depend.[2] The Clean Water Act of 1977 set up effluent standards on discharges to water.[3] The Comprehensive Environmental Response, Compensation, and Liability Act authorized cleanup of toxic contaminants.[4] Each of these measures compelled the oil industry to take into account impacts on the habitat when conducting its business.

This chapter describes the measures necessary to minimize the impacts on habitats before, during, and after an area is developed for oil production. Environmental planning before development can reduce both environmental impacts and expenses. The chapter reviews potential impacts of oil and gas development and the methods to minimize or mitigate the impacts in different habitat types.

### 3.2 Environmental Planning and Management

Activities of human populations and the support systems on which they depend have two types of impacts, loss of natural habitats and environmental contamination. Controlling and minimizing these environmental impacts is the focus of environmental planning and management activities in corporations and government agencies.

Native plant and animal populations are reduced, confined to small areas, or lost altogether as a result of disturbance, modification, and conversion of natural habitats. At some point, survival of species themselves may be in jeopardy. The reason that most species become endangered or threatened is not genetic or physiological but because their habitat is modified or eliminated or because the species is overharvested. It is not possible to achieve zero pollution or stop natural-habitat losses in the foreseeable future. This is as true in the mountains of southern California as it is in the forests of Brazil. Some important habitats can be protected in parks or reserves, but this strategy saves only habitat "islands" and, alone, cannot ensure maintenance of biodiversity over time. Environmental contamination reduces environmental quality and can interfere with ecosystem functions. Long-term solutions must emphasize ways to maintain ecological integrity and functions of "developed" areas. This approach is consistent with the concept of sustainable development.[5] The basic concepts presented in this section were published earlier.[6]

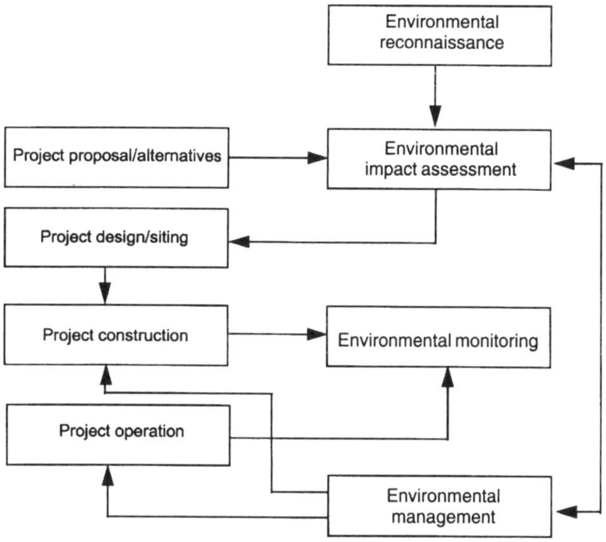


**Fig. 3.1—Environmental planning and management process.[6]**

**3.2.1 Development That Preserves Ecological Integrity.** Sustainable development is defined as development that is accomplished while preserving the ecological integrity and functions of the affected system. The following example of development resulted in preservation of ecological or biological integrity.

The Guadalupe Dunes oil field in Santa Barbara County, California, consists of oil wells, pipelines, roads, and oil storage and treatment facilities. Because most of this coastal property is maintained in open space and public access is restricted, it is one of the last places in the region where native dune vegetation survives, including some endangered species. Because of public use (dune buggies, dirt bikes), dunes adjacent to the oil field are damaged and no longer support native vegetation. Similarly, in Kern County, California, oil fields are some of the only remaining large tracts of land that have not been converted to agriculture or urbanization. Here, too, native plants and animals are present, including several endangered species. One working oil field was recently designated an ecological preserve and is being managed as such in cooperation with state agencies. This management will continue beyond the life of the oil field.[7]

For these examples, protection of natural ecosystems often was an unforeseen byproduct of the type of development that occurred. However, today these environments are protected by design, as part of managing the facilities. It is the latter, environmental protection by design, that must be encouraged and supported worldwide.

**3.2.2 Environmental Planning and Management Process.** The goals of environmental planning and management are to minimize adverse environmental impacts of development and to maintain or enhance the ecological integrity and functions of the natural system. The principles can be applied to development as diverse as an agricultural crop, housing tract, power plant, or oil production. The interdisciplinary process of environmental planning and management requires interaction of the environmental sciences, engineering, and operations. It can be applied to ongoing operations or new projects, although it has its greatest effect when applied to new projects early in their development, when siting, design, and engineering options are still open. **Fig. 3.1** illustrates the process.

As part of an environmental planning and management process, a project proposal is developed, including siting and design alternatives. An environmental reconnaissance study is conducted to identify major ecological features of each site. This need not be, and in most cases cannot be, a lengthy and expensive study. Much can be learned from a literature review, site visit, study of topographic maps, and aerial or satellite photos. On the basis of this information, an environmental impact assessment is prepared. The assessment includes analysis of potential impacts of the various alternatives as well as recommendations of alternatives and mitigation measures. This information is fed back into an environmental management plan for the project-siting and -design process. Site and project-design alternatives are selected, and permits and other approvals are obtained.

On-site environmental monitoring begins during the construction phase of a project. Monitoring can begin before construction if additional, site-specific environmental data are needed. Monitoring to establish a baseline against which to measure future impacts is usually not feasible because of the time, scope, and cost of a study required to define natural variation. Therefore, the monitoring program must focus on environmental parameters and populations likely to be affected during project construction and operation phases. Environmental impacts are assessed, and the information is fed back into an environmental management plan that becomes part of the project operations plan. Monitoring continues to some degree during operations so that impacts can be assessed continually and the management/operations plan can be modified as necessary.

Once this process is completed, the environmental and engineering information generated becomes extremely valuable to anyone planning similar developments or different projects in similar habitats. Therefore, dissemination of the results (including both successes and failures) is critical.

**3.2.3 Discussion.** A major effort is needed to strengthen the science base as well as to train and encourage scientists and engineers to practice the innovative thinking and collaboration required to implement sound environmental planning and management. Environmental scientists and engineers in the academic community must develop programs to promote this field. Industry scientists and engineers must be empowered to take risks and think unconventionally because they are the ones who generate most new ideas and apply them in the field. Governments need to develop regulatory frameworks that support, rather than impede, this process. Supporting and encouraging sound environmental planning and management offer some hope that the ecological integrity and functions of natural systems will be maintained when development occurs.

### 3.3 Oil- and Gas-Related Pressures on Habitat

Habitat, most simply described, is the place where a plant or animal lives. It is not necessarily synonymous with plant cover, which some may typically think of as being habitat. For example, an endangered San Joaquin kit fox usually lives in a burrow within desert saltbush scrub habitat comprising a variety of grasses, shrubs, and flowering plants. However, it can also live in a pipe beside a road or under a well pad in a densely developed oil field with ¼-acre or less spacing where no vegetation grows. That becomes habitat for the San Joaquin kit fox. Habitats can be terrestrial, aquatic, or both. Some species range throughout or require both types of habitat.

Habitat consists of a variety of biotic and abiotic factors, including air, water, soils, vegetation, and the wildlife and other interacting species.[8] Habitat must provide a combination of air, food, water, cover, and shelter for most wildlife species. Because a mixture of biotic and abiotic factors is involved, understanding the combined impacts and pressures of development can be very complex. Oil and gas activities can impact any of these habitat parameters. Plants and animals generally require air. Oil or gas activities can affect air quality, for example, by being directly toxic to the habitat or by reducing viability or affecting reproduction of species. These are some of the primary factors that must be recognized and integrated with oil and gas planning activities if plants, wildlife, and their habitat are to be conserved.

Adequate habitat is often the primary limiting factor for wildlife and plants. In southern California, cumulative habitat loss is the main cause for endangerment of many species in the central San Joaquin Valley, where almost 20 species are listed or under consideration for listing as threatened or endangered. Consequently, habitat considerations are of primary importance when planning and designing oil and gas facilities. Habitats are classified in a variety of ways to help in their management. Usually, different geographical areas have resources for identifying regional habitats. These resources can be used to assist with planning and design of oil and gas facilities.

**3.3.1 Habitat Loss.** One of the major impacts of human activities on ecosystems today is loss of habitat. Oil and gas exploration and development contribute to the overall cumulative impacts of habitat losses. However, when compared with many types of urban, industrial, and agricultural developments, the percentage of habitat affected by oil and gas activities is relatively small.

A notable advantage of many oil and gas developments and activities is that, in many cases, low and moderate levels of development can be compatible with certain plant and wildlife habitat uses when adequately mitigated. In some cases, such as moderately to lightly developed oil and gas fields, the presence of oil and gas activities has essentially preserved habitat from agricultural and urban development. Today, one such area, the Coles Levee Ecosystem Preserve, which encompasses the 25th largest oil field in the state of California, is a preserve for threatened and endangered species and wetlands.[9]

**3.3.2 Habitat Fragmentation.** Oil and gas activities also can contribute to habitat fragmentation. Fragmentation is the breaking up of large pieces of habitat into smaller pieces or islands. Fragmentation is of particular importance in the management of rare and endangered species. When species numbers are low, fragmentation increases the potential probability of extinction by leading to breeding depression, increasing predation and exposure to other losses and risks, such as diseases. Habitat fragmentation also can help contribute to population losses of common species.

**3.3.3 Habitat Degradation.** Oil and gas E&P activities degrade habitat in a wide variety of ways. Disturbance can increase edge effects, eroding away habitat quality of adjacent ecosystems. Disturbance can contribute to degradation by creating a setting for invasion of noxious weeds and early successional or disturbance species (such as coyotes, ravens, and blackbirds; other types of plants, and other wildlife pests, which frequently create adverse pressures on habitats and the species they support). Sometimes oil and gas activities directly pollute or contaminate habitats during upset conditions. Noise can frighten migrating species, or lights can interrupt feeding and movement patterns. Soil vibrations from drilling can affect the foraging of ground-dwelling species. Accidental fires from welding and related activities can burn adjacent undisturbed habitats, changing habitat quality. Small animals can die from eating uncontained garbage, such as plastic sandwich bags, a not uncommon occurrence.

**3.3.4 Plants and Wildlife.** Plants and wildlife are part of the habitat. They reflect the pressures of oil and gas development in many ways, from effects that are difficult to measure, such as reduced reproductive success because of human activity and stress, to genetic mutations from exposure to toxic materials.

Air quality is important for habitat quality. Adverse air quality impacts plant growth, development, and productivity. It can also have adverse impacts on wildlife species. Properly designed, maintained, and operated facilities can prevent adverse pressures on air.

Soils can be affected by construction activities that cause compaction, erosion, elimination of topsoil, and blowing dust and dirt. These can result directly and indirectly in on- and off-site habitat degradation. Spills of chemicals, oil, or condensate can result in direct contamination and degradation of habitat quality.

Water is essential in terrestrial and aquatic habitats, both as habitat and as a habitat requirement for many species. Oil and gas can affect water by seeping into ponds and springs from natural formations; however, significant impacts of oil and gas activities on water generally can be prevented. Mitigation is more difficult to design for impacts from spills and other catastrophic events.

Air, land, and water traffic creates different types of pressures on habitat, including noise stress, activity stress, interruptions of animal movements, or direct mortality. Habitat can increase noise effects created by oil and gas activities because of echoing or bouncing from geological or physical structures within the habitat. Habitat can also be used or manipulated to dampen noise effects. Habitat quality can be affected by the presence of light at the wrong time of day or night. Effects can range from changing animal movements, to impacting predation on species made visible by additional light, to disrupting natural-habitat aesthetics. Although vibrations that occur during drilling and geophysical activities usually are short term, their impacts can range from alteration of population movements to interruption of feeding patterns. They can occur in both terrestrial and aquatic habitats. Aesthetics may be in the eye of the beholder; however, it can be a significant concern in designing oil and gas developments. Air clarity, shapes, colors, and impacts on horizons are pressures that can be exerted on habitat from oil and gas developments.

Cultural resources also are effectively part of habitat. Even though they are not necessarily habitat qualities for plant and wildlife survival, they are protected by a variety of laws and can occur in, on, or under habitat. Consequently, they generally can be affected by any activity that affects soil or water.

**3.3.5 Addressing Habitat Pressures.** Even though oil and gas production activities have impacts on habitat, resource development can occur without leaving permanent adverse impacts. The pressures of oil and gas activities can be relieved, minimized, or eliminated by taking actions during the planning and design phase and by implementing awareness of environmental issues during operations and maintenance. A holistic ecosystem approach to planning and project design is one of the most cost-effective ways of developing methods to relieve, minimize, or eliminate these pressures. A project-by-project or species-by-species approach tends to result in expensive habitat micromanagement, sometimes with inconsistent and contradictory results. When dealing with rare species, significant funds often are spent attempting to protect wildlife and plant individuals rather than looking at the most effective means of ensuring the long-term continued existence of these species and the habitats that support them. Overall environmental protection measures are the best and most cost-effective approach to exploration and development activities. This approach integrates human needs with natural landscape in addition to protecting the environment.[10]

The first consideration in dealing with habitat pressures from oil and gas activities is to determine whether these impacts can be avoided by simple project design or alteration. Avoidance can eliminate habitat loss and fragmentation and minimize degradation. For example, building new facilities on existing disturbances accomplishes avoidance and usually reduces construction and maintenance costs at the same time. If pressures cannot be avoided, they generally can be minimized or reduced in size. Reduction in size by combining new facilities also reduces construction and maintenance costs. Planning can be used to ensure the presence of conservation corridors to offset natural fragmentation.[11]

Mitigation is implementation of an action designed to reduce adverse impacts or pressures. Because these cover a wide range of pressures and impacts, a detailed discussion is outside the scope of this chapter. Typical mitigation measures include reduced speed limits, watering roads for dust control, revegetation for erosion control, deflectors on lights, and employee environmental awareness training.

Habitat losses sometimes can be compensated for by permanent habitat acquisition and protection combined with enhancement and other management practices. Compensation generally is interpreted as "replacement" of developed habitat by conservation of undisturbed habitat in some ratio. As an example, in California, desert saltbush habitat typically is compensated for at a ratio of 1 : 3 acres.

Improving habitat is a strategy often used to increase carrying capacity, as is simply assisting restoration of degraded habitat. Enhancement can range from revegetation, to installation of special nesting or denning structures, to placement of additional watering sites. Restoration is used to refer to rehabilitation or reclamation of disturbed habitats. Precise restoration *per se* is impossible. However, habitats affected by oil and gas activities generally can be rehabilitated to an effectively designed, post-disturbance land use. Compared with many other types of disturbances, those created by oil and gas activities generally are relatively small and amenable to natural habitat successional patterns and implemented revegetation practices.

### 3.4 Sensitive Habitats

Worldwide, many different kinds of natural habitats are sensitive to the impact of human activities. Some are more resilient than others to environmental impacts and may recover over time. Others are less resilient and may require a great deal of recovery time. The following examples are some of the more general types of habitats, with a discussion of details that set them apart. The overall goal and guiding philosophy must be minimization of any environmental and social impacts of oil and gas operations.

**3.4.1 Tropical Rain Forests.** Tropical rain forests are complex environments that contain a greater abundance and diversity of plant and animal species than other terrestrial habitats. Relatively few of these species have been named, described, or studied scientifically. Tropical forests are some of the world's most sensitive environments. Increasing population pressures and rising expectations are giving rise to conversion of present prime forest ecosystems and will most likely intensify and will continue to do so well into the future. More than 27 million acres of mature tropical forests are converted to agriculture, pasture, or other uses yearly. The scientific community believes that deforestation is resulting in loss of plant and animal species at unprecedented rates. Mindful of these important issues and concerns, minimizing environmental impacts for the general, overall success of a project is essential. Major environmental and social issues are associated with any level of development in tropical forests because of the potential for significant impacts on the forest and resident indigenous people: deforestation and habitat alteration, pollution, and social impacts. These impacts should be minimized during all phases of a project. The economic and environmental future of the area after the field is decommissioned must be considered.

***Function and Value.*** Tropical rain forests are rich in substances, both medicinal and industrial. These substances are necessities to their indigenous residents and may be as useful to society at large as are the minerals beneath them.

***Permits and Approvals.*** Most areas of the world where oil and gas activities take place are owned by federal, state, local, or private entities and require permits for access and use. Identifying land ownership and governing regulations early in project planning is imperative. Visit the site, and include owners in the planning process to expedite approvals and permitting.

***Permitting Process.*** The permitting process varies according to the requirements of the particular country, state, local, or private regulations and policies. The process of gaining access to a petroleum resource can differ significantly from area to area. The permitting process usually can be accelerated by honest cooperation, by demonstration of knowledge of local environmental and social issues, and by having a good environmental management system to offer in addressing project issues.

***Special Planning.*** Those doing work in tropical rain forests must be committed to doing it the right way at all times. Early environmental planning is critical to reaching this goal. Although mistakes have been made in the past in relation to the environment and to indigenous people during petroleum activities in many rain forests, petroleum can be explored for and produced in an environmentally safe manner with minimum impacts on rain-forest ecology and on resident indigenous people.

***Discussion.*** The performance guidelines mentioned previously should be followed in all aspects of the program, from initial discussion and planning stages through exploration, development, production, and site restoration. These guidelines outline the standards of environmental management and performance, many of which are new to petroleum development in the rain forest. Examples of specific practices that can significantly reduce footprints and environmental impacts include the following.

1. Minimize building of roads, maximize helicopter use.
2. Minimize size of clearings for pathways and facilities; open the smallest footprint possible.
3. Minimize use of heavy equipment, maximize use of hand clearing.
4. Establish community-development programs.
5. Establish on-site waste-water-treatment and waste-minimization practices.
6. Fund archaeological surveys; collect and clear artifacts before any construction.
7. Plan for and use forest-saving practices, including selecting trees necessary for lumber at spaced intervals to minimize both size of openings and natural free fall, eliminating clear cutting, selecting trees approved by locals, leaving tree litter in place rather than burning it, and maximizing use of any trees cut. Promoting the use of foresters maximizes the amount of lumber cut from each tree.
8. Salvage topsoil, stockpile for site restoration, and protect from weathering and leaching.
9. Establish on-site nurseries for propagation of seedlings used in site restoration.
10. Topsoil bare areas; seed and plant even during production where possible.
11. Inject waste drilling muds and cuttings; use no surface disposal.
12. Train all personnel in environmental and social issues.
13. Audit and monitor environment during all phases of activity.

In short, exploration, development, operation, and maintenance should be planned with needs of site restoration in mind.

***Mitigations.*** Scientific resources cannot be used effectively to develop and implement an environmental plan without the understanding and support of management, field personnel, the host government, and the local population. Design and use of scientific resources to protect and sustain the environment and the local population will fail unless these procedures are communicated to, understood by, and followed by all active parties. This information should be made an integral part of a project to guide planning and operations during site preparation, well drilling, construction, operation, and subsequent restoration activities.

To conduct operations in accordance with established guidelines, social and ecological knowledge is necessary. These data are needed for planning, for conducting impact assessments for daily operations, and for proper site restoration. Environmental assessments include searches for relevant publications, discussions with experts, and environmental impact evaluations.

Site visits are indispensable for gaining environmental knowledge for planning and for learning specific details about the proposed site. Both aerial and ground assessments should be made by company personnel and consultants to obtain first-hand knowledge of the proposed development area and to aid in finalizing operational guidelines and planning for construction, drilling, and restoration phases. During these visits, initiate communication with local indigenous people. Apprise them of the project, and use their vast knowledge for proper planning and management of the project.

Site-specific biodiversity studies addressing water quality, fisheries, forest fauna and flora, and soil morphology and chemistry produce additional information to aid in planning, construction, operations, and restoration. During drilling operations, water quality must be monitored. Waste water can be reinjected into the well annulus or treated to water quality standards and released. Geofabric can be used on the soil surface under construction (e.g., drilling rig platform), providing a stable mud-free work surface.

Where it is necessary to temporarily remove topsoil during site preparation, cover and store this soil in a pile for future restoration work to protect it from the deteriorating influence of solar radiation and from leaching by rainfall. If such protection is not provided, physical and chemical changes reduce soil quality. If the topsoil and the contained root systems are left in place and protected from significant disturbance when a site is cleared of vegetation down to the surface of the topsoil, regrowth through natural means is dramatic. Where the topsoil and root systems are disturbed, regrowth is significantly retarded.

To accelerate the restoration process in cleared areas, experience has shown and consultation has confirmed that leaching of bare areas must be prevented and that mulching of bare soil to retain moisture and to provide shade and nutrients for seedlings is essential to successful restoration. Also, reseeding and seedling planting in disturbed sites must be at a greater density than normally done to accelerate recovery. Although it appears paradoxical to the abundance of rich tropical vegetation, the most important characteristic of all rain forest soils is their low nutrient content for plants. However, in an undisturbed rain forest, there is a closed cycle of nutrients from vegetation to soils to vegetation. The soil beneath its natural forest cover reaches a state of equilibrium where soil nutrients are replaced as they are used up by dead vegetation falling on the soil surface and decaying and the decay products entering the soil. Outside this cycle, loss of nutrients is very slight and is made up by fresh plant nutrients set free by decomposition of rock at lower depth. Disturbance of the natural vegetative cover breaks this nutrient cycle.

Most tropical forests are not pristine in the sense of never having been inhabited by humans. Most forests have been inhabited prehistorically and historically, and what may look like a pristine forest is actually a mosaic of forest patches that have recovered from past disturbances. Natural creation of gaps in rain forests and their recovery is a major reason for the great diversity and health of a "natural" rain forest. Acquisition of lumber and other materials when necessary should closely resemble the natural creation of forest gaps by felling isolated trees and not clear cutting. This greatly aids regrowth and restoration of the forest. Cutting of trees must be coordinated with the local indigenous people to avoid destroying vegetation that is important to them. Disturbed forest areas recover more successfully if damage to seedlings, adolescent trees, the soil surface, root systems in the upper soil layers, and water-drainage patterns can be minimized.

Central to developing the resources of the rain forest is sustainability and minimization of destruction to the environment and the indigenous people. Careless development of petroleum resources in rain forests can lead to short-term prosperity for segments of the local population, long-term destructive changes to the local sociocultural structure, and dependence of the local indigenous population on a very limited resource that will be depleted in a few decades. Traditional socioeconomic and cultural aspects of life in many rain forest areas already have been affected significantly by earlier and ongoing contact with outsiders. Historically, impoverished indigenous and colonial populations have resorted to sustainable agricultural and silvicultural practices to provide income for survival. The goal should be to ensure that the local rain forest indigenous population has viable options for maintaining and/or improving their quality of life without destroying the forest ecosystem through sustainable resource utilization.

**3.4.2 Wetlands.** Approximately 71% of the Earth's surface is under the ocean. Most information about the ocean relates to the shallow shelf areas at the margins of the continents. The ocean has met many human economic, food, and recreational needs for centuries. As the population increased, the role of the ocean in meeting these needs also increased, and this increased use has impacted the natural ecosystems of the ocean. Some, such as commercial fisheries and water quality near coastal population centers, are affected more than others. The ocean has a tremendous capacity to deal with man-made pollution; however, this capacity may be finite. It is important to protect coastal areas as habitat for fish and other wildlife. These are areas where feeding and spawning grounds are located, and fish, mollusk, and crustacean populations will disappear if the areas are harmed or destroyed. Yearly, a relatively large amount of habitat of these species is lost to pollution, unwise development, and other human activities.

Other problems associated with human activities include invasions of exotic plants in coastal wetlands, impacts from energy and mineral extraction, and recreation. Contamination in bays and harbors is far higher than in areas of ocean outfalls and open coastal systems. However, there is very good evidence that negative pollution trends can be reversed. Monitoring data show that levels of contamination have declined in the southern California area where water, sediment, and marine life were heavily contaminated 10 to 20 years ago. This is a direct result of source control and treatment. Pollution trends in harbors and bays are less encouraging.

It is important that oil- and gas-related activities do not add significantly to the loss of ocean resources and habitats. The ocean environment is much more stable than land environments. Because of this stability, organisms that live in the ocean generally have not developed highly specialized regulatory systems to combat sudden changes that might occur within their environment. Therefore, they are affected by relatively small changes.

The Natl. Research Council's Committee on Characterization of Wetlands recently defined wetlands: "A wetland is an ecosystem that depends on constant or recurrent, shallow inundation or saturation at or near the surface of the substrate. The minimum essential characteristics of a wetland are recurrent, sustained inundation or saturation at or near the surface and the presence of physical, chemical, and biological features reflective of recurrent, sustained inundation or saturation. Common diagnostic features of wetlands are hydric soils and hydrophytic vegetation. These features will be present except where specific physicochemical, biotic, or anthropogenic factors have removed them or prevented their development."[12] This definition probably will make its way into official manuals to define wetlands and into regulations and federal and state wetlands policies.

Wetlands are, at times, difficult to define exactly because of their geographical extent and the wide variety of hydrologic conditions in which they are found. Wetlands include bogs, swamps, marshes, mires, fens, and other wet habitats usually between more terrestrial habitats and open waters of lakes, streams, and the ocean. Wetlands are found throughout the world under many different names. They are found on every continent except Antarctica and in every climate regime from the tropics to the Arctic tundra, even arid areas. To be a wetland, however, they must have the three main components described in the previous definition: wetland hydrology, wetland soils, and wetland vegetation. In the U.S., an Army Corps of Engineers wetlands delineation manual is the official reference for identifying and describing wetlands.[13] However, the Committee on Characterization of Wetlands has recommended a new manual be produced because of new wetlands knowledge available. When planning oil and gas activities in the U.S., the U.S. national policy of "no net wetlands loss" must be considered. Permitting decisions are driven by regulatory agencies. More than half of U.S. wetlands have been altered, and there is much worldwide concern about this issue.

***Function and Value.*** It is well-known that the oceans influence the weather and the climate of a good part of the Earth. The oceans are a means of transportation and a source of food both for humans and for the plants and animals that live in and around them. As mentioned earlier, the ocean, long a source of human food from fishing and farming the sea, also may be a source of medicinal products. It is a source of energy from oil and gas production and a potential source of thermal energy. Wave and tide action is already used to generate electricity in certain areas. Ocean waters are used to cool power plants and industries and as a medium to assimilate and dispose of sewage, dredge spoils, and other industrial wastes.

Wetlands values are also well-known. Wetlands play an extremely important role in nature as nursery areas for animals and plants, in flood control, in recharging groundwater, as sediment traps, in filtering pollutants, in providing habitat for wildlife, and as recreational and educational areas, to name a few major uses. Recognizing the value of wetlands, legislators have translated this into wetland protection laws, regulations, management plans, and policy.

***Permits and Approvals.*** In the U.S. and its protectorates, the adjacent ocean is owned by the coastal state out to 3 miles (or in the case of Texas, Louisiana, and Florida, out to 3 leagues). Beyond this lie federal waters, under federal government control; these extend to the edge of the U.S. economic zone and are restricted to U.S. commercial use.

In most of these areas, permits are required for commercial and private uses of wetlands and ocean resources, including commercial and recreational fishing, oil and gas E&P, mineral E&P, ocean dumping, dredging and filling, and other forms of construction. Permits are obtained from various federal and state agencies, depending on jurisdiction. These agencies include state lands offices, state departments of fish and game, state natural resources offices, the U.S. Bureau of Land Management, U.S. Marine Fisheries, and the U.S. Fish and Wildlife Service.

***Permitting Process.*** Obtaining permits to use ocean resources usually is done through a governmental agency that controls the use of the various resources in the ocean. Agencies involved need to be identified early in the project and invited to be a part of project planning. U.S. regulations governing ocean usage include the Marine Mammal Protection Act,[14] the federal Endangered Species Act of 1973,[2] the Oil Pollution Act of 1990,[15] the Clean Water Act,[3] and the National Environmental Policy Act.[1] Those applying for permits should become familiar with the provisions of these regulations.

The primary regulations protecting wetlands and controlling their usage in the U.S. are the Clean Water Act and the Food Security Act[16] (agricultural wetlands). Some states have wetlands regulations that require landowners to know the boundaries of the wetlands on their property. Nationwide permits can authorize filling of relatively small areas if the permitted activity is consistent with the Clean Water Act. However, compensation for any wetlands loss most likely will be required on a 1 : 1 or greater basis. In the U.S., permits for alteration of wetlands are mandatory and many state and federal agencies usually are involved. These permits typically require an environmental impact assessment to determine possible risks to the wetlands area under consideration, to look for alternatives with less impact, and to provide scientific data about the specific wetland. It is important to identify and become familiar with the regulations and regulatory agencies applicable to the wetland areas being considered for oil and gas use as early in the planning process as possible. Sec. 404 of the Clean Water Act is the main permit to consider. It is administered by the U.S. Army Corps of Engineers; but the ultimate authority for determining the reach of the act lies with the U.S. Environmental Protection Agency. A company contemplating use of wetlands must contact this agency's office closest to the project site and any similar state office.

***Special Planning.*** Working in any sensitive area like the ocean and wetlands also requires proper and early environmental planning. As discussed earlier, attempts to minimize environmental impacts of the proposed activities must be addressed early in the planning process. Again, the process by which this is done is environmental planning and management. The goals of environmental planning and management are to minimize adverse environmental impacts of development and to maintain or enhance the ecological integrity and function of the natural ecological system. Constant monitoring for impacts during all phases of petroleum operations (exploration, development, production, and site closure) and establishment of a good restoration plan are essential to mitigation of environmental impacts. In the case of wetlands (especially) and other sensitive areas, the best solution is to try to avoid development in these sensitive environments whenever possible.

***Mitigations.*** Scientists have concluded that the single largest long-term threat to the viability of the ocean fisheries of the U.S. is the increasing loss of fish habitat to pollution, unwise development, and other human activities (such as overfishing). Fish habitat is anywhere fish are found. Habitat loss is occurring in spite of many regulations, policies, and management activities. It is imperative that oil- and gas-related activities do not add significantly to this habitat loss. Personnel working in wetlands and ocean environments should receive training on the sensitive plants and animals and on the habitats of these areas. Critical areas, such as wetlands, reefs, and saltwater marshes, should be protected from any development impacts. Wetland restoration and research should be supported as well as research to evaluate long-term impacts of oil- and gas-related activities. As mentioned earlier, the U.S. national goal is no net wetlands loss. The best way to accomplish this is to avoid activities in wetlands wherever possible. When unavoidable and use is permitted by a regulatory agency, wetlands taken must be replaced at a 1 : 1 or greater ratio. This is done by restoring, creating, or buying wetlands. It is not a good principle to destroy a natural wetland for a project and compensate for it by creating or restoring a wetland. Although the restoration acreage is 1 : 1 or greater, many consider this practice to be a net loss of wetlands.

**3.4.3 Arid Regions.** The word desert usually is applied to arid regions of very low rainfall unsuitable for human habitation, even though desert areas are inhabited by humans and some support a surprising number of plants and animals. The extent of the world's deserts depends on the definition used. Estimates based on vegetation adapted to arid conditions gives about 35% of the Earth's land area. Deserts are subject to the most extreme temperature fluctuations of any habitat type. During the day, they are exposed to intense sunlight where the temperature can rise to 130°F in the shade; at night, temperatures can fall to below freezing. Low humidity is the reason for the extreme temperature fluctuations in deserts. Some deserts, such as parts of the Sahara, are almost devoid of life; but more commonly, there are drought-resistant shrubs, such as creosote bush, sagebrush, mesquite, greasewood, succulent plants, cactus, and the euphorbias. Also, following desert rains, many small rapid-growing annual herbs germinate from seeds present in the ground, then grow, flower, produce the next crop of seeds, and die within a relatively short time.

Desert vegetation seldom forms a canopy over the ground and is usually separated by open ground. The ground between desert shrubs, bushes, and trees is usually wind-swept fine-texture soil, gravel, or rock and always has very little humus. The vegetation usually has shallow, wide-spreading, many-branched root systems adapted to absorb any moisture on the ground surface quickly.

Many deserts have abundant animal life that is active primarily at night or during the brief periods of early morning or late afternoon when the heat is not so intense. During the day, the animals remain in cool underground burrows, in cavities of plants (e.g., spiders and insects), or in the shade of plants. Examples of these types of animals are rodents (kangaroo rat, ground squirrel), snakes (sidewinder), lizards (blunt-nosed leopard lizard), turtles (desert tortoise), jackrabbits, insects, and spiders. A number of desert animals sleep during the hot summer months (aestivation).

***Function and Value.*** Beyond plant and animal natural resources, deserts provide recreational value (e.g., nature viewing, camping, off-road use, hunting) and natural resources, such as oil, gas, and borax.

***Permits and Approvals.*** Most desert land in the U.S. is owned by the federal government (e.g., Bureau of Land Management), with considerable sections set aside for special uses, such as national monuments, national recreational areas, wildlife refuges, national forests, state parks, critical habitat areas for endangered species, Indian reservations, and military reservations. Areas of the desert are also privately owned. Much of U.S. desert is managed under one or more regulations, both federal and state, including the Desert Protection Act,[17] federal and state endangered-species acts, and other such legislation.

***Permitting Process.*** Because desert land is owned by public or private entities, it is necessary to determine ownership and make contact with the owner to arrange for use by lease or lease sale, easement, rental, or other means early in the planning process. Use of land (e.g., petroleum E&P), if it is not purchased, usually is controlled by permit, especially if environmentally sensitive areas are present. In the case of endangered or threatened species in an area where activity is proposed, a permit to "take" a protected species must be obtained if there is a chance that the organism may be affected. This is done under Sec. 7(a) of the Endangered Species Act.[2] These permits must be obtained from the proper authority before the planned activity begins. Some environmental permits can take a considerable time to obtain; therefore, this time period must be a part of the planning process. It is usually necessary, or at least expedient, to involve the regulatory agencies early in the planning process. In the long run, it saves time and trouble to work closely and cooperatively with them.

***Special Planning.*** Working in any sensitive area like the desert requires early and good environmental planning. As discussed in Sec. 3.2, attempts to minimize environmental impacts of proposed activities must be addressed early in the planning process. This is done through environmental planning and management, with the goals of minimizing adverse environmental impacts of development and maintaining or enhancing the ecological integrity and function of the natural desert system. Constant monitoring for impacts during all phases of petroleum operations (exploration, development, production, and site closure) and establishment of a good restoration plan are essential to mitigation of environmental impacts.

***Mitigations.*** A preactivity survey of the project area should be conducted early in the process by a qualified biologist. All signs of sensitive species and their habitat should be marked for exclusion from oil and gas activities. Establishing buffer zones around animal dens, burrows, and sensitive plants is a good idea.

### 3.5 Habitat Restoration

The objective of habitat restoration is to reproduce the original vegetation of a site. Restored vegetation must be self-sustaining and functional; that is, nutrient cycling and other ecosystem processes must be established. Ecosystem function depends on exacting site conditions and on soil microbiology; thus, soil preparation and introduction of key microorganisms can be as important as the introduction of plants.

It is important to realize that restoration is not landscaping: restoration requires establishment of ecosystem function, while much of landscaping practice is compensation for the missing ecosystem

**TABLE 3.1—SOURCE OF HABITAT DESTRUCTION AND PROGNOSIS FOR RESTORATION**

| Causative Agent | Effects: Primary | Effects: Secondary | Prognosis |
|---|---|---|---|
| Damage to above-ground vegetation | Loss of photosynthetic tissue and reproductive structures. | Reduction or loss of important soil microorganisms. | Good if disturbance is removed or reduced. |
| Damage to above-ground vegetation and below-ground components of vegetation | Loss of photosynthetic tissue and reproductive structures; removal of primary producers; loss of plant propagules. | Reduction or loss of important soil microorganisms; sealing of soil surface; modification of moisture, shade, and soil surface conditions; loss of organic input. | Good with restoration of soil-surface conditions, reintroduction of plant propagules and soil microbiota and with control of invasive plants. |
| Modification of natural hydrology | Alteration of moisture conditions in soil and subsoil; change in suitability of site for original vegetation. | Changes in decomposition and other microbial processes; changes in soil fauna. | Very poor for original vegetation; alternative vegetation type may be feasible. Much more important for riparian than upland vegetation. |
| Soil toxicity | Direct inhibition of plant growth. | Loss of soil microbiota. | Depends on toxin; may be poor without detoxification or capping with favorable soil. |
| Soil compaction | Loss of soil permeability to air and water; creation of adverse conditions for root growth. | Creation of inhospitable conditions for soil microbiota. | Very poor without decompaction. |

function. This brief overview of restoration defines the most important factors for successful restoration, with a reference to the mechanisms by which those factors operate.

**3.5.1 Sources of Destruction. Table 3.1** summarizes the kinds of activities that result in a need for habitat restoration, generally arranged from easier to more difficult restoration problems.

**3.5.2 Primary Biological Factors in Restoration.** Soil biota include all biological components of the soil taken together. Most soil biota are destroyed during mechanical disturbance, and the most important components are very slow to reinvade after destruction. All groups in the soil biota are reintroduced with properly salvaged topsoil, and some groups are available from commercial sources.

Root systems of wild plants almost always are colonized by beneficial fungi; the combination of root and fungus is called mycorrhiza. Mycorrhizal fungi extend into the soil where they are far more effective than unaided plant roots in absorbing nutrients. Mycorrhizal fungi have been shown to favor improved species diversity.[18,19] Forest trees often are ectomycorrhizal (usually in association with mushroom-forming fungi); while grasses, shrubs, and other plants usually are endomycorrhizal or arbuscular mycorrhizal (in association with microscopic soil-borne fungi). Plant communities that are largely arbuscular mycorrhizal usually can be served by one or a few species of mycorrhizal fungi. Mycorrhizal fungi are present in almost all native soils but are easily killed or rendered ineffective by the kinds of disturbances listed in Table 3.1.

Other key soil microorganisms include decomposers, which contribute to the breakdown and mineralization of plant and animal debris; pathogen antagonists, which are bacteria and fungi that counteract the effects of root disease; and plant-growth-promoting rhizobacteria, which are root-surface bacteria that greatly stimulate plant growth and health. The surface of natural soils is protected from erosion and sealing by a layer of algae and other primitive plants, collectively called the cryptogamic crust.[20]

Small and microscopic animals, the soil fauna, play a key role in soil processes. The smaller soil animals return microbial nutrients to the plant-available pool and maintain soil pore space. Medium to large soil animals, including some rodents, disperse mycorrhizal fungi.

**3.5.3 Restoration Alternatives.** ***Cost vs. Time.*** Often, the most costly steps in restoration are procedures that require heavy equipment used for decompaction or creation of favorable hydrological conditions. While some earthworks may be unavoidable, other aspects of the project may allow tradeoffs between cost and time. Examples include use of seed rather than container plants and use of commercial mycorrhizal inoculum rather than salvaged topsoil.

***Amelioration of Unfavorable Conditions.*** **Table 3.2** summarizes restoration methods over a range of site conditions. Obviously, the first step, and sometimes the only step necessary, is removal of the disturbance. There are many cases of mechanically impacted land that have been able to return steadily to the predisturbance state. This is most likely when no serious soil compaction has taken place and when surrounding areas can provide a steady natural input of native plants and microorganisms.

Soil compaction inhibits root growth and water infiltration[21] and is a primary concern in any project that involves construction or vehicle traffic. There is little point in going forward on such sites without first ripping or otherwise relieving compaction to the expected rooting depth. Sealing of the soil surface, which limits water and air exchange, can be prevented with surface heterogeneity, a variety of microtopography and debris that roughens and opens the soil. The regular pattern of microcatchments and ridges created by a land imprinter (discussed later) often is an effective solution.

Soil biota are best introduced with quality topsoil. Commercial preparations are available for mycorrhizal fungi and some other organisms and can be used when no topsoil is available.

**3.5.4 Project Design.** ***Habitat Restoration Industry.*** Typical arrangements in the restoration industry call for consultants, suppliers, and contractors who are independent of each other. Although general orientation can come from this and other technical works,[22,23] restoration is complex and draws on several technical disciplines. An experienced consultant can evaluate the site, prepare the restoration plan, negotiate the approach with regulatory agencies, and draw up technical specifications for the field work. Plant materials are provided by professional seed dealers and by native-plant nurseries. It may be necessary to contract for seeds and plants a year or more in advance. Plant materials usually are installed by a landscape contractor.

***Control of Hydrologic Conditions.*** If grading results in drainage of a wetland soil or redirection of groundwater flow, the site can become unable to support the original vegetation. Irrigating such sites in perpetuity is rarely practical, and resource agencies wisely decline these offers in environmental mitigation. More realistic solutions involve changing expectations for the project or relocating the restoration effort to a more suitable site. In some cases, it is possible to lower the grade within reach of the water table or spread out groundwater through construction of a partially buried weir.

**3.5.5 Method.** ***Topsoil Application.*** Topsoil salvage is a very effective step in habitat restoration.[20,24] Quality topsoil brings with it soil heterogeneity, propagules of native plants and soil organisms, and important soil properties (such as cation exchange capacity) that may be minimal in subgrade. Topsoil application often results in

| TABLE 3.2—RESTORATION PROCEDURES FOR MATRIX OF PROJECT CONDITIONS | | | | |
|---|---|---|---|---|
| | Land Type | | | |
| Project Objectives | Level and Well-Worked Ground | Level But Rough and Unworked or Compacted | Steep and Rough But Accessible | Very Steep or Inaccessible to Equipment |
| Long-term recovery, low to medium cost | Remove disturbance; imprint with seeds of facultative mycotrophs or scarify, broadcast seed, and cover. | Remove disturbance; decompact by ripping if necessary; imprint with seeds of facultative mycotrophs; or scarify, broadcast seed, and cover. | Remove disturbance; decompact by ripping; imprint with seeds of facultative mycotrophs or scarify, broadcast seed, and cover. | Remove disturbance; decompact by hand labor; apply seeds of facultative mycotrophs by hand or hydraulically. |
| Medium-term recovery, medium cost | Perform all steps given above; plant island areas with topsoil, surface heterogeneity, and salvage or container plants; weed control in early stages. | | | |
| Fast results, high cost | Perform all steps given above; plant cuttings, salvaged, and container plants; continuing weed control; phased planting and remedial planting in subsequent years. | | | |

greatly reduced weed problems and higher species diversity than other kinds of treatment.

The "topsoil" sold in the nursery trade is not the topsoil necessary for restoration. Topsoil for restoration must originate with healthy native vegetation of the type intended for the project. The suitable soil layer usually is defined by the presence of fine roots, which support most microbial activity. Stockpiling is best done when soil organisms are dormant, as in a dry or cold season. Dormant organisms survive handling much better than active organisms.

Topsoil collection usually is done with a bulldozer and should include rocks and plant material to provide plant propagules and soil heterogeneity. Store topsoil in conditions that minimize handling, preferably near the intended site of application. Avoid moving or turning the stockpile unnecessarily. Topsoil stockpiles should be more than 3 ft deep if the soil contains a large fraction of clay and as much as 6 ft deep if predominantly sand.

Soil stored up to 3 years retains considerable biological activity under most conditions. Mycorrhizal fungi decrease considerably with 2 years or more of storage, and supplemental inoculation is advisable at that point. Nevertheless, even very old topsoil has considerable value for its physical properties and a few long-lived organisms.

Care must be taken not to compact the topsoil during handling. Transport by scraper almost always leads to compaction, which can result in nearly complete loss of microbial activity. The soil should be carried by truck, then moved with a bulldozer before the next truck drives over the earlier load.

***Mycorrhizal Inoculation.*** Microbial inoculation is an important component of restoration projects that do not reapply recently salvaged topsoil. Mycorrhizal inoculation is least important in wetlands, but upland restorations fail without inoculation unless mycorrhizal fungi are already present in the soil. Mycorrhizal fungi are unlikely to be sufficient on sites that are occupied by annual weeds or are unvegetated.

Some plant species are more mycotrophic than others; that is, they depend more on mycorrhiza for uptake of soil nutrients. Facultative mycotrophs have an important role in habitat restoration because they help increase any introduced inoculum. Over time, facultative mycotrophs make the site suitable for the original native species, most of which are obligate mycotrophs. Facultative mycotrophs exist in most native floras and can be identified with fair confidence by noting which native species volunteer on roadsides or other seriously disturbed sites. Short-lived perennial grasses and composites are often facultative mycotrophs and are of some value in this context. Chenopods and crucifers usually do not become mycorrhizal at all[25] and generally have little value in restoration.

Other beneficial organisms, especially pathogen antagonists, are sometimes available from commercial sources. Some soil-binding algae are commercially available, or native cryptogamic crust can be spread over approximately 10 times the original area to hasten recovery of these important organisms.[26]

***Land Imprinting.*** A method with wide applicability, especially in arid climates, is land imprinting.[27] A toothed, tractor-drawn roller forms a pattern of depressions and ridges that trap water and allow exhaustion of soil air. This valuable soil-surface heterogeneity is of great benefit to the restoration project. Most commercial machines are fitted with seed-dispensing mechanisms, allowing the roller to replace seed-application methods that would otherwise require a separate costly step. After a rain, water infiltrates the impressions and provides capillary rewetting for several days. This is more likely to bring up slow-germinating native seeds than application methods that leave the seeds on the surface or only loosely incorporated. Some specially designed machines provide a means of injecting mycorrhizal inoculum as part of a one-pass operation.[19] This is not a trivial accomplishment because the inoculum must be delivered into the root zone of the new seedlings.

***Sources of Plants.*** There is accumulating evidence that the best plant material for each project originates close to the intended site.[28] The use of site-specific material requires advance arrangements with native-seed and container-plant suppliers, who may have to collect seed or start crops of plants 1 year or more in advance of the planting date. It is important to contact suppliers at an early stage to schedule production.

***Seed Introduction.*** Most facultative mycotrophs and other species with a functional role are easily introduced by seed. Late-successional species, usually the main objectives of habitat restoration, may be difficult to introduce by seed because of complex dormancy mechanisms. Seeding rates must take into account the amount of pure live seed (PLS) in the lot. This is calculated by multiplying the fractions representing purity and germination by the weight of the bulk seed. Seed count (number of seeds per unit weight) is required for each species to calculate the number of plants.

The final number of plants is influenced not only by purity and germination, but by survivorship after germination, a number that is very difficult to determine except by trial and error. Because it is difficult to anticipate survivorship for each species, most specifiers plan for the summed rates of all seeds, then allocate numbers in proportion to the desired final composition of the vegetation. A reasonable total seed rate might fall in the range of 10 to 50 lbm seed/acre, depending upon habitat, number of species, and preferences of the specifier. Another rule of thumb is to use 5,380 PLS/$ft^2$ (all species combined) if broadcasting or half that number if applying by seed drill or imprinter. Allocating density by number of seeds is decidedly superior to allocating by weight because of the great variability in seed size among plant species.

Hand broadcasting seed rarely gives satisfactory results unless the seeds are raked into the soil. Spot seeding is an efficient use of scarce seed; it is applied under shrubs, between rocks, or in other microsites that provide good prospects of seedling success. Hydraulic seeding, "hydroseeding," is popular in landscaping and has become a significant method in restoration. It is more expensive and generally less effective than imprinting or drill seeding, but is often used because of its ready availability and because project contractors are familiar with its use. Seeds can be applied during the dry or dormant season if protected from seed-eating animals. Large seeds, including acorns, readily fall prey to animals and usually must be protected.

| TABLE 3.3—QUALITY-CONTROL CONSIDERATIONS | |
|---|---|
| Topsoil salvage | Correct identification of topsoil; collection during dormant season; minimal transport and handling; storage for shortest possible time; no compaction. |
| Seeds | PLS certified by supplier on basis of lab test; seeds originate from similar habitat near job site. |
| Seed application | Uniformity of coverage; no damage to seeds; no displacement of seeds by rain or irrigation; good imprints if using land imprinting. |
| Container plants | Hardened against frost; not root bound or spiraling in containers; root systems fill containers; tops not oversized compared with roots; no serious pests or diseases; ready for planting by agreed date; not broken or desiccated during transport. |
| Plant introduction | Careful handling of plants; correct siting of plants; firming of soil around roots. |
| Microbial inoculum | Certification by supplier; recommendation of supplier by recognized authorities. |
| Microbial inoculation of plants | Convincing statement of facilities and personnel; certification by independent laboratory; sample root preparations provided. |
| Inoculation in the field | Evidence of inoculum in the root zone after application; careful handling and storage of inoculum. |
| Reporting to resource agencies | Reports meet agency's formatting and content criteria; reports accurately show success or remedial needs of project . |

***Containers.*** Containers provide a means of introducing plants that are difficult to introduce from seed and offer a way to start the project with some larger plants. The cost of container plants can be high, especially with close spacing. A useful strategy is to approximate the natural spacing of full-size plants.

Some plant species that are difficult to grow from seed can be brought to the site by animals over a period of years. Bird perches and clumped vegetation, which attracts rodents, can sometimes replace or supplement container plants.

Small containers are less expensive and often result in better plants than large containers; however, they are subject to rapid drying if not properly handled. Narrow, deep containers give the best performance in arid conditions.[29] A disadvantage is that small containers are closely-spaced during production. Native shrubs are often brittle, and closely spaced containers lead to breakage during handling. The ideal container size for each species provides sufficient horizontal space in the nursery while allowing a relatively deep root system. Small container plants, especially oaks, may require protection from deer, rabbits, and rodents. Various protective devices are available from forestry dealers. Some protection is afforded by planting very large land areas because some herbivores are unwilling to stray far from pre-existing vegetative cover.

Direct use of cuttings at the site works well with willows and some other riparian and wetland species. Cuttings generally are not a reliable way to plant upland sites. Salvage of wild plants can provide inexpensive material if the plants are moved immediately to the new site. If repotting and storage are required, salvage may prove more costly than purchasing new plants. The best candidates for salvage include wetland and riparian species, succulents, and perennial grasses. A peculiarity of plant salvage is that the plant must be oriented at the new site in its original compass direction to prevent sun scald. It is important not to destroy an off-site location for plant or topsoil salvage. Such a practice contradicts the intent of restoration and, in many cases, is illegal.

***Planting Strategies.*** It is very important that planting be timed with natural seasonal cycles. Soil moisture, desiccating influences, and temperature are all seasonal factors to which plants are strongly tuned. Some species fail to produce roots or otherwise adapt to the new site if planted out of season. The most favorable season for planting usually is the season of greatest root growth, often late fall, unless the plants have not been hardened for the coming winter conditions. In that case, the best time is the earliest possible spring planting date. In the tropics, the rainy season is preferred to the dry season.

Ideally, planting is carried out in a series of phases. A quick cover of functional species, including facultative mycotrophs, is introduced by seed during the first phase, with late successional species introduced later. In practice, few projects have the luxury of years of phased planting. "Island planting" is establishment of small, high-input "islands" within a larger, mostly untreated matrix. A good intermediate strategy is imprinting the whole area with facultative mycotrophs, then establishing smaller patches that include topsoil and container plants or late successional species. The islands provide a source of plants that spread outward over time.

***Maintenance.*** Most restoration projects require a certain amount of post-planting care, including weed control. Ruderals (a group that includes weeds) realize their very high potential growth rates in favorable conditions.[30] Because of weed competition, fertilization is a grave error in any but the most nutrient-deficient conditions (i.e., coarse subgrade material derived from granitic rocks). Most fertilizer formulations have the potential to prevent plants from becoming mycorrhizal, often leading to serious imbalance between top and root systems. Low-phosphorus, slow-release formulations are most likely to be compatible with restoration objectives.

Irrigation should be avoided whenever possible, in part because it can be the highest single cost of the project. Erosion, disease problems, and increased weed growth are predictable outcomes of heavy irrigation. Shallow irrigation prevents root systems from extending into deeper soil, which is necessary for self-sufficiency. Irrigation does the least damage in riparian or wetland situations; there the only real disadvantage is cost. Bainbridge *et al.*[29] considered irrigation to be nearly indispensable in desert plantings of container stock and described inexpensive methods suited to small or noncommercial projects.

**3.5.6 Evaluation Techniques.** Restoration sites need frequent evaluation to warn restoration personnel of the need for remedial action and to ensure resource agencies that a real ecosystem has been created. Evaluation for the second purpose usually requires formalized quantitative criteria. Previously used criteria include the survival and height or stem diameter of container plants and the percent cover and other measures of above-ground vegetation. These criteria can unintentionally force use of high-input maintenance methods, seriously delaying establishment of ecosystem function.

In recent years, restoration monitoring has become more sophisticated, with increasing attempts to evaluate ecosystem function. These methods rely on the presence of insect indicators[31,32] or such below-ground processes as root proliferation and mycorrhizal fungi. The degree of soil aggregation is a promising measure that integrates root and microbial processes.[33]

**3.5.7 Quality Control.** Most failures result from profitable "shortcuts" routinely taken by each segment of the restoration industry. Examples include nearly verbatim reuse of restoration plans by consultants and unloading of outdated seed or overgrown plants by suppliers. Landscape contractors are perhaps the most likely to use unwise shortcuts. Specifications must anticipate and exclude every possible corner-cutting strategy, especially when policy requires acceptance of the lowest bid. A substantial fraction of the payment should be retained until the work meets specifications.

Quality control begins with careful choice of the restoration consultant. The best way to find a good consultant is referral, preferably by several sources in the restoration industry. Interlocking business affiliations are very common and do not result in sound advice to the client. Ask to see examples of the consultant's work or obtain examples from public documents. Good specifications use site evaluation to tailor specifications to the conditions at the project site. Ask how the consultant plans to deal with soil compaction, lack of soil microorganisms, and other matters emphasized in this section. The specifications should provide standards for acceptance of seed, container plants, and installation work. **Table 3.3** provides a brief review of the factors that are important to the quality of each segment of the restoration project.

## References


1. Natl. Environmental Policy Act (NEPA), 42 USC 4321–4347 (1969).
2. Endangered Species Act (ESA), 7 USC 136; 16 USC 460 *et seq.* (1973).
3. Clean Water Act (CWA) 33 USC s/s 121 *et seq.* (1977).
4. Comprehensive Environmental Response, Compensation, and Liability Act (CERCLA or Superfund) 42 USC s/s 9601 et seq. (1980).
5. Lovejoy, T.E.: "People and Biodiversity," *Nature Conservancy* (January/February 1994) 29.
6. Lindstedt-Siva, J. *et al.*: "Engineering for Development in Environmentally Sensitive Areas: Oil Operations in a Rain Forest," *Engineering within Ecological Constraints*, P.C. Schulze (ed.), Natl. Academy Press, Washington, DC (1996) 141–61.
7. Loll, S. and Steinhauer, S.: "Coles Levee and North Dauphin Island Win Arco Environmental Award in AOGC 'Sweep'," *Environmental News* (1993) **12,** No. 3, 1.
8. Caughly, M. and Sinclair, R.E.: *Wildlife Ecology and Management*, Blackwell Scientific Publications, Boston, Massachusetts (1994) 334.
9. Wolfe, M.H. and Gelbs, T.R.: "Making the Most of Endangered Species Compliance: Development of the Coles Levee Ecosystem Preserve," paper SPE 25969 presented at the 1993 SPE/EPA Exploration and Production Environmental Conference, San Antonio, Texas, 7–10 March.
10. Cooperrider, A.: "Reintegrating Humans and Nature," *Landscape Linkages and Biodiversity*, W.E. Hudson (ed.), Island Press, Washington, DC (1991) 194.
11. Soule, M.E.: "Theory and Strategy," *Landscape Linkages and Biodiversity*, W.E. Hudson (ed.), Island Press, Washington, DC (1991) 196.
12. *Wetlands: Characteristics and Boundaries*, Natl. Academy Press, Washington, DC (1995) 308.
13. *Corps of Engineers Wetlands Delineation Manual*, Natl. Technical Information Service, U.S. Dept. of Commerce, Vicksburg, Mississippi (1987).
14. Marine Mammals Protection Act, 16 USC 1361 *et seq.*, 1401–1407, 1538, 4107 (1972).
15. Oil Pollution Act (OPA), 33 USC 2702–2761 (1990).
16. Food Security Act, 16 USC 3801–3862 (1985).
17. California Desert Protection Act, 16 USC 410 aaa (1994).
18. Grange, A.C., Brown, V.K., and Farmer, L.M.: "A Test of Mycorrhizal Benefit in an Early Successional Plant Community," *New Phytologist* (1990) **115,** No. 1, 85.
19. St. John, T.V.: "Mycorrhizal Inoculation with a Modified Land Imprinter," *Restoration & Management Notes* (1996) **14,** No. 1, 84.
20. St. John, T.V.: "The Importance of Mycorrhizal Fungi and Other Biological Beneficial Organisms in Biodiversity Projects," U.S. Dept. of Agriculture (USDA) Forest Service General Technical Report RM-22199-105, *Proc.*, T.D. Landis (ed.), Western Forest Nursery Assn., Davis, California (1993) 191.
21. Alexander, E.B. and Poff, R.: *Soil Disturbance and Compaction in Wild Land Management*, Earth Resources Monograph, USDA Forest Service, Pacific Southwest Region, San Francisco (1985) 80.
22. Gray, D.H. and Leiser, A.T.: *Biotechnical Slope Protection and Erosion Control*, Van Nostrand Reinhold Co. Inc., New York City (1982).
23. Schiechtl, H.M.: *Bioengineering for Land Reclamation and Conservation*, U. of Alberta Press, Edmonton, Alta., Canada (1980) 404.
24. Hargis, N.E. and Redente, E.F.: "Soil Handling for Surface Mine Reclamation," *J. Soil Water Conservation* (June 1984) **39,** 300.
25. Gerdemann, J.W.: "Vesicular-Arbuscular Mycorrhizae and Plant Growth," *Ann. Rev. Phytopathology* (December 1968) **6,** 397.
26. St. Clair, L.L., Johansen, J.R., and Webb, B.L.: "Rapid Stabilization of Fire-Disturbed Sites Using a Soil Crust Slurry: Inoculation Studies," *Reclamation, Revegetation, Restoration* (1986) **4,** No. 4, 261.
27. Dixon, R.M. and Carr, A.B.: "Land Imprinting for Low Cost Revegetation," *Erosion Control* (April 1994) **1,** 38.
28. Millar, C.I. and Libby, W.J.: "Disneyland or Native Ecosystem: Genetics and the Restorationist," *Restoration & Management Notes* (August 1989) **7,** 18.
29. Bainbridge, D.A., Sorensen, N., and Virginia, R.A.: "Revegetating Desert Plant Communities," General Technical Report RM-221, USDA Forest Service, *Proc.*, Western Forest Nursery Assn., Davis, California (1993) 21.
30. St. John, T.V.: "Soil Disturbance and the Mineral Nutrition of Native Plants," *Proc.*, Soc. for Ecological Restoration and Management Second Native Plant Revegetation Symposium, San Diego, California (1988) 34.
31. Anderson, A.N.: "Ants as Indicators of Restoration Success at a Uranium Mine in Tropical Australia," *Restoration Ecology* (September 1993) **1,** 156.
32. Williams, K.S.: "Use of Terrestrial Arthropods to Evaluate Restored Riparian Woodlands," *Restoration Ecology* (November 1993) **1,** 107.
33. Bethlenfalvay, G.J. and Schüepp, H.: "Arbuscular Mycorrhizae and Agrosystem Stability," *Impact of Arbuscular Mycorrhizas on Sustainable Agriculture and Natural Ecosystems,* S. Gianinazzi and H. Schüepp (eds.), Birkhäuser-Verlag, Basel, Switzerland (1994) 82.


## General References


Barker, G.W. and Steele, E.J.: "Exploration and Production Operations in an Environmentally Sensitive Area," paper SPE 28737 presented at the 1994 SPE International Petroleum Conference and Exhibition in Mexico, Veracruz, Mexico, 10–13 October.

*Ecological Aspects of Development in the Humid Tropics,* Natl. Academy Press, Washington, DC (1982).

"Environmental Guidelines for Worldwide Geophysical Operations," Intl. Assn. Geophysical Contractors, Houston (1993) 42.

Kendeigh, S.C.: *Animal Ecology*, Prentice-Hall Inc., Englewood Cliffs, New Jersey (1961) 468.

Lewis, S.: *The Rainforest Book: How You Can Save the World's Rainforests*, Living Plant Press, Los Angeles, California (1990).

"Loss of Biological Diversity: A Global Crisis Requiring International Solutions," Publication NSB-89-171, Natl. Science Foundation, Washington, DC (1989).

Mitsch, W.J. and Gosselink, J.G.: *Wetlands*, second edition, Van Nostrand Reinhold, New York City (1993) 722.

Niering, W.E.: *Wetlands*, Audubon Soc. Nature Guide, Alfred A. Knopf, New York City (1985) 638.

*Oil Exploration in the Tropics: Guidelines for Environmental Protection*, Intl. Union for Conservation of Nature, Gland, Switzerland (1991) 30.

*Oil Industry Operating Guidelines for Tropical Rainforests*, E&P Forum, London (1991) 18.

Raven, P.H.: "Defining Biodiversity," *Nature Conservancy* (January/February 1994) 10.

*Research Priorities in Tropical Biology*, Natl. Academy Press, Washington, DC (1980).

Schmidet-Nielsen, K.: *Desert Animals: Physiological Problems of Heat and Water*, Oxford U. Press, New York City (1964) 277.

"Stemming the Tide of Coastal Fish Habitat Loss," *Proc.*, Natl. Coalition for Marine Conservation Symposium on Conservation of Coastal Fish Habitat, Baltimore, Maryland (March 1991) 258.

Thurman, H.V.: *Essentials of Oceanography*, second edition, Merrill Publishing Co., Columbus, Ohio (1987) 370.

Wilson, E.O.: *Biodiversity*, Natl. Academy Press, Washington, DC (1988).


## SI Metric Conversion Factors

| | | | |
|---|---|---|---|
| acre | × 4.046 856 | E − 01 | = ha |
| ft | × 3.048* | E − 01 | = m |
| $ft^2$ | × 9.290 304* | E − 02 | = $m^2$ |
| °F | (°F − 32)/1.8 | | = °C |
| lbm | × 4.535 924 | E − 01 | = kg |

*Conversion factor is exact.

# Chapter 4
# Impact Analysis

**Jesse D. Frederick,** WZI Inc.
**Fred Simon,** Consultant
**Michael T. Cotton,** Exxon Co. Intl.
**Rodney E. Horton,** Oklahoma Dept. of Environmental Quality
**Robin S. Smith,** Munro Garrett Intl.

## 4.1 Overview

It has often been stated that a U.S. Natl. Environmental Policy Act[1] (NEPA) would not have been enacted in 1969 were it not for the Apollo Program. The space experience convinced U.S. citizens that the Earth is a closed ecosystem (sometimes called "spaceship Earth") that requires stewardship. Our increased mobility allowed us to view and perhaps realize the fragility of nature. Regardless of the cause, a gradual shift occurred in concern for the environment that resulted in broad-reaching environmental policies and an industry shift toward stewardship and sound environmental practices. Although conservation existed in the past [the British Parliament tortured and hanged a man for selling "sea coal" (peat) in London during the 13th Century[2]], diverse objectives precluded any comprehensive policy making.

The first attempt at federal legislation in the U.S. took place in 1959. It sought to "declare a national policy on conservation, development, and utilization of natural resources ... to meet human, economic, and national defense requirements, including recreational, wildlife, scenic and scientific values, and the enhancement of the national heritage for future generations."[1] This bill was endorsed by the Democratic party for the 1960 election platform but was abandoned by President Kennedy. The Vietnam War and the Great Society consumed President Johnson's time and energy. Eventually the U.S. Congress took the lead and, in 1968, the first pass at NEPA was proposed. President Nixon signed the bill into law in 1970,[3] four presidents and 2 decades after the first attempt. Other developed nations followed similar paths to implementation of environmental-management programs, and the World Bank developed its own standard for environmental-impact analysis. Ultimately, the Intl. Standards Organization (ISO) developed the ISO 14000 series to govern environmental-management practices.[4]

The following are part of NEPA.[1]

1. Declares that the policy of the U.S. federal government is to use all practicable means to create and maintain conditions under which man and nature can exist in productive harmony and fulfill the social, economic, and other requirements of present and future generations of Americans.
2. Establishes within the Executive Office of the U.S. President a full-time, three-man Council on Environmental Quality (CEQ).
3. Requires the President to submit environmental-quality reports to Congress annually.
4. Defines the duties of the CEQ: (a) to assist the President in the preparation of the annual report, (b) to develop and recommend to the President national policies that promote environmental quality, and (c) to accumulate necessary data for continuing analysis of changes or trends in the national environment.

The specific wording given here regarding the requirement for environmental-impact statements was a last-minute provision.[3]

"All agencies of the federal government shall include in every recommendation or report on proposals for legislation and other major federal actions significantly affecting the quality of the human environment, a detailed statement by the responsible official on:

(i) The environmental impact of the proposed action,

(ii) Any adverse environmental effects which cannot be avoided should the proposal be implemented,

(iii) Alternatives to the proposed action,

(iv) The relationship between local short-term uses of man's environment and the maintenance and enhancement of long-term productivity, and

(v) Any irreversible and irretrievable commitments of resources which would be involved in the proposed action should it be implemented.

Prior to making any detailed statement, the responsible federal official shall consult with and obtain comments of any federal agency which has jurisdiction or social expertise... ."

This section, as applied, is the basis for the U.S. government's use of NEPA reviews for projects involving federal agencies.[1]

From these humble beginnings, more rigorous review procedures have been established, such as the California Environmental Quality Act (CEQA), which is widely considered the most thorough review process in the U.S. All environmental-impact analyses (EIA's) have common elements that stem from the principles stated in NEPA.

In addition to the previously mentioned regulatory-mandated impact analysis, industry and commercial standards and practices of the oil and gas industry dictate that a presale review of environmental liabilities be performed to establish a body of information upon which potential buyers may rely pursuant to their "due-diligence" review and to identify issues that may be raised by the buyers after the sale of the property.

The following sections of this chapter address the current range of tools for assessing and managing environmental issues in E&P settings and provide a focused discussion regarding practical considerations in the application of these tools. The final sections of this chapter are devoted to environmental-impact studies.

## 4.2 Tools for Environmental-Management-Based Assessments in E&P

Over the last several decades, environmental-management tools have been developed that fit a variety of informational requirements. Many of the techniques used in the construction of these tools and the related vernacular are similar. Misuse of these terms is not uncommon, resulting in needless confusion. The following relate to the transfer of oilfield property in general.

1. Oilfield properties often, if not always, have environmental liabilities that result from activities inherent in the exploration, production, and processing of oil and gas.

2. Liability for practices and activities during the period of ownership remain with any substantial oil company or entity in perpetuity because of the deep-pocket theory, regardless of prior or subsequent ownership.

3. A property transfer proceeds more smoothly, hence, more quickly, if the needs of all stakeholders are anticipated and adequately addressed in advance of the due-diligence review. Stakeholders include sellers, buyers, lenders, the public, and governmental agencies that grant entitlements.

4. Disclosure of the specific condition of the property at the time of the transaction is prudent because of risk management as well as because of specific federal, state, and common-law requirements.

5. Misuse of environmental-management tools and terms (such as site assessment and due diligence) can lead to unnecessary confusion or delay and, in extreme cases, termination of the transaction.

The review of acquisitions and divestitures is one of the most important responsibilities of today's environmental managers and environmental counsel. Most corporate entities and lending institutions have developed policies and procedures to cover property-transfer activities. The following discussion delineates the similarities and differences between due-diligence reviews for acquisitions and disclosures for divestitures, environmental audits, site assessments, and EIA's—e.g., CEQA and NEPA. Contrary to popular usage, the terms and the specific tools that they represent are not interchangeable. Each addresses a unique set of stakeholder expectations, and the related report contains unique contents.

**4.2.1 Due Diligence.** Relative to the sale of real property, a due-diligence review is the act of exercising reasonable care in determining whether an acquisition should be made and, if so, under what conditions (e.g., representations and warranties, indemnifications, or reserve funds for cleanup). A due-diligence review addresses the expectations of the purchaser, the purchaser's partners, stockholders, and lenders.

The term due diligence has recently been interpreted to represent the process of assessing environmental issues involved in a property transfer. The term actually refers to the review process that serves to define whether a property seller or an entity being examined has been duly diligent in meeting obligations stemming from regulatory and legal requirements that define and allow their operations to exist or take place. This extremely broad review usually is done to satisfy special inquiries and presale screening. A prudent purchaser normally requires a seller to warrant that he/she has been duly diligent in meeting all obligations and requirements set forth in regulation and law and that there are no undisclosed pending actions or other encumbrances that could result in liability for the purchaser.

The complete due-diligence review typically includes an analysis and discussion regarding many, if not all, of the following items: site assessments, audits, related benchmark analyses, environmental-management-program compliance plans, permit certifications, environmental-impact studies, decrees, and past or pending enforcement actions. Such information creates a record for the purchaser that he or she, too, was diligent in protecting shareholder, investor, and partner interests.

**4.2.2. Disclosure.** Disclosures are representations made by the seller to a buyer regarding conditions or situations that may result in post-sale liabilities. These disclosures take the form of representations and warranties, certifications, written responses, studies, reports, and interviews. They are the basis of future environmental claims and should address residual liability.

In the case of a property sale, environmental disclosures are made by the seller to the buyer and are presumed to be an affirmation of fact. Inaccurate representations or warranties may provide the buyer with a contractual basis for suing the divesting party, their partners, and contractors regarding environmental issues. Some disclosures concerning the presence, release, or threatened release of hazardous substances are required by governmental mandates; in the U.S., federal and state laws [e.g., the Comprehensive Environmental Response, Compensation and Liability Act (CERCLA),[5] the Resource Conservation and Recovery Act (RCRA),[6] and state health and safety codes] define these requirements. Some disclosures can be of such a sensitive nature that material harm may be incurred from undue communication with enforcement agencies or the divulgence of business confidential information; in such situations, a confidentiality agreement usually is required before the disclosures.

Generally, a buyer does not rely solely on the disclosures made by the seller. A seller should request that the buyer warrant that it was provided adequate opportunity to investigate the property and records related to its condition before transfer and is satisfied with its condition. Disclosures should be limited to what is actually known. While an "as-is" provision is sometimes used, it probably will not relieve residual liability for cleanup costs or fines for past activities imposed by such regulatory agencies as the U.S. Environmental Protection Agency; however, a buyer's risk can be covered by an indemnification from the seller to cover the residual liabilities. Without the indemnification, the as-is provision poses an exceptional liability for an uneducated buyer.

**4.2.3 Environmental Audits.** An environmental audit is a review of environmental programs and activities to determine compliance with requirements stemming from licenses, contracts, rules, regulations, laws, and/or company policy. Audits frequently are used to assist a company in organizing and managing effective environmental programs, to meet U.S. Securities and Exchange Commission (SEC) disclosure requirements, and for legislative and regulatory strategy development.

Some believe that all audits should be exhaustive in nature, with large teams, extensive checklists, and record inspections. Others believe that specific audits should focus on management systems and major flaws in compliance activities, relying on the fact that once a facility understands its basic environmental mission, an audit merely crosschecks that the systems are performing properly.

Standards for the performance of environmental, health, and safety audits have been developed by the Environmental Auditing Roundtable.[7] These general standards can be applied to a wide variety of complex legal and regulatory requirements, depending on the circumstances of the individual audit.

1. The auditors must have adequate qualifications, technical knowledge, training, and proficiency.

2. Professional care must be exercised to ensure accuracy, consistency, and objectivity in the audit performance.

3. Auditors must be objective and independent, free of conflict of interest, and not subject to internal or external pressure.

4. Audits should have clear and explicit objectives based on systematic plans and procedures.

5. Quality-control and -assurance measures must be incorporated.

6. A formal report, consistent with the audit scope and objectives should be prepared.

**4.2.4 Site Assessments.** SEC *Regulation S-K*[8] provides reporting requirements for Form 10-K, Items 101 (description of business), 103, (Legal Proceedings), and 303 (Management Discussion and Analysis). In addition, the U.S. Financial Accounting Standards Board (FASB) provides guidance for reporting contingent liability in *FASB Statement No. 5*.[9]

A site assessment is a site-specific investigation, either intrusive or more limited in scope, used to delineate the extent of contamination and related environmental liabilities. Site assessments serve specific purposes that are determined by the initiating party (the buyer for inclusion in a due-diligence review, the seller for utilization in disclosure, or the owner for release cleanup activities).

There are three levels of site assessments, commonly referred to as Phases 1 through 3. Phase 1 is a nonintrusive investigation, usually conducted to determine whether there is a need for Phase 2 sampling and suggests locations for such sampling. Phase 2 describes the intrusive (sampling) investigation, which typically takes place after Phase 1 is complete. A Phase 3 assessment occurs when a contaminated area has been identified that requires remedial activities. Phase 3 completes the investigation and details the remediation plans. Phase 3 is rarely performed in relation to a property transfer.

American Soc. for Testing and Materials (ASTM) *E-1527-97*,[10] defines the practices necessary to complete a Phase 1 site assessment. ASTM *E-1527-97* suggested practices have to be weighed vs. practical considerations during implementation. It is typically referenced in corporate guidelines as a starting point to establish potential liability and plans for further study to ensure reasonable care to protect the shareholder. As this standard shows, the Phase 1 site assessment focuses on contamination-related liability and does not concern itself with any other potential source of liability, such as violations, fines, and consent decrees.

**4.2.5 EIA's.** An EIA is the analysis of the environmental impacts related to a specific action being proposed, generally for the purpose of soliciting public participation in the project design and selection of mitigation measures. Impact assessments most often address the information that the public and governmental entities require to grant discretionary permits, licenses, and other entitlements.

In the U.S., federal agencies are required under NEPA to do the following.

1. Develop methods and procedures that ensure that environmental resources are given appropriate consideration in decision-making, along with economic and technical consideration.
2. Use a systematic and interdisciplinary approach.
3. Make diligent efforts to involve the public and provide for public disclosure of proposed actions on public land, including analysis of the proposed action and any alternatives involved in the decision-making process.

Fulfillment of the NEPA process is accomplished through availability and participation in the construction of documents, such as environmental assessments, environmental-impact statements (EIS's), records of decisions, and categorical exclusion reviews. Each of these reviews and their documentation has specific requirements in accordance with law, rule, and regulation.

## 4.3 Practical Considerations Related to Use of Assessment Tools for E&P Property Transfers

**4.3.1 Facility-Related Transfers vs. Oil Fields.** The types of assets involved in a transfer can differ significantly. They include facility sales (typically covering a relatively small area); small property sales; and fee-simple, large-interest sales among others. The size and extent of some sales are so great that they encompass several types of possible sale transactions. Maintenance areas, administration buildings, gas-handling facilities, and utility facilities might be limited in size and easily subjected to a thorough, classic, nonintrusive Phase 1 study as detailed in ASTM *E-1527-97*; however, any Phase 1 assessment performed on oilfield assets usually needs to be tempered by practical considerations because of property size.

Potential buyers typically are given a defined period to accomplish due-diligence reviews. During this time, the review team has to look at the available information, usually in a data room, and determine the extent of the information needed to fulfill any corporate directive and to establish the extent of environmentally related liability. Once the environmental team is allowed to enter the data room, they have to act quickly to determine compliance requirements, operations, CERCLA and RCRA issues, and any ongoing remediation. At this time, the team will probably initiate a review with the regulatory agencies to establish independent verification of enforcement and remediation actions and to establish the body of required permits, licenses, and approvals.

The team also contacts and interviews appropriate and knowledgeable personnel to fill in any gaps that they feel exist and to define the body of information better. Once this is done, the team performs a visual evaluation (including photographs) of the wells, sites, or items in question. Also, the team looks for and investigates undocumented problem areas. After this is completed, the team should write a report defining the type and extent of liability and provide recommendations related to the sale and post-sale treatment of the liabilities. The negotiating team frequently uses this report on environmental liabilities to set a price or to make and substantiate a decision to abandon the sale.

All work must be accomplished in the fixed time allotted in the presale agreements. As such, the team has to weigh time vs. thoroughness carefully and has to rely on experience when electing to forego a more complete review. Concerns related to insufficient information and poorly defined liabilities usually result in requests by the review team for broader indemnification or a higher premium attached to the value for acquired risk. The risk valuation can be used in the price negotiations.

The adequacy of the information in a data room determines the extent of field activity and other related investigations. Review teams will be larger and more intrusive if the buyers view the data as insufficient for satisfying their due-diligence requirements.

**4.3.2 What Buyers Expect To See in a Data Room.** Once the data rooms are opened, potential buyers will seek information that helps to define and place a value on their risk stemming from the acquisition. Potential buyers are looking for a complete and cataloged set of documents for each operating piece of equipment, including all permits, licenses, and approvals.

They also want to see any previous site assessments, consent decrees, and documents that address prior, ongoing, or potential enforcement actions. This information can be audits, benchmark studies, environmental-management programs, and compliance plans.

The seller frequently takes the position that the data (as provided), access to the facility, and all records provided to the buyer after the bid acceptance are adequate to inform a prudent buyer. However, from a practical standpoint, issues that surface during the due-diligence review that were not disclosed in the data room have a tendency to delay the closing of property transfers, decrease the price received, or perhaps create a situation in which the sale is not consummated because of time pressures.

## 4.4 EIA and Considerations Related to E&P

The EIA, often called an EIS, should provide the project proponent, responsible agencies, and the affected public a chance to assess the project scientifically. The process should promote sustainable development by allowing growth to occur while tempering its adverse impact. The analysis process can be either a state- or federal (and in some rare instances local) -mandated review and typically involves a period of study followed by a report that is submitted to the public for comment. The EIA takes into account the information found in Chaps. 2 through 8.

The EIA represents an attempt on the part of responsible agencies to deal with the social contract that exists between industry, government, and society in general. The EIA reflects the concept that industry is allowed to act only with permission from the public; it is in this vein that the analysis process attempts to assist all affected parties. Concepts critical to the process are that the process should involve an interdisciplinary approach, that the process should result in integrated management of the project's impacts, and that the standards of significance for impacts should be based on practical constraints as opposed to unachievable no-growth-type thresholds.

An EIA has four basic purposes that are best reflected in the CEQA guidelines.[11]

1. Inform public decision-makers and the public about potential significant environmental impacts of proposed activities.
2. Identify ways to avoid or significantly reduce these significant environmental impacts.
3. Prevent significant, avoidable damage to the environment by requiring project changes, alternatives, or mitigation measures

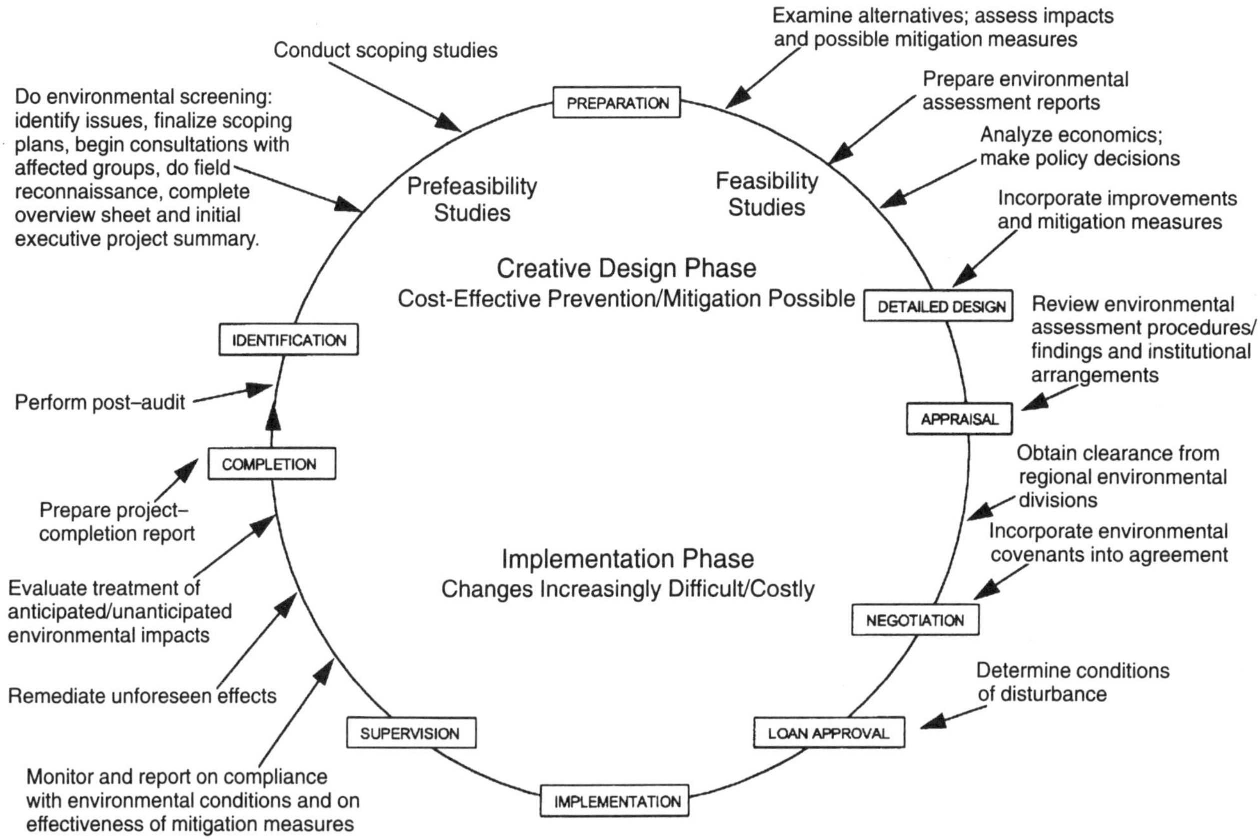


**Fig. 4.1—Environmental assessment and the project cycle.**

that are technically, legally, economically, socially, and environmentally feasible.

4. Disclose to the public the reasons why a public agency should or should not approve a project, particularly if the project will cause significant environmental impacts.

For simplicity, CEQA is cited throughout this chapter as an excellent example of assessment laws and regulations used by many states in the U.S.

Before initiating an EIA, the project proponent should review the general project with the various permitting agencies to determine who will take the lead in the process; what exemptions, if any, exist; how the review will proceed (flow chart); and whether any overlapping reviews exist that may require resolution.

The impact analysis focuses on four levels of assessment.

1. Pre-existing condition of the affected area.
2. Project-related impacts.
3. Project mitigations.
4. Residual impacts after project mitigation.

In addition to these levels of assessment, a monitoring program is put into effect to ensure that the project proceeds in a manner that both reflects the EIS and follows sound environmental practices.

The EIA addresses the overall effect of a project on the environment. The analysis involves a concerted effort on the part of numerous responsible agencies having primary oversight over particular environmental or other special programs. Each potential area of impact (discussed in Sec. 4.4.1) is assessed on the basis of specific forms of disturbance to the pre-existing/natural setting, transport of the disturbing agents, and additional synergistic impacts resulting from the combination of the disturbing agents. Chaps. 3 and 5 through 7 provide more detailed discussions. A "no-build" alternative is frequently reviewed as a basis for reference.

**4.4.1 Environmental Impacts.** ***Air.*** Air impacts result from emissions from a variety of point sources. These emissions can take the form of criteria pollutants [volatile organic compounds (VOC's), nitrogen oxides ($NO_x$), sulfur oxides ($SO_x$), carbon monoxide (CO), and particulate matter with a < 10-mm diameter ($PM_{10}$)] and those hazardous air pollutants listed under Title III of the U.S. Clean Air Act Amendments[12] or other regionally related, toxicity-related guidelines. The analysis typically involves incorporation of all data that have been submitted to the agencies for permit review as well as additional emissions and risk modeling.

***Water.*** Water-related impacts stem from discharges to surface water and/or groundwater as well as storm-water-runoff-related discharges. The emissions can take the form of inorganic or organic contamination, heavy-metal salts, or suspended solids. The analysis typically involves incorporation of all permit-related data as well as studies on migration and defusion.

***Soil.*** Soil impacts are related to construction activities, spills that become perched or result in transport to the groundwater, or both. Contamination can be from inorganic or organic chemicals, heavy-metal salts, or suspended solids or can take the form of road-construction materials, construction-equipment fluids, drilling muds, or defoliants. The analysis typically involves a nonintrusive Phase 1 site assessment and, if needed, an intrusive Phase 2 site assessment. Phase 2 site assessments typically are required for projects proposed for previously disturbed settings.

***Biological.*** Habitat impacts are related to activities that have the potential to disturb the flora and fauna in their natural setting. These disturbances can result from activities related to preconstruction, construction, operating, and demolition. Habitat impacts can be light, noise, soil disturbances, water and food contamination, and any impediment to natural movement and reproduction. The analysis typically involves incorporation of any laws related to endangered species, such as the U.S. Endangered Species Act,[13] or state programs.

***Archeological.*** Archeological impacts typically are the result of construction disturbances. The impacts can take the form of excavations related to road building and site preparation as well as foundation excavations that disturb sites containing artifacts of historical

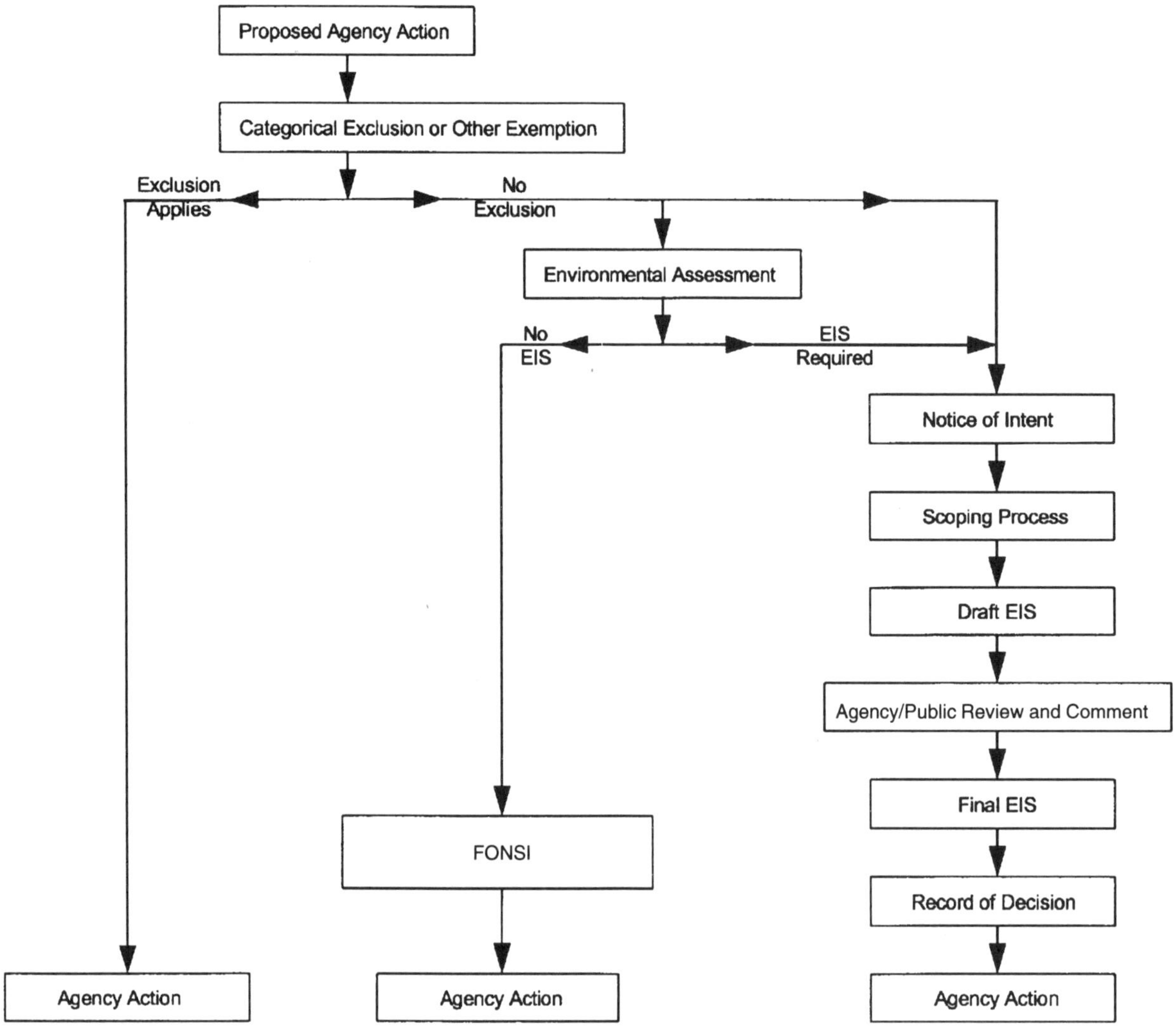


**Fig. 4.2—Overview of NEPA environmental review.**

or archeological value. The analysis typically involves historical reviews and field surveys.

***Societal.*** Societal impacts are those that stem from modifications of social structures, employment structures, property values, support services, recreation, and transportation. The analysis typically involves both socioeconomic and transportation studies.

***Health and Safety.*** Health and safety impacts involve both employees and the surrounding community and stem from preconstruction-, constructional-, operational-, and abandonment-related activities. These impacts can be airborne pathogens released through soil disturbance, spills of toxic chemicals, catastrophic releases of both toxic and explosive compounds, accidents, and noise. The analysis typically involves risk-analysis-type studies, analysis of socioeconomic-based considerations, and reviews of emergency-response plans.

**4.4.2 EIA.** There are three basic types of analysis.

1. An abbreviated impact analysis that indicates no significant impacts. NEPA refers to this as a finding of no significant impact (FONSI); CEQA refers to this as a negative declaration.

2. An abbreviated impact analysis that indicates no significant impacts after inclusion of mitigations as design features to reduce the degree of impact from the proposed project.

3. A detailed impact analysis that includes project mitigations and residual impacts.

World Bank, NEPA, and CEQA process flow charts (**Figs. 4.1 through 4.3**) show how the overseeing agencies view a given project and how they proceed with their review.

The World Bank focuses on projects related to developing nations. These projects often involve infrastructure and growth inducement. The pivotal approval is the loan approval, not the issuance of permits, licenses, or authorizations that may not exist because of the absence of adequate environmental laws and regulations in the subject country or locales.[14] The review process is less structured than a NEPA or CEQA review and allows (owing to its financial nature) the proponent to negotiate mitigations, which are ultimately incorporated into the loan agreement as covenants.

The NEPA review focuses on actions taken by a U.S. federal agency, such as sale or lease of federal lands, legislative changes that compel an agency to take broad actions (e.g., deregulation of gas and electricity), or federal funding of major projects. The pivotal approvals for projects that are not exempted or approved are either the FONSI or the record of decision for the EIS. Although not granting a right, these approvals allow the agency to proceed with its intended action. The agency, for reasons not related to NEPA, could decide to deny approval or elect not to proceed with the proposed action.

The NEPA review structure does not have a distinct negotiation phase; however, as a practical matter, the project proponent (if a private party is involved) can negotiate easily during the writing of the environmental assessment, during the writing of the draft EIS, and during the public-comment period. Once the agency has responded to public comments and is writing the final EIS, incorporating major modifications to the project without another review period is difficult and will delay issuance of the final EIS.

The CEQA process focuses on projects that require discretionary permits and are not categorically exempted from CEQA by statute.

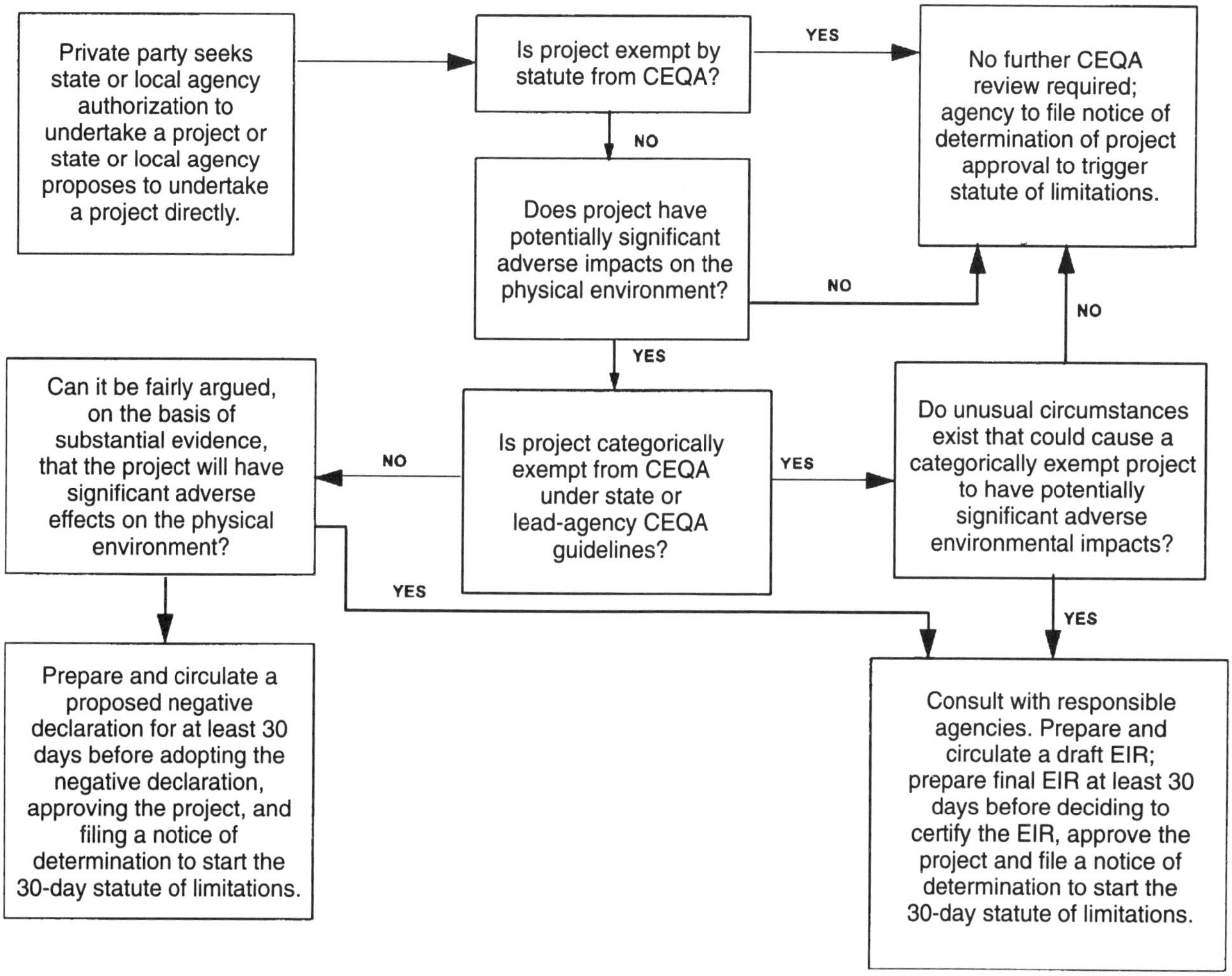


**Fig. 4.3—Overview of CEQA compliance process.**

CEQA is written so that it applies to agency actions (local and state) as well as private parties seeking approvals (i.e., licenses, permits, or authorizations). The pivotal approvals are the negative declaration (stating that the project will have no significant adverse impacts) and the notice of determination [certifying the final environmental-impact report (EIR) as being adequate].

Certification of the EIR grants no right to the related discretionary permits. The lead agency or a responsible agency could in fact rely on a certified EIR, and the discretionary board could vote to deny the issuance of the related permit. This is often the case with many other state programs.

As a practical matter, a project proponent can negotiate mitigations with the agencies up to the time the issuance of the permit is approved as long as the mitigation does not create an unassessed impact that may be significant.

***Relationship to Permits.*** An impact analysis is usually necessary for discretionary permits and licenses or other other actions involving governing bodies. A discretionary permit is one that requires review by a public body and for which there are no pre-established minimum design criteria for approval. Ministerial permits can be granted before completion of the impact analysis or without consideration of an impact analysis; however, many programs do not allow agencies to grant any permits until the impact assessment is completed.

***Overlapping Reviews.*** In some instances, an impact assessment is required by both state and federal agencies. In some rare instances, the impact assessment could be identical for each. However, some states' impact-assessment programs are more stringent than others or focus on certain issues.

***Preproject Setting.*** This element should focus on the preproject (undisturbed) setting for the area where the proposed project is to be located. It should address those areas of concern where any potential project-related activities might result in impacts, positive or negative. **Table 4.1** lists the types of information that should be provided in the section of the impact analysis that deals with preproject settings.

**TABLE 4.1—INFORMATION FOR PREPROJECT-SETTING IMPACT ANALYSIS**

| Potential Area of Impact | Information Necessary |
|---|---|
| Air | Ambient air quality, emission sources of interest, receptor locations, weather data. |
| Water | Water usage, locations and types of waterways (artificial and natural), recreation areas, rainfall data. |
| Soil/geological | Soil types, soil analysis, locations of disturbed soil, possible migration paths, geologic structures, seismicity. |
| Biological | Biota survey. |
| Archeological | Historical survey, existing archeological sites. |
| Societal | Geopolitical data, demographic data, census data, employment information, anthropological studies. |
| Health and safety | Reference data for risk analysis, description of sensitive receptors and their location. |

**TABLE 4.2—SOURCES OF AIR POLLUTANTS IN OIL AND GAS PRODUCTION**

| | Tanks | Produced Water | Glycol Unit | Combustion Equipment | |
|---|---|---|---|---|---|
| | | | | Liquid Fuel | Natural Gas/Propane |
| Criteria pollutants | | | | | |
| VOC | X | X | X | X | X |
| $NO_x$ | | | X | X | X |
| $SO_x$ | | | P | P | P |
| CO | | | X | X | X |
| $PM_{10}$ | | | X | X | X |
| Hazardous air pollutants | | | | | |
| Acetaldehyde | | | X | X | X |
| Acrolein | | | X | X | X |
| Benzene | X | X | X | X | X |
| Beryllium | | | | P | |
| Cadmium | | | | P | |
| Chlorine | | X | | | |
| Chromium | | | | P | |
| Creosol | X | | | | |
| Cumene | X | | | | |
| Ethyl benzene | X | X | X | X | X |
| Ethylene glycol | | | X | | |
| Formaldehyde | | | | X | X |
| Glycol ethers | | | X | | |
| Hexane | X | | | | |
| Hydrogen sulfide | P | P | | | |
| Lead | | | | P | |
| Manganese | | | | P | |
| Mercury | | | | P | |
| Naphthalene | X | P | X | X | X |
| Nickel | | | | P | |
| Phosphorus | | | | P | |
| Polycyclic organic matter | | | | P | X |
| Selenium | | | | P | |
| Toluene | X | X | X | X | X |
| Xylenes | X | X | X | X | X |

X = present and P = particulate.

***Project Implementation.*** *Project Description.* This element should focus on the design features of the project alternatives, including the no-build option, and should address both the magnitude and duration of any potential activities that might result in some degree of impact, positive or negative. The following information should agree with any other permit applications and design data.

1. Site location.
2. Air emissions.
3. Discharges to surrounding waters.
4. Noise.
5. Solid-waste discharges.
6. Clearing, excavation, and other disturbances.
7. Transportation and construction activities.
8. Health and safety.

Oil- and gas-related projects and activities that might warrant consideration in an EIS are seismic logging; exploratory, offshore, and production drilling; field development; support facilities and treatment facilities; storage and transfer; transmission piping; and road, platform, and power-plant construction.

For the oil and gas industry, stationary equipment that represents potential sources of impact that should be considered in the EIS if associated with the proposed project includes stock, test, and skim tanks; valves and flanges; boilers; gas-driven-pumping, water-treatment, and lease-automated custody-transfer units; flotation devices; steam generators; separators; heater treaters; free-water knockouts; glycol dehydrators; stuffing-box emissions; sumps; injection pumps; casing vents; compressors; internal-combustion engines; flares; and indirect heaters.

Mobile equipment that represents potential sources of impact and should be considered in the EIS if associated with the proposed project includes mobile and offshore drilling rigs, earth-moving equipment, transport trucks, supply boats, aircraft, and personal transportation.

Impacts related to a given project must be addressed without inclusion of mitigations. Misapplied mitigations can create an undesired secondary impact that could otherwise have been avoided (e.g., safety-related problems, excessive property disturbance, or misuse of water resources). During this phase of review, the responsible agency might establish to its satisfaction that the proposed project is relatively innocuous and has no significant impacts warranting further review.

***Air.*** Several types of pollutants emitted from oil- and gas-production facilities could impact the air. The extent of the impact related to these emissions is estimated on the basis of the amount of criteria pollutants or hazardous air pollutants emitted by a facility and the proximity to sensitive receptors, such as schools, residences, hospi-

| TABLE 4.3—POSSIBLE POLLUTANTS IN PRODUCED WATER | |
|---|---|
| Chlorides | Other volatile organics |
| Total dissolved solids | Carbon disulfide |
| Total phenols | Chloromethane |
| Priority metals | *m*-xylene |
| Antimony | *o*- and *p*-xylene |
| Arsenic | Vinyl acetate |
| Beryllium | 2-butanone |
| Cadmium | 2-hexanone |
| Chromium | 2-propanone |
| Copper | Priority pollutant semivolatile organics |
| Lead | *bis*-(2-ethylhexy)phthalate |
| Mercury | di-*n*-butylphthalate |
| Nickel | Naphthalene |
| Selenium | Phenol |
| Silver | Other semivolatile organics |
| Thallium | Benzoic acid |
| Zinc | Benzyl alcohol |
| Other metals | Hexanoic acid |
| Aluminum | *n*-decane |
| Barium | *n*-docosane |
| Boron | *n*-dodecane |
| Calcium | *n*-eicosane |
| Cobalt | *n*-hexosane |
| Iron | *n*-hexadecane |
| Magnesium | *n*-octacosane |
| Manganese | *n*-octadecane |
| Molybdenum | *n*-tetracosane |
| Strontium | *n*-tetradecene |
| Sulfur | *n*-triacontane |
| Tim | *o*-creosol |
| Titanium | *p*-creosol |
| Vanadium | 1,2 and 3,4-diepoxybutane |
| Yttrium | 2-methyl naphthalene |
| Zinc | 2,4-dimethyl phenol |
| Priority polluant volatile organics | Radionuclides |
| Benzene | Gross alpha |
| Ethyl benzene | Gross beta |
| Methylene chloride | Lead-210 |
| Toluene | Radium-226 |
| Trichlorofluoromethane | Radium-228 |

tals, recreation areas, and protected or sensitive habitats. Chaps. 6 and 7 detail potential emissions and treatment of potential impacts.

The following are the criteria pollutants.

1. VOC's are caused by evaporation whenever a hydrocarbon has contact with air. VOC emissions are also present in the exhaust stream of combustion equipment such as boilers, heaters, and internal-combustion engines.

2. $NO_x$ are emitted from fuel-burning equipment as part of the combustion process.

3. $SO_x$, typically referred to in the form of sulfur dioxide ($SO_2$), are emitted from fuel-burning equipment. The amount of $SO_2$ emitted depends on the amount of sulfur in the fuel.

4. CO is emitted from fuel-burning equipment.

5. $PM_{10}$ are emitted by fuel burning and any fugitive-dust source.

6. Lead can be emitted from burning coal, fuel oil, and crude oil.

Title III of the 1990 Amendments to the U.S. Clean Air Act contains a list of 189 hazardous air pollutants. Several of the listed hazardous air pollutants occur naturally in crude oil or natural gas; therefore, oil and gas production results in emissions of certain listed hazardous air pollutants. **Table 4.2** gives a partial list of potential air contaminants from some oil- and gas-related equipment.

***Water.*** Oil- and gas-production facilities emit several types of pollutants that might impact water. The extent of the impacts related to these emissions is estimated on the basis of the amount and concentration of the pollutants discharged by a facility, the transport mechanisms of the pollutant from the point of discharge to the receiving waters, the dilution effect of the receiving waters, and the proximity to sensitive receptors (e.g., drinking-water sources, recreational areas, and protected or sensitive habitats). Chaps. 5 through 7 provide details on the potential discharges and treatment of potential impacts.

Possible sources of water-related contamination are produced water, unconfined spills, drilling mud, equipment lubrication, equipment-cooling water, surface runoff, casing failures, and processing wastes. **Table 4.3** shows possible components found in produced waters, and Chap. 7 provides further details.

***Soil.*** Soil-related contamination stems from the same chemicals that affect groundwater; however, the more volatile chemicals tend to evaporate over time. Soil contamination typically is considered in conjunction with groundwater contamination because of the migration of the chemicals through the soil to the groundwater. Possible sources of water-related contamination are gas-processing wastes, produced water, unconfined spills, drilling mud, equipment lubrication, equipment-cooling water, surface runoff, and casing failures.

***Biological.*** Biological impacts tend to result from activities (1) that result in violation of any environmental law or regulation designed to protect wildlife, fisheries, plant species, or habitat areas; (2) that directly harm a sensitive species or cause a net loss of habitat; (3) that interfere with movement, migration, or other essential behavioral patterns; (4) that cause a loss of riparian lands, wetlands, or other sensitive habitats; and (5) that cause a sensitive species population to drop below self-sustaining levels. For the purposes of this discussion, the term sensitive species is any species listed as threatened or endangered according to the U.S. Endangered Species Act or as defined or listed in any national or local law or regulation.

Sec. 9 of the Endangered Species Act prohibits the taking of listed species without special exemption. Taking is defined as harassing, harming, pursuing, hunting, shooting, wounding, killing, trapping, capturing, collecting, or attempting to engage in any such conduct. Harm is further defined to include significant habitat modification or degradation that results in death or injury to listed species by significantly impairing essential behavioral patterns, including breeding, feeding, or sheltering.

Possible sources of impact are construction activities, new roads, unconfined spills, noise, lighting, soil contamination, trash and debris, and encroachment.

***Archeological.*** Archeological impacts are those (1) that violate any federal, state, or local laws, regulations, or programs; (2) that disturb or prevent future access to a prehistoric, historic, or cultural site that is listed or is eligible for listing on the U.S. Natl. Historical Register or other such listings of historic resources; (3) that disturb or prevent future access to a unique paleontological site; and (4) that disturb the site of human remains. Possible sources of impact are construction activities, new roads, property development, and encroachment.

***Socioeconomic, Societal, and Cultural.*** Socioeconomic, societal and cultural impacts are those (1) that make demands on public services (fire and police protection, schools, parks, and recreation), (2) that modify traffic and transportation patterns, (3) that affect employment patterns, (4) that affect property values, and (5) that disturb or render inaccessible any site of cultural importance to an ethnic or social group. In the U.S., this section is rarely necessary for projects that involve the oil and gas industry because the projects generally are located in remote regions where socioeconomic, societal, and cultural impacts are minimal. The exception for NEPA projects would be those on or near native American reservations. However, outside of the U.S., the impacts on indigenous populations can pose a serious issue that warrants considerable attention.

Possible sources of impact are construction activities, new roads, unconfined spills, noise, lighting, soil contamination, trash and debris, encroachment, aesthetic obstructions, property development, and production activities.

***Health and Safety.*** Health and safety impacts are those that violate any health- and/or safety-related federal, state, or local laws; regulations; or programs. In the U.S., this section is rarely necessary for projects that involve the oil and gas industry because the industry relies heavily on trained professionals and because the projects usually are located in remote regions where uncontrolled health and safety impacts are minimal; the exception is projects on or near native American reservations. Chap. 7 provides further details for projects close to sensitive receptors or for which the lead agency requires a risk analysis.

***Residual Impacts.*** Residual impacts are those impacts that remain after mitigations have been put into place. These impacts usually are not significant (the lead agency usually seeks to reduce all significant impacts through mitigation). In some instances, the residual impact still may be significant but not prohibitive. The final decision on the acceptability of residual risk is left to the governing body holding the discretionary responsibility for final project approval, licensing, or permit issuance. Possible sources of residual impacts are construction and production activities, new roads, confined spills, noise, lighting, soil contamination, trash and debris, encroachment, aesthetic obstructions, property development, and air emissions.

***Impacts After Closure and/or Abandonment.*** Impacts after closure and/or abandonment exist after the project life has ended and are artifacts of project impacts over its lifetime and/or result from abandonment activities. Many residual impacts fall in this category. Possible sources of closure/abandonment impacts are demolition activities, road damage, spill sites, soil contamination, water contamination, trash and debris, encroachment, aesthetic obstructions, property development, air emissions, and well abandonment.

***Project Mitigations.*** Mitigations are those measures or design features that help to ameliorate or reduce adverse impacts of a proposed project or action. These measures usually are imposed by the reviewing agency after proposed-project impacts have been compared with the original setting. The level and type of project mitigations typically are negotiated between the project proponent, the reviewing agency, and other stakeholders. Eventually, the mitigations themselves become a part of the project, voluntarily, by contract, or by agency-specified obligation. Compliance with laws and regulatory requirements should not be part of the mitigations because the obligation to follow laws and governing regulations is mandatory.

Mitigations can be incorporated into the design and operational features by the project proponent before review by the agency. This can sometimes benefit the project by reducing the initial impact of the facility to such a degree that the need for a full impact analysis is waived. On the other hand, the project proponent may want to wait for the public and the agencies to have an opportunity to comment before adopting or agreeing to mitigations. This typically is done for one of two reasons: either the project proponent doesn't want to volunteer costly mitigations or the agency or proponent feels that public participation is essential to the project's success.

Timing of the submission of possible mitigations should be based on a clear understanding of mitigation costs, the cultural approach to negotiation and the environment, and the process that the governing body wants to follow.

## 4.5 Summary

This chapter has attempted to cover the broad range of issues facing the environmental and nonenvironmental professionals engaged in E&P. Some projects that require impact analysis are exploration, production, facility installation, closure, and remediation.

Step-by-step procedures for assessment programs, such as NEPA, CEQA, World Bank, ISO 14000 series, and locally used programs, can be obtained from those agencies to establish the current regulatory and statutory requirements and schedules for activities (e.g., public review and comment).

In addition to the references cited, the general references provide information on the issues involved and on where to look for guidance to ensure the success of a project or acquisition in a manner that reflects sound environmental management.

## References


1. "Natl. Environmental Policy Act," Public Law 91-190, 42 USC§4321-4347, Government Printing Office, Washington, DC (1 January 1970).
2. Wark, K. and Warner, C.F.: *Air Pollution: Its Origin and Control*, Harper and Row, New York City (1976) 2.
3. Bass, R.E. and Herson, A.J.: *Mastering NEPA: A Step-By-Step Guide*, Solano Press Books, Point Arena, California (1996) 60–69.
4. *Series 14000, Environmental Management*, ISO, ISO Central Secretariat, Geneva, Switzerland (1997).
5. "The Comprehensive Environmental Response, Compensation, and Liability Act (CERCLA of Superfund)," 42 USC 9601 *et seq.* (1980).
6. "The Resource Conservation and Recovery Act," 40 CFR (1976) Part 280, Subtitle I.
7. Friedman, F.D.: *Practical Guide to Environmental Management*, Environmental Law Inst., Washington, DC (1993).
8. The Securities Exchange Act 17 CFR (1934) 229.
9. *Statement No. 5, Accounting for Contingencies*, FASB, Norwalk, Connecticut (March 1975).
10. *Std. E-1527-97, Standard Practice for Environmental Site Assessments: Phase I Environmental Site Assessment Process*, ASTM, West Conshohocken, Pennsylvania (1997).
11. "The California Environmental Quality Act," amended, 14 CFR (29 March 1999) Chap. 3, Secs. 15000 to 15387.
12. "Federal Clean Air Act Amendments," 42 USC 7401 *et seq.* (1990).
13. "The Endangered Species Act (ESA)," 7 USC 136; 16 USC 460 *et seq.* (1973).
14. Boekel, C.P.: "Regulations to Industry Compliance: Building Institutional Capabilities," Technical Report No. 11, United Natl. Environment Program, Paris (1992).


## General References


Cairncross, F.: *Costing the Earth*, Harvard Business School Press, Boston, Massachusetts (1991).

Leitman, J.: *Energy Environment-Linkages in the Urban Sector*, Urban Management Program, World Bank, New York City (1996).

Porter, L. *et al.*: *A Guidebook for Technology Assessment and Impact Analysis*, Elsevier North Holland Inc., New York City (1980).

Smith, A.L.: "Securing the Future: In Partnership With the Environment," *JPT* (June 1996) 526.

Smith, R.M.: "Environmental Issues and Solutions for Exploratory Drilling in Sensitive Areas," paper SPE 29704 presented at the 1995 SPE/EPA Exploration and Production Environmental Conference, Houston, 27–29 March.

# Chapter 5
# Terrestrial and Ocean Environments

**Don L. Warner,** U. of Missouri
**Terence R. Moore,** Arco E&P Technology
**A. Bonny Crews,** Arco E&P Technology
**James P. Ray,** Shell Chemical Co.

## 5.1 Overview

This chapter describes terrestrial and ocean environments; E&P impacts that can occur, and have occurred, to those environments; and the fate and transport of E&P contaminants in those environments.

**5.1.1 Soils.** The first section of the chapter describes soils as an environment. E&P activities can impact soils mechanically, but the concern in this chapter is with contaminants introduced from E&P activities, particularly saline water, hydrocarbons, and metals. Damages to the soil environment from these contaminants include changes to soil physical, chemical, and biological properties and to soil agricultural capability. Some E&P contaminants are held up in the soil profile where they may remain or, in the case of hydrocarbons, may be lost by volatilization or physical, chemical, and biological degradation. Oilfield waters can have a range of total-dissolved-solids (TDS) concentrations but are most commonly more saline than seawater. The principal salts of oilfield waters are the anions chloride (Cl) and sulfate (S) and the cations sodium (Na) and calcium (Ca), with most waters dominated by sodium chloride (NaCl).

The soluble salts that occur in oilfield waters are very mobile in soils and are transported at about the same rate as water. Transport is predominantly vertical, downward toward the groundwater table. Anions, particularly chloride, are transported with little or no retardation. Cations are subject to retardation by ion exchange.

Transport of petroleum hydrocarbons through soils at E&P sites depends on the characteristics of both the soil and the hydrocarbon. Low- and medium-gravity oils (g/cm$^3$) are very viscous compared with water and do not flow readily through soil. When spilled on the soil surface, they are more likely to flow along the surface rather than penetrate deeply into the soil.

High-oil-gravity crude oils and refined products contained in fuels, lean oil, or oil-based drilling fluids have the potential to move more readily through soils because of their lower viscosity. In cases where significant volumes are applied to the soil over time, these oils can pass through the vadose zone to form nonaqueous-phase layers at the interface of the vadose and saturated zones.

The rate of hydrocarbon migration in soils depends strongly on soil physical properties. Hydraulic conductivity to petroleum is much higher in sandy soils than in clay-rich soils. The solubility of most hydrocarbon constituents is relatively low. Compounds, such as benzene, however, can partition into the water phase and be transported in a dissolved phase. These hydrocarbons are present in a state that is optimum for biodegradation or adsorption to organic matter in soils.

Hydrocarbons in soil can be attenuated by a number of physical, chemical, or biological processes. These mechanisms combine to provide an effective barrier that often prevents hydrocarbons from interacting with the groundwater system.

**5.1.2 Vadose Zone and Groundwater.** The vadose zone and groundwater are considered in the second section. From a strictly technical view, the vadose, or unsaturated, zone that occurs above the groundwater table includes soils. However, because soils are an important medium and not just a pathway for infiltrating water, they have been treated separately.

Groundwater enters the subsurface by infiltrating vertically downward from the surface, where it is the result of rainfall or snow melt. If present, water-soluble E&P contaminants move with the infiltrating water and, if unattenuated in the vadose zone, enter the groundwater system. As mentioned earlier, liquid petroleum is not miscible with water and travels through the vadose zone as a separate phase.

Causes of E&P-related groundwater contamination result primarily from such historical practices as use of unlined drilling and production pits; improperly constructed or abandoned wells; and spills or leaks of produced fluids, oilfield chemicals, or wastes.

The primary groundwater impact is from highly saline waters that result from the very high TDS and chlorine content of many petroleum reservoir waters. Groundwater contamination by petroleum and natural gas from E&P occurs but is less common. Crude petroleum varies in viscosity, and its viscosity frequently determines its mobility in the subsurface. A close correlation exists between oil gravity and viscosity. The higher the oil gravity, the lower the specific gravity and the viscosity and, therefore, the greater the mobility. In addition, higher-oil-gravity petroleum is more likely to contain aromatic compounds, which are soluble in groundwater.

The average velocity of groundwater flow and of unretarded dissolved groundwater contaminants is calculated with a variation of Darcy's law. Groundwater velocities in unconfined formations average less than 0.3 m/d even in very permeable media. Exceptions include areas of very high permeability or steep gradients close to streams or other sources of discharge. Groundwater in deeper formations tends to travel more slowly than near-surface groundwater.

The groundwater velocity calculated with Darcy's law is an average value for molecules of groundwater or ions or for molecules of a contaminant unretarded by chemical reaction or sorption. Chlo-

| TABLE 5.1—AREAS WHERE E&P ACTIVITY AFFECTS THE ENVIRONMENT |
|---|
| Acoustics |
| Geophysical surveys |
| Other industrially generated sounds |
| Physical bottom disturbances |
| Structural impacts |
| Burials |
| Debris |
| Offshore discharges |
| Drilling muds |
| Produced water |
| Platform removals |
| Spills |

ride is such a contaminant. On the other hand, sodium is not because it can be subject to ion exchange for other common cations, particularly calcium or magnesium. Thus, if an aquifer were contaminated by an NaCl reservoir brine, the chloride ions would be expected to travel down the groundwater potentiometric gradient at the same velocity as water molecules, whereas the sodium ions could travel at a slower average velocity.

In addition to the possibility of retardation during travel in an aquifer, contaminants are subject to hydrodynamic dispersion, which is the combined processes of diffusion and mechanical mixing. Through dispersion, contaminants are diluted and spread out during travel. As a result of possible retardation and through dispersion, the concentration of a dissolved groundwater contaminant eventually is reduced to the point where it is not distinguishable as a contaminant. For example, condensate (or similar hydrocarbons) released to the subsurface can partition into the following four phases: (1) vapor phase, (2) residual phase (adsorbed and trapped), (3) free phase (mobile liquid hydrocarbons), and (4) dissolved phase. The vapor phase exists only in the unsaturated zone. Most condensate constituents have relatively high vapor pressures, which means they evaporate rather quickly when exposed to air at 1 atm of pressure.

If condensate or petroleum products are released to the subsurface, a certain amount is adsorbed as residual in pore spaces in the unsaturated zone or as a coating surrounding soil or formation particles. This residual tends to release contamination in the form of vapors and as dissolved hydrocarbons if recharge water percolates through it. If the unsaturated zone is very thick or if a petroleum release is small, the entire volume of a release may be consumed as residual. Residual also tends to be trapped in the saturated zone as small globules that are unable to pass from large pores through restricted water-wet pore throats. Residual hydrocarbon in the saturated zone is a continuing source of dissolved contamination.

Like crude oil, condensate is a mixture of different compounds. Crude-oil constituents are distilled into the gasoline fraction, diesel fraction, and other such products. Condensate is very similar to refined gasoline and generally is considered immiscible in water. However, many compounds in condensate dissolve on contact with water. The most commonly used marker compounds for dissolved condensate constituents are aromatic compounds, such as benzene, toluene, ethylbenzene, and xylenes, collectively referred to as BTEX. BTEX constituents most commonly are listed by state agencies in the regulation of releases of light hydrocarbons because they are the most water soluble of hydrocarbon-condensate compounds.

**5.1.3 Oceans.** Oceans are the most dynamic and difficult domains for man to work in and to explore for minerals. Since the beginning of the 20th Century, the oil and gas industry has ventured seaward from the coastal boundaries: from drilling piers extending from the beach, to floating barges, to fixed structures, to semisubmersible platforms, to today's modern drillships and tension-leg platforms. This evolution was slow at first because explorationists had to adapt their knowledge of oil and gas drilling onshore to an entirely new environment that posed many formidable operational challenges.

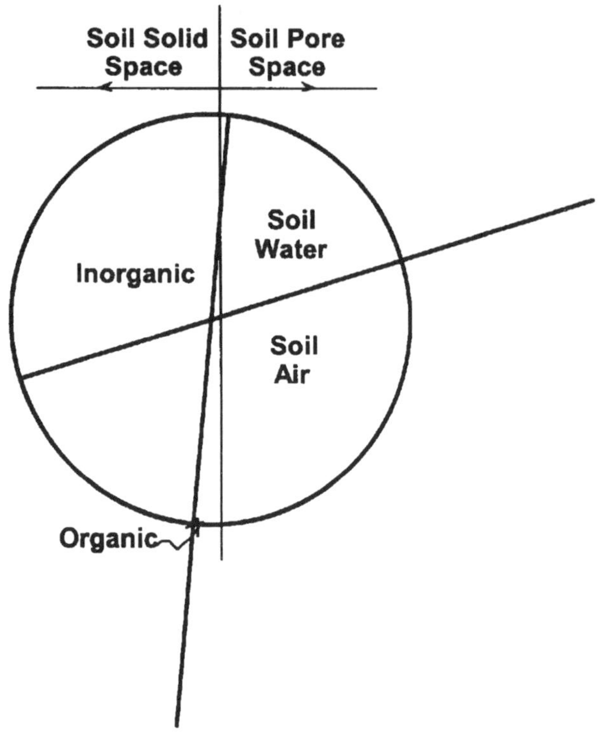


**Fig. 5.1—Four phases of soil on a volumetric basis (from Ref. 3).**

As a user of the ocean, the petroleum industry interacts with and affects the environment in myriad ways. **Table 5.1** defines the areas of E&P activity where environmental impacts occur. These activities result in physical, chemical, and biological impacts of varying magnitude and duration. However, effluent discharges in the outer continental shelf (OCS) have minimal impact, partly because of their relatively nontoxic nature and partly because rapid natural dispersion occurs.

Biological impacts and chemical contamination have been predominantly limited to an area within 100 to 200 m of most platforms. Beyond these distances, it is virtually impossible to document significant biological effects and the still detectable chemical signal is very low and not related to toxic components (or levels). Most impacts appear to be more related to direct burial beneath the platforms. Impacts resulting from seismic surveys appear to be minimal (if they occur at all), and marine-mammal response to OCS activity appears to be small and primarily of local duration.

## 5.2 Soils

Civil engineers consider soil to include all material, organic and inorganic, overlying bedrock.[1] Soil scientists, on the other hand, describe soil as "the collection of natural bodies on the Earth's surface, in places modified or even made by man of earthy materials, containing living matter and supporting or capable of supporting plants out-of-doors."[2] The emphasis for soil scientists is on understanding the morphology and physical, chemical, and biological processes important in soil formation (genesis). Understanding of the natural processes involved in soil genesis is important to managing and maintaining the productivity of the soil resource.

The process of exploring, producing, and refining hydrocarbons can disrupt the properties of soils significantly. The objective of reclaiming these sites is to restore soils, as nearly as possible, to their original productivity. With this goal in mind, we chose a soil-science approach to the discussion of the effects of petroleum E&P on soil properties and methods for reclamation.

**5.2.1 Properties of Soils.** ***Soil Physical/Chemical Properties.*** An in-depth discussion of soil properties is beyond the scope of this section; however, many excellent texts can serve as references concerning soil properties in general,[3] soil genesis and taxonomy,[4,5] nutrient cycling and soil fertility,[6-9] soil chemistry,[10] saline and sodic soils,[11-13] soil

| TABLE 5.2—USDA SYSTEM OF PARTICLE-SIZE FRACTIONS | |
|---|---|
| Particle-Size Fraction | Range of Diameter (mm) |
| Gravel | >2 |
| Very coarse sand | 1 to 2 |
| Coarse sand | 0.5 to 1 |
| Medium sand | 0.25 to 0.5 |
| Fine sand | 0.1 to 0.25 |
| Very fine sand | 0.05 to 0.1 |
| Silt | 0.002 to 0.05 |
| Clay | <0.002 |

mineralogy,[14-16] soil microbiology,[17,18] and soil physics and water relationships.[19,20] The following discussion of soils focuses on the properties that are most affected by E&P activities or that have the greatest effect on soil reclamation following E&P.

*Four Phases.* The most basic understanding of soils concerns the four phases present in soils: inorganic, organic, soil water, and soil air (**Fig. 5.1**). The inorganic phase generally forms the greatest mass of the soil and consists of mineral particles derived from unconsolidated sediments or sediments weathered from underlying rock. The organic phase may range from <0.1 to >90%, with a common range of 0.5 to 5% organic matter. The soil air and soil water phases occupy the pore space of the soil and are therefore inversely related to one another. Soil air differs from ambient air, typically by having higher concentrations of $CO_2$ and lower concentrations of $O_2$. As soil-water content increases, soil air necessarily decreases, and vice versa. Total soil porosity typically ranges from 35 to 60 vol% and is composed of a mixture of soil water and soil air.

*Particle Size.* Particle size refers to the actual diameter of a soil particle and is one of the most fundamental of all soil properties. Particles can be classified in many ways on the basis of their size. Soil scientists typically use the U.S. Dept. of Agriculture (USDA) system for particle-size description (**Table 5.2**).[21] Size of individual particles has no meaning when discussing a bulk soil sample. Particle-size information for soil samples is summarized by use of particle-size-distribution classes.

*Particle-Size-Distribution Classes.* Several systems exist for classifying particle-size distribution, but the most common are included in the Unified Soil Classification System[22] and the USDA system. The Unified System is based on particle size and physical behavior (such as plasticity index and liquid limit). This system is most useful for predicting soil behavior for construction purposes and is used by engineers in many environmental applications. The USDA system is based solely on particle-size distribution (summarized in **Fig. 5.2**). The particle-size classes included on the textural triangle can provide important clues about the chemistry, aeration, and water relationships of the soil.

***Clay Properties.*** The term "clay" can have multiple meanings, depending upon its context. Clay can refer to a particle with a diameter of >2 mm, a soil with >40% clay-sized particles, or a group of minerals that commonly comprises the clay-sized fraction of soils. The clay-sized particles in soil play a key role in determining a soil's physical and chemical properties. Important properties of clays include size, shape, charge, and dispersion character.

*Size.* In the USDA system for classification of soils, clay particles are those >2 mm in diameter. All analytical tests for particle size are based on a rate of particle settling from suspension.

*Shape.* Clay-sized particles may have many shapes, but plate-shaped morphology is predominant. This reflects the minerals that dominate the clay-sized fraction. Micaceous types of minerals (2:1 and 1:1 phylosilicates) have a cleavage pattern that produces very thin flakes or plates. The small size and plate-shaped morphology of clays results in a large amount of surface area per unit mass. This large surface area and the accompanying charge characteristics cause the clay-sized fraction to dominate the chemical and physical properties of soils.

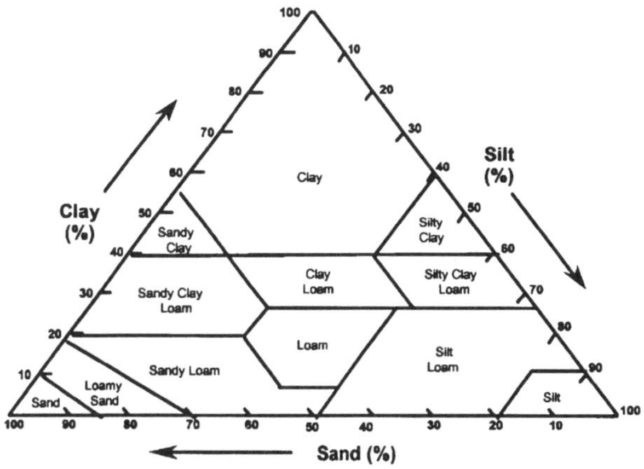


**Fig. 5.2—USDA system of particle-size-distribution classes (from Ref. 21).**

*Charge Character.* Clay particles have a net negative charge that results from substitution of metals in the crystalline lattice or from a pH-dependent charge at crystal edges. This negative charge is balanced by loosely held aqueous cations that reside at or near the particle surface. These cations are held loosely enough so that they can be replaced easily by other cations in the soil solution (**Fig. 5.3**). The ability of soils to exchange cations is called the cation-exchange capacity (CEC) of soils. CEC is responsible for much of the chemical behavior of soils and their ability to attenuate inorganic pollutants.

**5.2.2 Effects of E&P Activities on Soils.** The effects of E&P activities on soils may be manifested in many ways, depending on the type of E&P process, the contaminant, and the characteristics of the site. It is easiest to discuss these effects in terms of the types of contaminants that may be present. The vast majority of environmental remediation actions are focused on three types of contaminants: salts, hydrocarbons, and metals.

***Soluble Salts.*** Throughout the life of a typical oil field, more salt water is produced than oil. The salt content of produced waters may vary widely, but the majority have high concentrations of NaCl salts. Salts from produced waters may produce both chemical and physical effects in soils.

*Chemical Effects: Exchangeable Sodium Percentage (ESP).* Soluble salts contained in produced water interact with clay particles through the cation-exchange mechanism. As the $Na^+$ content of the soil solution is increased from produced water, the number of cation-exchange sites occupied by $Na^+$ increases (**Fig. 5.4**). The percentage of exchange sites occupied by $Na^+$ can affect soil physical and chemical properties dramatically. A fundamental property of soil commonly measured in salt-affected soils is ESP. Fig. 5.4 also shows the formula for ESP calculations.

*Chemical Effects: Sodium Adsorption Ratio (SAR).* A distinct relationship exists between the cation concentrations in solution and those on the cation-exchange sites. The SAR is used to express the relative amount of sodium compared with the other cations in the solution. The mathematical expression for SAR, $R_{a\mathrm{Na}}$, is

$$R_{a\mathrm{Na}} = \mathrm{Na}/[(\mathrm{Ca} + \mathrm{Mg})/2]^{0.5}, \quad\quad (1)$$

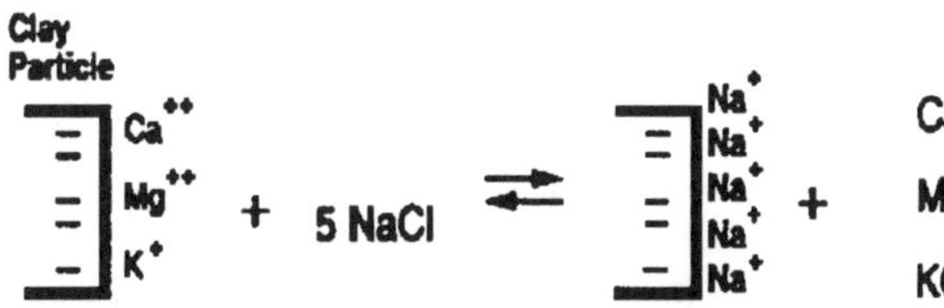


**Fig. 5.3—Schematic of the cation-exchange mechanism, soil clays reacting with salts from produced water: units of 1 meq charge/100 g mass.**

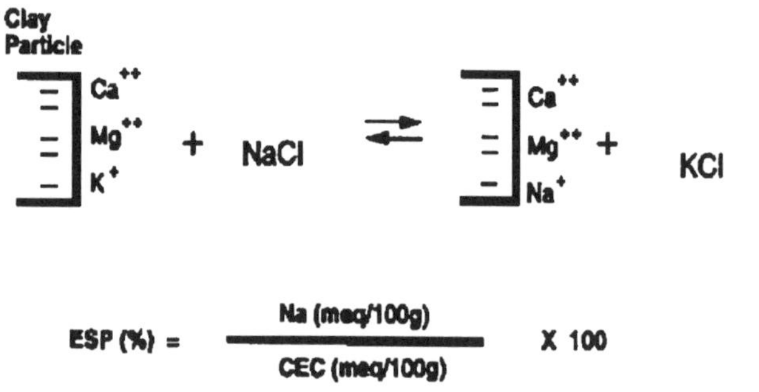


**Fig. 5.4—Cation-exchange mechanisms for soil clay and calculation of the ESP.**

where Na, Ca, and magnesium (Mg) are in milliequivalent concentrations. Ref. 21 provides an empirical equation that relates sodium in solution (SAR) to sodium on cation-exchange sites (ESP); **Fig. 5.5** presents this graphically.

The utility of SAR is twofold. The ESP of a saline soil is often difficult to determine because of excess salts in the soil sample. SAR can be measured easily and used in conjunction with estimates of CEC to determine calcium (gypsum) requirements for soil remediation. The SAR can also be used to evaluate the effects of irrigation water on soil properties. Over time, soils equilibrate to an ESP that corresponds to the SAR of irrigation water (Fig. 5.5). SAR can be used in this manner to evaluate irrigation waters that may be used in remediation or potentially to evaluate low-salinity produced waters as a source of irrigation water.

*Chemical Effects: pH.* Calcareous soils [those that contain calcium carbonate ($CaCO_3$)] typically have pH values that range from 7.4 to 8.2. Calcareous soils impacted by produced water typically increase one pH unit to approximately 9.2. Although the pH increase does not affect plants directly, other more subtle effects can occur. The higher soil pH may increase dispersion of clays and affect physical properties (discussed in the next section) and may also cause minor changes in the availability of nutrients to plants. Optimum soil pH range for plant-nutrient uptake is from 6 to 7 for most nutrients. Increased pH may make such nutrients as phosphorus and iron less available to plants. Remediation techniques typically correct the situation by supplying additional nutrients and lowering pH.

*Physical Effects.* The most deleterious and difficult–to-correct produced-water impacts are caused by soil physical properties. Osmotic effects are often visible immediately following salt impacts.

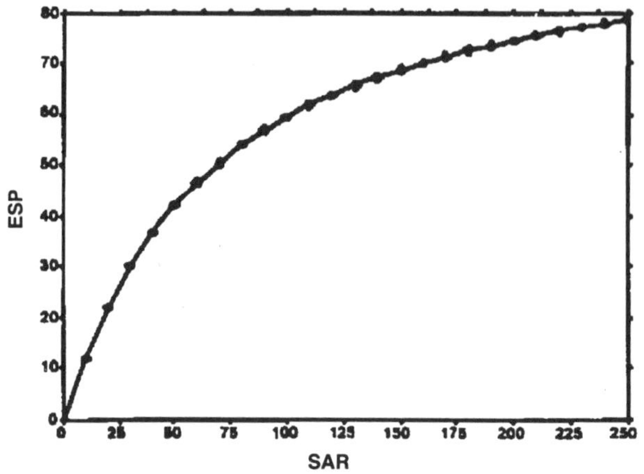


**Fig. 5.5—Relationship between ESP and SAR, relating sodium in solution to sodium on soil-exchange sites (from Ref. 21).**

An effect that occurs when water leaches salts from soils is dispersion; dispersion can affect important soil properties, such as infiltration rate, permeability, and erodibility.

*Physical Effects: Osmotic.* Osmotic effects typically are the first signs of produced-water damage. These effects are manifest in the health of plants in the spill area. **Fig. 5.6** illustrates the basic theory behind osmosis. Water tends to flow from areas of low salt concentration to areas of high concentration when separated by a semipermeable membrane (water can pass, but salts cannot). In a normal setting, evapotranspiration from leaves causes salt concentrations to increase in a plant. Subsequently, water flows from the soil solution into plant roots to equilibrate the system.

Introduction of soluble salts to the soil solution decreases the amount of water available to plants. Even though the soil may be very wet, plants cannot take up water because of the high osmotic potential in the soil solution (**Fig. 5.7**). Correction of this problem requires removal of soluble salts from the plant-rooting zone.

*Physical Effects: Dispersion.* Soil particles placed in water settle out at a rate proportional to their size. Clay particles ( < 2 mm) may stay suspended in water almost indefinitely under the right conditions. A suspension is said to be "dispersed" when clay particles re-

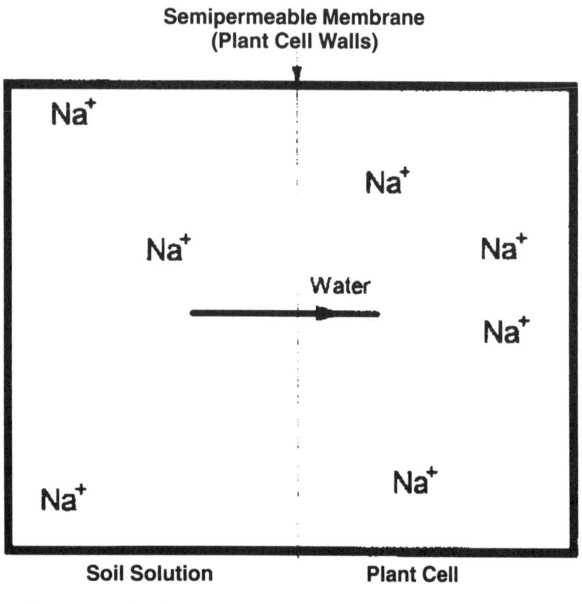


**Fig. 5.6—Plant/soil osmotic relationships encountered in typical soils. Water moves from lower salt concentrations in the soil to the plant, which has higher salt concentrations resulting from evapotranspiration at the leaves.**

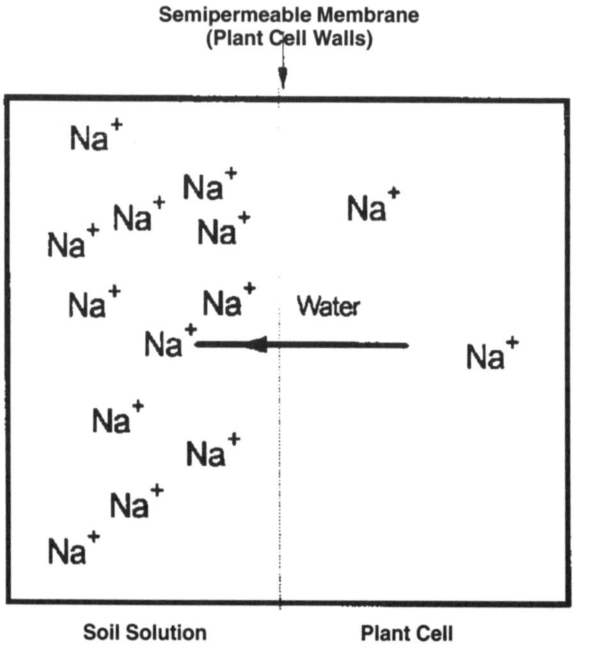


**Fig. 5.7—High salt concentrations in the soil solution prevent plant water uptake. Very high concentrations can result in plant stress or death.**

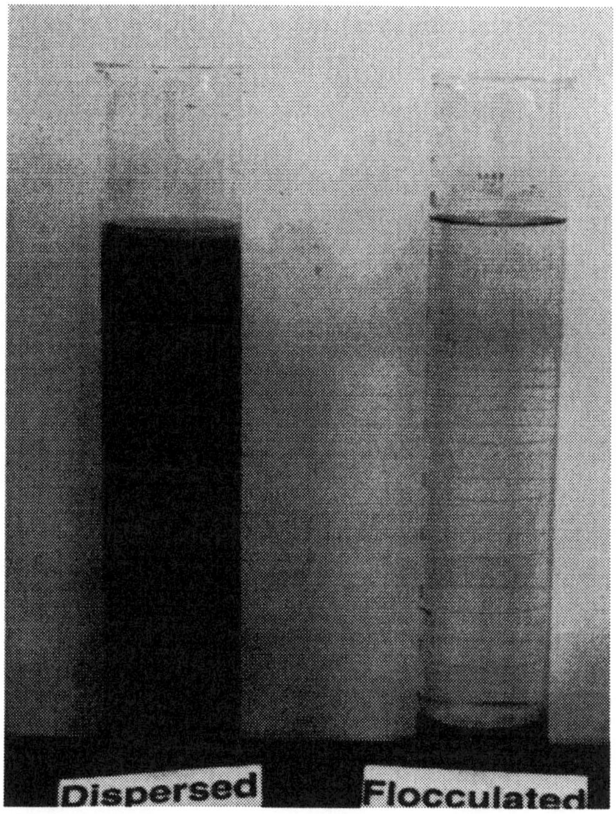


**Fig. 5.8—Contrast of a dispersed (left) vs. a flocculated (right) soil suspension.**

main suspended (**Fig. 5.8**). This condition can be contrasted to a "flocculated" suspension, where clays settle out rapidly.

Suspensions become dispersed when clay particles act independently. In dispersed suspensions, hydration shells develop around soil clay particles and prevent them from interacting, forming larger particles (flocculating), and settling out (**Fig. 5.9**). Size of the hydration shell is dependent on many factors, but the type of exchangeable cation is very important. Sodium has a very large hydrated radius in solution. Clays with a high ESP can, therefore, develop large hydrated shells that repel other particles. Large hydrated shells result in dispersed clays that can stay in suspension almost indefinitely. In contrast, most other common cations that saturate the CEC of clays have a smaller hydrated radius than sodium. Calcium-saturated clays do not have sufficient hydration shells to prevent interac-

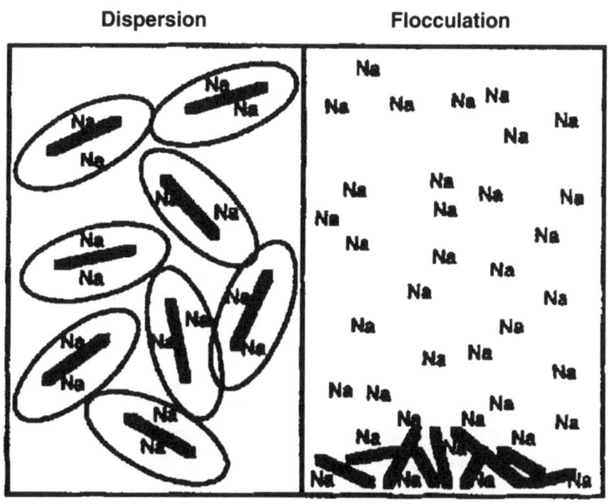


**Fig. 5.10—Simplified drawing of the effects of salt content on flocculation and dispersion of clay suspensions.**

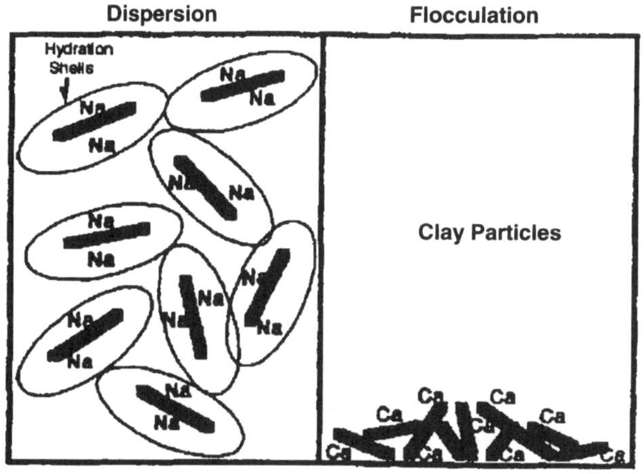


**Fig. 5.9—Simplified drawing of the effects of saturating cations on flocculation and dispersion of clay suspensions.**

tion between particles and commonly flocculate and settle out of suspension (Fig. 5.9).

An additional chemical factor that determines whether a clay flocculates or disperses is the salt concentration. At very high salt concentrations, sodium ions compete for water molecules to hydrate themselves (**Fig. 5.10**). At high salt content, the hydrated shell around sodium-saturated clays may be small enough for the particles to interact (flocculate) and settle out of suspension. The conditions that favor dispersion of clays are high ESP and low salt concentrations (leaching of salts from the system).

*Physical Effects: Permeability.* Water added to leach salts from soils moves through a maze of pores on its way down through the soil profile. This is accomplished readily in flocculated soils because of low concentrations of suspended solids in the water. A dispersed soil, however, has high concentrations of suspended solids and rapidly plugs pore space and reduces permeability. **Fig. 5.11** illustrates an example of permeability reduction as a soil becomes dispersed. In field situations, permeability may be reduced by several orders of magnitude.

**Fig. 5.12** plots a hypothetical example of electrical conductivity (EC)—which is related to salt content—with depth in a produced-water-impacted soil. Salt concentrations are highest at the surface and decrease with depth. Initially, hydraulic conductivity remains high because of the flocculating effect of high salt content. As salts

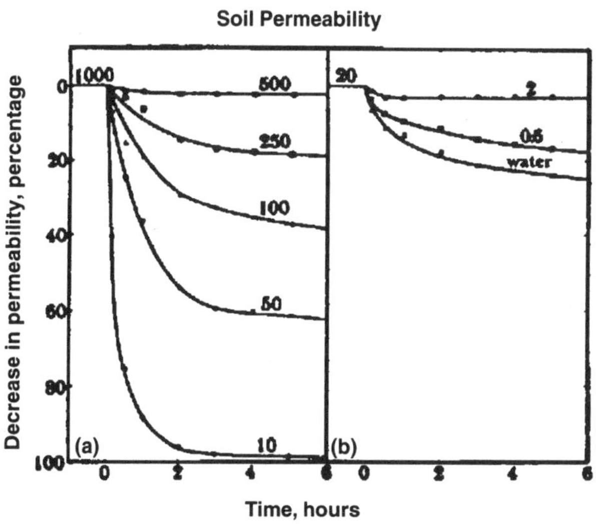


**Fig. 5.11—Permeability reduction as a function of time for a soil saturated with (a) sodium and (b) calcium. Numbers above lines refer to the salt content of leaching solutions (meq/L). The soil is most affected by sodium saturation and leaching with low-salinity water (from Ref. 23).**

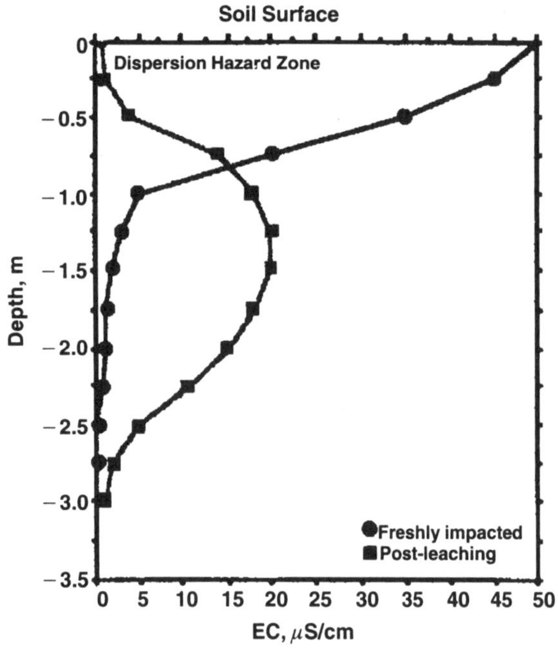


**Fig. 5.12—Hypothetical depth profile of salts in a soil impacted by produced water; dispersion hazard zone is at the surface.**

leach naturally from the system, clays may disperse and adversely affect the physical properties of the surface soil. The upper portion of the soil is leached with relatively pure water during rainfall. In the early stages of leaching, a profile like the post-leaching line may develop (Fig. 5.12). Low salt content in the upper few inches can result in dispersion and "crusting."

Crusting results from dispersion and the mechanical action of raindrop impact. Crusting reduces the infiltration rate of soils and produces two problems: erosion and low leaching potential. Leaching of salts from the rooting zone requires that high-quality water pass through the soil. Crusting diminishes infiltration and maintains the salinity, slowing the natural process of leaching and remediation.

*Physical Effects: Erosion.* Accelerated soil erosion is a common symptom of produced-water-impacted soils. Erosion occurs when clays are dispersed, soil is disaggregated by raindrop impact, rainwater primarily runs off (low infiltration rate), and surface runoff is focused. Erosion is one of the most visible signs of produced-water-impacted soil but can be remediated with methods of saline-soil reclamation and soil-conservation practices.

***Hydrocarbons.*** Petroleum hydrocarbons from oil fields are produced as crudes that vary in oil gravity as well as in the ratio of saturates, aromatics, and nitrogen/sulfur/oxygen compounds. The lighter, more soluble constituents of the oils can affect plants and animals (including microorganisms) because the compounds are highly bioavailable and readily pass across cell membranes.

Separate-phase hydrocarbons in soil can occlude pore spaces, inhibiting the exchange of gases and liquids. Thus, hydrocarbon-saturated soil exhibits decreased effective permeability and inhibited oxygen exchange. This is particularly evident with heavier fractions, such as crude.

*Biological.* E&P hydrocarbons in soils influence soil biota through two separate mechanisms. The first effect is toxicity of the petroleum constituents to biological systems. This is of concern particularly with petroleum mixtures that have not been subject to weathering and biological degradation and therefore contain volatile, soluble constituents. The toxicity effect is most evident with fresh spills or exposures and becomes less of a concern as soluble compounds dissipate and degrade. The second mechanism is the overall ecological impact of microbial degradation of hydrocarbons. Biodegradation results in decreased available oxygen, increased carbon dioxide ($CO_2$), and decreased soluble nutrients (particularly nitrogen). The deficit of oxygen and soluble nutrients is addressed in land-based bioremediation activities where replenishment of such factors as oxygen and nitrogen is paramount.

**TABLE 5.3—CONCENTRATIONS OF METALS IN SOILS AMENDED WITH E&P WASTE**

| Element | Metal Concentration (mg/kg) |
|---|---|
| Arsenic | 41 |
| Barium | 180 000 |
| Boron | 2* |
| Cadmium | 26 |
| Chromium | 1 500 |
| Copper | 750 |
| Lead | 300 |
| Mercury | 17 |
| Nickel | 210 |
| Zinc | 1 400 |

*Measured in mg/L.

*Biological: Oxygen.* Oxygen (as a constituent of air) in soil diffuses through the larger pore spaces and then through the water film, where it is used for aerobic respiration by soil biota and microbiota. Oxygen limitations occur in soils impacted by hydrocarbons because of intense microbiological activity involved in petroleum biodegradation. The presence of large quantities of degradable petroleum hydrocarbons (compared with background concentrations of native-soil organic matter) typically results in increased microbial respiration to the extent that oxygen becomes limiting in such soils. Therefore, decreased concentrations of soil oxygen are available for respiration because oxygen is used as soon as it diffuses into the water film and the normal oxygen flux of a hydrocarbon-contaminated soil is disrupted.

*Biological: Metals.* Metals can be added to soils through several E&P activities, but the most common is through land application of drilling waste. Most drilling wastes are disposed of on site by land spreading, land treating, land farming, or burial. To address the issue of land application of E&P waste, the American Petroleum Inst. (API) performed a risk-based evaluation of the techniques.[24] Metals and the mineralogical forms used in drilling fluids were the focus. The study included 14 pathways of land-applied E&P waste exposure to plants, animals, and humans. **Table 5.3** summarizes the study with a list of maximum concentrations of metals in soils amended with E&P waste.

**5.2.3 Characterizing Soils Affected by E&P Activities.** ***Salts.*** Soluble salts are measured in water extracts of soils, but the water/soil ratio may vary widely. Ratios commonly used for water extracts of soils include saturated paste, 1 : 1, 2 : 1, and 5 : 1. It is very important to specify the extraction ratio when ordering analyses or at least receive the ratio with sample results. Similarly, the soil water content at saturated paste should always be requested when a saturated-paste analysis is performed.

Saturated-paste extract is the best estimator of salinity encountered by plants grown in affected soil. Most agronomic research concerning salt tolerance of plants is performed with saturated-paste-extract data. Laboratories commonly want to use ratios of 2 : 1 or higher because of the ease of analysis. Saturated-paste extracts must be obtained by vacuum filtration, whereas 2 : 1 (or higher-ratio) extracts often can be separated by decantation.

*EC.* EC of water extracts is the simplest and most cost-effective method to obtain estimates of salt content. Salt content in solution and EC are fairly well correlated over the range of salt content in soils. The empirical relationship developed for USDA Handbook 60[21] is EC (in μ℧/cm ) × 640 = mg salt/L. EC measurements are the primary diagnostic key for assessing soil salinity. Salinity-hazard classes and salt tolerance in plants are based primarily on EC measurements.

*Specific Ions: Chloride.* Measurement of chlorides in aqueous soil extracts is common because of regulatory requirements. In reality, measurement of cationic components (e.g., Na, Ca, and Mg) provides much more information about the potential behavior of an affected soil. Chlorides are viewed more as an estimate of total salt content. EC provides similar information at a fraction of the cost.

Anionic constituents typically are used for fingerprinting sources of salts. Anions are particularly good for this exercise because they do not interact as readily with soils, as do cations. Measurement of chlorides is not sufficient to perform a fingerprinting exercise. Measurement of all major anions (including bromide) is suggested if fingerprinting is a data requirement.

*Specific Ions: Boron.* Some produced waters may contain sufficient concentrations of boron to affect plant growth in produced-water-impacted soils. High levels of boron in produced waters are a basinwide effect; therefore, boron analysis is not required unless the basin is prone to high boron or unless boron concentrations in the field are unknown.

The range of boron concentration that causes specific ion toxicity to plants varies widely, depending on the species and the method of analysis.[25] A value of $< 2$ mg/L is commonly used as an evaluation criterion; this value also can be used to evaluate produced waters with respect to boron hazard.

*ESP.* ESP is a fundamental soil property when determining a soil's propensity to disperse. Refs. 21 and 26 provide information on methods for ESP analysis. Highly saline soils require modification of some analytical techniques; Refs. 21 and 26 include these modifications.

In some cases, the effect of soluble salts on ESP measurements can be significant. When measuring exchangeable sodium (ammonium acetate extractable) of saline soil, a soluble-sodium analysis (water extractable) must always be performed. Results of the water extract must be subtracted from the ammonium acetate extract to determine exchangeable sodium. The individual requesting this analysis must understand the potential for error, because experience has shown that many laboratories do not perform these corrections unless requested to do so. Errors result in artificially high ESP.

*SAR.* SAR is a property of fluids and not soil solids. SAR determination may be performed on water extracts of soils or on potential irrigation waters (including beneficial use of produced waters). Fig. 5.5 showed the relationship between ESP and SAR. SAR determination is a relatively inexpensive analysis compared with ESP analysis and can yield information that can be used in a similar manner.

***Hydrocarbons.*** The choice of techniques for analysis of hydrocarbons in soil requires careful attention to the source of the original material if known (e.g., crude oil or light or heavy refined products) and to any changes that might have occurred to the hydrocarbon in the soil. The two general categories of hydrocarbon analyses commonly used by analytical laboratories are total petroleum hydrocarbons and oil and grease. Total petroleum hydrocarbons measured by gas chromatography can provide more accurate results if the laboratory uses internal standards and baseline integration.

*Total Petroleum Hydrocarbons.* Total petroleum hydrocarbons is the term that encompasses analysis of total extractable hydrocarbons or total recoverable hydrocarbons in soil. A commonly applied (or misapplied) method is U.S. Environmental Protection Agency (EPA) Method 418.1, total petroleum hydrocarbons by freon extraction, silica or alumina cleanup, and infrared detection.[27] Several aspects of this method can be modified to improve accuracy.

Cleanup of the extract is performed to remove nonhydrocarbon constituents that might otherwise show up in the analysis. These constituents comprise polar compounds, such as fats, plant waxes, some of the microbial biomass, and partially oxidized intermediates of microbial degradation. The prescribed mass of silica (or alumina) specified in the U.S. EPA method is often overwhelmed by the biodegradation byproducts in a soil, and the silica becomes saturated before removal of all polar compounds is complete. The amount of silica used should be sufficient to prevent saturation with polar intermediates.

Infrared spectrophotometry responds differently to different classes of hydrocarbons and therefore can greatly under- or overestimate the concentration of hydrocarbons in the soil. Its use is most appropriate for petroleum compounds that contain primarily fully saturated aliphatics, with few aromatics or branched and cyclic aliphatics.

*Oil and Grease.* Oil and grease (U.S. EPA 413.1) is a gravimetric analytical technique that does not include a cleanup step to remove polar constituents.[27] In general, gravimetric methods are more applicable to hydrocarbon mixtures that have few volatile constituents. The hydrocarbon concentration in a highly biodegraded oily soil may be overestimated with this method because polar-degradation intermediates will be quantified as hydrocarbons.

***Metals.*** Analytical techniques for metals in soil may vary, depending on the relevant government regulatory agency. The difference is usually in the extracting solution used. The most common technique is the U.S. EPA 3050 extraction.[27] Louisiana or states following the format of Louisiana 29-B regulations[26] use different metal extracts. The primary metal of exception is barium (Ba). The metal extraction for Louisiana 29-B regulations does not extract the majority of barium from samples containing barium sulfate ($BaSO_4$). An alternative extraction was devised for barium and is referred to as "true total barium."[13,26]

**5.2.4 Fate and Transport of E&P Constituents in Soils.** Soil is a complex system of porous material with a wide range and distribution of pore sizes. This distribution of pore sizes is responsible for the hydraulic conductivity, infiltration rate, aeration status, and water-holding capacity of soils. These physical properties in turn govern the physical transport of E&P contaminants through the soil. Chemical properties of soils also affect the solubility and mobility of E&P contaminants in soils.

A discussion of the fate and transport of all E&P constituents in soils is beyond the scope of this section. The following discussions briefly illustrate some of the important concepts concerning fate and transport of the most common classes of E&P contaminants: salts, metals, and hydrocarbons.

***Soluble Salts.*** *Source.* Produced waters are the source of most soluble-salt contamination around E&P operations. As their name denotes, these salts are highly water soluble and the most mobile of potential E&P contaminants. Produced waters most often are highly saline and dominated by NaCl. Produced waters on the opposite end of the salinity range, however, may have low salt concentrations and be dominated by sodium carbonate salts. The Gas Research Inst. (GRI) recently published a convenient survey of gas-related produced-water compositions across the U.S.[28]

Chemical composition of produced waters may differ regionally or stratigraphically. Chemical composition of waters may provide clues about the origin of produced waters if the source is unknown.[29-31]

*Attenuation.* Soluble salts are very mobile in soils and groundwaters and may be transported at roughly the same rate as water. Anions do not interact significantly with soils and often are used as tracers of water migration in field tests. On the other hand, cations are subject to cation-exchange reactions with the soil. Cation exchange may cause a minor change in the ratio of sodium to other cations in the soil solution. The ionic strength of produced waters typically is very high, however, and the effects of cation-exchange reactions on soil pore water are very difficult to measure.

Because soluble salts are transported readily in water, the most effective attenuation mechanism is to limit the water that enters the soil. This can be achieved by capping with a low-permeability material or by use of plants that extract and evapotranspire water from the soil.

Most soils affected by produced water eventually disperse at the surface. This dispersion results in plugging of soil pores and development of a thin crust or low-permeability zone at the surface (Fig. 5.12). This action limits water infiltration and minimizes transport of salts through the soils.

*Transport.* Release of produced water may be through minor leaks, catastrophic leaks, or past waste-management techniques. The type of release governs the volume and rate of the release and therefore the mechanism of soluble-salt transport.

*Transport: Overland Flow.* When produced water is released in quantities that exceed the infiltration rate of soils, overland flow down ditches or natural surface drains may occur. Although this mechanism is capable of moving large amounts of water and soluble salts, the release is visible at the surface. The visible nature of the flow allows spill response plans to be implemented and water to be physically removed from the surface using temporary containment structures and vacuum tank trucks.

*Transport: Saturated Flow.* Soils are normally part of the vadose zone; therefore, unsaturated-flow conditions dominate most of the time. However, saturated-flow conditions may occur, at least seasonally, in upland and wetlands soils. One example of saturated flow in soils concerns past produced-water-management practices. Before the innovation of produced-water-injection disposal, a common management technique was produced-water evaporation ponds. The hydraulic head provided by ponding resulted in saturated-flow conditions in these surface soils. Transport of soluble salts under these saturated conditions potentially was many orders of magnitude greater than transport resulting from typical unsaturated-flow conditions currently experienced.

*Transport: Unsaturated Flow.* Discussions of unsaturated flow in soils generally resort to complex mathematical formulas even in the initial stages. We attempt to summarize some of these concepts. For further information, soil-physics texts are available.[19,20] Unsaturated flow in soils is a complex phenomenon by which water movement is driven by gravitational and capillary forces. Capillary forces result from the attraction of water molecules to the hydrophilic surfaces of soil particles. Soils of finer-particle-size distribution classes (e.g., clays) usually exhibit higher capillary forces because of larger surface area and smaller-pore-size distribution. Under unsaturated conditions, capillary forces cause water to move from areas of high water content to areas of low water content. Strong capillary forces can often dominate gravitational forces and result in upward flow of water and soluble salts.

Because of evapotranspiration or evaporation from the soil surface, the surface of most soils in arid or semiarid areas is very dry for most of the year. In these cases, it is common for water to move upward in response to capillary forces. As water moves upward, so do the salts dissolved in the water. When waters are evaporated at the surface or extracted from the soil by plants, salts are precipitated at or near the soil surface. Even when rainfall and infiltration do occur, the negative net water balance for the area continues to result in cycling of the salts back to the surface. These effects are most pronounced in areas with high negative water balance, with relatively fine-textured soils, and with water tables close to the surface.

Water balance is critical to predicting the fate and transport of salts in soil. Where a net negative water balance (effective precipitation-evaporation/evapotranspiration) exists, the potential for capillary rise of salts to the surface exists. Estimates of how long soils remain dry in the rooting zone are used to classify soils. In the U.S., soil-survey reports may be used to obtain soil-moisture classification of soils on a county-by-county basis.

Water balance of a site is critical in remediation of produced-water-impacted soils. Covering the impacted soil or pit surface with clean topsoil may provide temporary low salt conditions for revegetation. However, a negative water balance ensures that salts eventually return to the surface. In arid, semiarid, or even some temperate areas, capillary barriers, irrigation, or excavation removal may have to be considered as remediation alternatives.

*Enhancing Transport.* Remediation of soils impacted by produced water often involves leaching of soluble salts to depths greater than the rooting zone of the soil. Careful evaluation of impacts to groundwater is always a prerequisite to this remediation strategy. Tools, such as the fate and transport program VADSAT, may be used to evaluate the risk of using leaching as a form of soil remediation.[32]

If leaching of salts is the remediation alternative selected, optimization of soluble-salt transport is desired. The agricultural research community has been developing management techniques for saline soils for many years. Techniques to prevent or reverse dispersion, maintain optimal infiltration rates, provide an adequate water balance, prevent erosion, and optimize revegetation are available but are site specific. The following are some aspects of saline-soil remediation that all sites would have in common.

1. Prevent or reverse dispersion of the soil through the use of soil amendments, such as gypsum or calcium nitrate.

2. Protect the soil surface (i.e., infiltration) and minimize runoff through the use of mulches.

3. Protect the soil surface from erosion by mulching and shaping to avoid focusing of surface water.

4. Promote vegetation by responsible use of soil amendments (e.g., fertilizers) and use of salt-tolerant species.

Soil scientists can be consulted to develop saline-soil remediation protocols that are region specific and that can be implemented by E&P operations personnel.

***Hydrocarbons.*** *Source.* Crude oil is the most obvious petroleum hydrocarbon that might enter the soil zone during E&P activities. Crude oils have a wide range of physical and chemical properties that affect fate and transport in the soil. Refined petroleum hydrocarbon fractions (products) may also be released during E&P activities. The most common refined products are those contained in fuels, lean oil from gas plants, and oil-based drilling muds.

*Transport.* Transport of petroleum hydrocarbons through soils at E&P sites depends on the characteristics of both the soil and the hydrocarbon material. Low- to medium-gravity oils have very high viscosity compared with water and do not flow readily through soils. When spilled on the soil surface, they are more likely to flow along the surface rather than penetrate deeply into the soil. Typical minor crude-oil spills produce shallow contamination that is readily degraded by natural biological processes or enhanced bioremediation.

High-gravity crude oils and refined products contained in fuels, lean oil, or oil-based drilling fluids have the potential to move more readily through soils because of their lower viscosity. In cases where significant volumes are applied to the soil over time, these oils can pass through the vadose zone to form nonaqueous-phase layers at the interface of the vadose and saturated zones. Underground-storage-tank programs have generated a significant body of literature that describes migration of lower-viscosity hydrocarbons through soils.[33]

Crude oil and refined petroleum products (fuels, lean oil, or oil-based drilling-mud constituents) are not miscible with water and travel through the soil as a separate, continuous phase. These hydrocarbons move only in response to hydraulic head provided by the hydrocarbon phase and do not enter pore spaces occupied by water. Hydrocarbons move into soil pore spaces until gravitational forces exceed the capillary forces that hold hydrocarbons in the pore space. Then, the hydrocarbon phase moves downward to occupy deeper pore space in the soil. The quantity of hydrocarbon held in the pore space by capillary forces vs. gravity is called the residual hydrocarbon saturation. The residual hydrocarbon saturation of a soil depends on pore-size distribution, water content of the soil, and viscosity of the hydrocarbon. A rule of thumb for residual concentration of low-viscosity hydrocarbons is one-third of the water-holding capacity of the soil.[34] Values for kerosene retention in coarse materials range from 1.25 $g/m^3$ (stony sediments) to 10 $g/m^3$ for fine sand and silt.[35]

Rate of hydrocarbon migration in soils is strongly dependent on soil physical properties. Hydraulic conductivity to petroleum is much higher in sandy soils than in clay-rich soils. This higher conductivity stems from the larger pore-size distribution and lower water-holding capacity of sandy soils. Water contained in finer-textured soils decreases the pore space available for migration of hydrocarbons. In finer-textured soil, migration of hydrocarbons may be restricted to large structural voids in the soil and be faster than predicted from soil texture alone. Soil structure is an important aspect of hydrocarbon migration in fine-textured soil and should always be documented when evaluating the potential of a soil to transmit hydrocarbons.

The solubility of most hydrocarbon constituents is relatively low. Compounds such as benzene, however, can partition into the water phase and be transported in a dissolved phase. These hydrocarbons are present in a state that is optimum for biodegradation or adsorption to organic matter in soils.

*Attenuation.* Hydrocarbons in soil can be attenuated by a number of physical, chemical, or biological processes. These mechanisms combine to provide an effective barrier that often prevents hydrocarbons from interacting with the groundwater system.

*Attenuation: Physical and Chemical Mechanisms.* The simplest and an often overlooked mechanism of hydrocarbon attenuation in soils is volatilization of the lighter petroleum fractions. Although a small amount of hydrocarbons may be lost to the atmosphere, residual hydrocarbons have higher viscosity and often develop a tar-like consistency at or near the soil surface. These residual hydrocarbons typically have high viscous-, low volatile- or soluble-hydrocarbon concentrations and are not mobile in the soil environment. Losses of small amounts of crude oil over time near wellheads tend to cause accumulations of residual hydrocarbons.

Hydrocarbons migrating through soils often are trapped when the hydraulic head of hydrocarbons is removed. Residual hydrocarbons are trapped in pore spaces and are not readily displaced by water. Residual hydrocarbons contained in these soil pore spaces may partition into their more soluble constituents in soil water and be subject to biological degradation.

Dissolved hydrocarbon constituents often do not migrate significantly in soils because of adsorption and biodegradation processes. Dissolved petroleum hydrocarbons in soil water have a great affinity for organic matter and mineral surfaces of soils. The partitioning of hydrocarbons from soil solution to soil solids often is described by partitioning coefficients. Adsorption usually is correlated with organic-matter content of the soil, and partitioning coefficients, $K_{oc}$, are normalized to the organic-matter content of the soil. A wide range of adsorption coefficients for soils exists, depending on the specific organic compound. The literature contains compilations of $K_{oc}$ values.[36] Detailed treatment of partitioning coefficients and adsorption is beyond the scope of this text.

*Attenuation: Biological Mechanisms.* Microbial communities in soil comprise bacteria, fungi, algae, and microinvertebrates. These communities function in extremely complex consortia to degrade organic matter and cycle inorganic nutrients. Petroleum hydrocarbons are readily degradable in most soils by the indigenous microbial communities under the proper concentrations of inorganic nutrients and oxygen (or alternative electron acceptor).

The natural rates of biodegradation that occur in soil typically are slow because of the limitation of one of several factors required for ongoing biodegradation. The factor in soil that most often becomes limiting to petroleum biodegradation is oxygen because the rate of oxygen diffusion into the soil pore spaces is slower than the rate of microbial utilization of the oxygen.

Petroleum-hydrocarbon biodegradation is facilitated by solubilization of the hydrocarbon into the soil aqueous phase. Thus, highly soluble petroleum constituents are more readily degradable than less soluble ones. Size and structure of hydrocarbons are also factors that affect their biodegradability. In general, biodegradability decreases with increasing carbon-chain length. Structure complexities, such as branching, multiple bonds, and cyclic structures, also impair biodegradation.

Biological attenuation of hydrocarbons can occur under aerobic or anaerobic conditions in the saturated and unsaturated zones. The overall effect of petroleum biodegradation is to make the remaining hydrocarbons less bioavailable and therefore less mobile and toxic.

***Metals.*** *Source.* Drilling fluids contain clays, barite, formation cuttings, and other naturally occurring materials that may have a range of metal concentrations. These compounds, in addition to bases, corrosion inhibitors, and other drilling-fluid constituents, determine the metal content of E&P wastes. The fate and transport of metals depends strongly on the form of the metal, geochemical behavior of the metal, drilling-waste matrix, leaching regime, soil matrix, and geochemical behavior of the vadose and saturated zones. The most effective way to approach fate and transport issues is through risk-based analysis of metals in E&P waste.

*Risk-Based Approach to Metals Management.* API published metals-management guidelines for land application of E&P waste to soils.[24] Development of these guidelines involved a risk-based analysis of exposure of humans and the environment to E&P wastes. The method closely paralleled development of metals criteria for land application of municipal waste.[37,38] The API study considered 14 pathways by which exposure of metals could occur. Table 5.3, which contains allowable metals concentrations at sites where E&P wastes have been land applied, summarizes the results of this analysis. For the purposes of this study, E&P wastes were equivalent to drilling wastes.

The fate and transport of barium was an important aspect of developing metals guidance values. Barite ($BaSO_4$) is the form of barium used in E&P applications. Its use as a weighting agent in drilling-fluid systems greatly decreases the risk of blowouts during drilling. Barite solubility is extremely low compared with the aqueous concentrations required to affect human health and the environment adversely. Barium also interacts strongly with the cation-exchange sites of soil clays. Considering the geochemical behavior of barite in upland soils, a barium concentration of 180 000 mg/kg (30% $BaSO_4$) is considered protective of human health and the environment.

*Other Metals.* Cooling-water systems are necessary for heat-exchange equipment used in gas-plant operations. Corrosion inhibitors are necessary in these cooling-water systems to extend the life of heat exchangers and other equipment. Chromate corrosion inhibitors are very effective and were in common use until recently. Loss of treated cooling water from these systems led to chromium migration at a number of operating and former gas-plant sites.

Chromate is an anionic $Cr^{+6}$ compound that is relatively mobile in soils. $Cr^{+6}$ is readily reduced to $Cr^{+3}$ by metallic or ferrous iron. Attenuation of chromate migration in soils is commonly accomplished with ferrous sulfate or organic compounds that result in a low reduction/oxidation potential and subsequent release of indigenous ferrous iron from the soil.

## 5.3 Vadose Zone and Groundwater

Geologists, engineers, and other environmental specialists are increasingly being called on to address the unwanted presence of E&P-related contaminants released into groundwater. This section describes the groundwater environment and discusses the occurrence and fate of E&P-related groundwater contaminants.

**5.3.1 Hydrologic Cycle.** Groundwater constitutes the portion of the Earth's water-circulation system known as the hydrologic cycle (**Fig. 5.13**). Inflow to the Earth's hydrologic system arrives as rainfall or snow melt. A portion of the inflow infiltrates into the subsurface and, when it reaches the saturated subsurface, it is called groundwater. Groundwater flows laterally; in response to gravity; and then again to the surface as springs, discharge to streams, lakes or the ocean or removal by evapotranspiration (evaporation plus transpiration by plants).

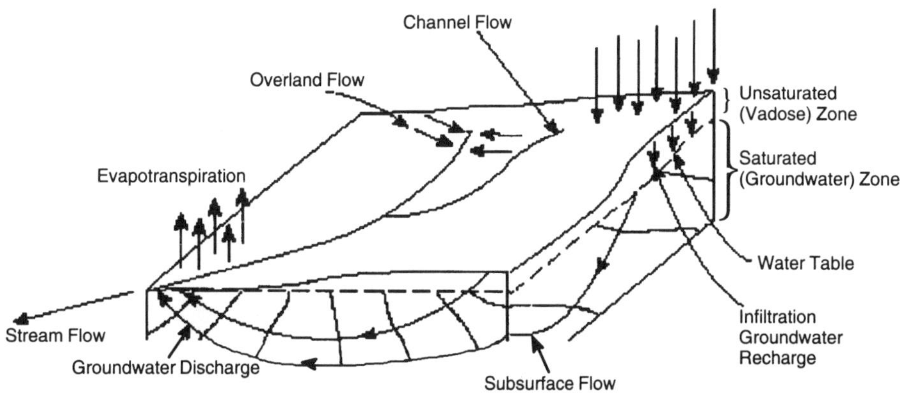


**Fig. 5.13—Schematic representation of hydrologic cycle (from Ref. 39).**

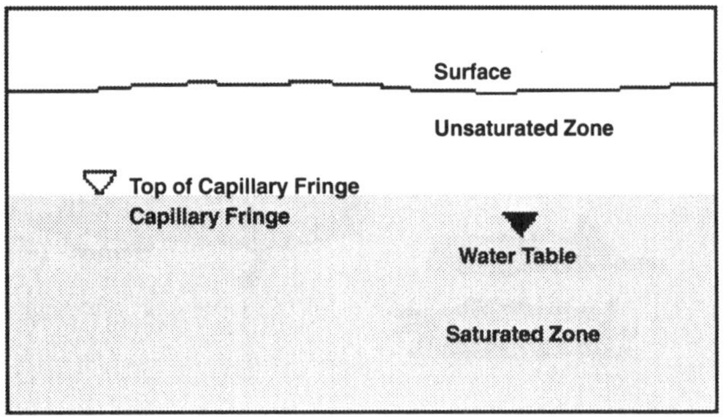


Fig. 5.14—Three primary shallow subsurface hydrologic environments.

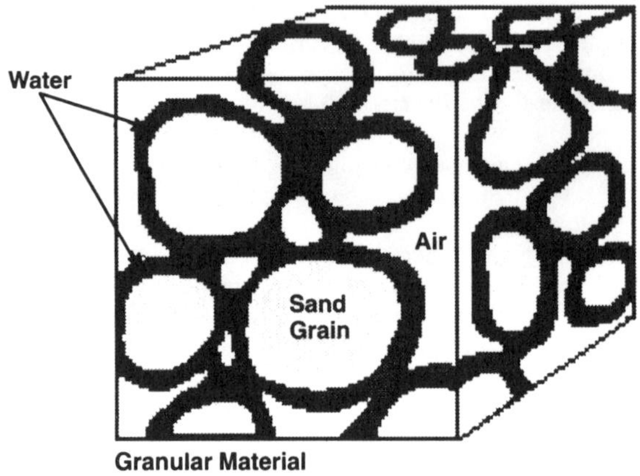


Fig. 5.15—Field capacity: water retained as a film on rock surfaces and in capillary-sized openings after drainage by gravity (from Ref. 40).

**5.3.2 Subsurface Environment.** Three primary hydrologic environments exist in the shallow subsurface: (1) the unsaturated (vadose) zone, (2) the capillary fringe or zone, and (3) the saturated (groundwater) zone (**Fig. 5.14**).

***Unsaturated Zone.*** The unsaturated zone is the portion of the subsurface that lies above the water table. For our purposes, the unsaturated zone is the area above the capillary fringe. Technically, the capillary fringe is part of the unsaturated zone; however, we describe it separately because it has enough of its own unique characteristics.

Above the capillary fringe, soils (or formations) contain a volume of pore water equal to or less than field capacity after gravity drainage has ceased. Field capacity is defined as the amount of water held by soil or formation after drainage by gravity and is expressed as the percentage of water retained by weight vs. the weight of dry soil. This is water left behind as coatings on grains and as globules held in pore throats by surface tension (**Fig. 5.15**). The actual percentage varies according to grain size or type of material, with coarser material having lower field capacity. Field capacities typically range from approximately 3 to 5% in coarse sand to nearly 50% in clay.

Soil may contain less moisture than field capacity because of evaporation or plant transpiration. Generally, if a soil contains less moisture than field capacity, this moisture deficiency must be replaced before water can percolate downward as groundwater recharge. Exceptions to this are when desiccation cracks and fractures and when root and wormholes act as direct conduits during a heavy rain.

***Capillary Fringe.*** The capillary fringe (**Fig. 5.16**) is saturated, but it has pore-water pressure that is less than atmospheric pressure. In this zone, capillary pressures draw water from the water table by molecular attraction between the soil and water and by surface tension of the water. These forces create a negative pore-water pressure with respect to atmospheric pressure, and water is drawn up from the water table below. A capillary fringe can exist only above a water table. Water tables exist only in unconfined conditions. If a formation has permeable material from the water table to land surface, it is unconfined.

In fine-grained formations, the capillary fringe can saturate a significant height of soil above the water table. This occurs because abundant small-diameter pore throats represent closely packed small-diameter capillary tubes. In general, coarse-grained formations have thinner capillary zones than fine-grained formations because the larger-diameter pore throats sustain a lower capillary height.

***Saturated Zone.*** By definition, the saturated zone begins at the water table. If a monitoring well has a screen (or perforations) across the portion of the subsurface that changes from an unsaturated to a saturated condition, the water level in the well marks the water table. This water level, however, is really the level where pore-water pressure equals atmospheric pressure, not necessarily the point where the formation changes from a saturated to an unsaturated condition. Therefore, the water table is really a plane that is defined by pore-water pressure. This is also called a potentiometric surface, although there are other mappable entities that also are called potentiometric surfaces. Water levels measured in piezometers or wells cased into confined or artesian aquifers represent potentiometric surfaces but are not considered water tables.

People usually associate water tables with aquifers. In hydrogeology, an aquifer is a geologic unit that can store and transmit water at rates fast enough to supply reasonable amounts to wells. This is a function of permeability. Water tables, however, frequently cross geologic boundaries and exist in low-permeability soils and rocks. Tight formations may therefore display a water table but do not yield water at high rates. In low-permeability soils or formations, the water table is difficult to determine within a reasonable time after pen-

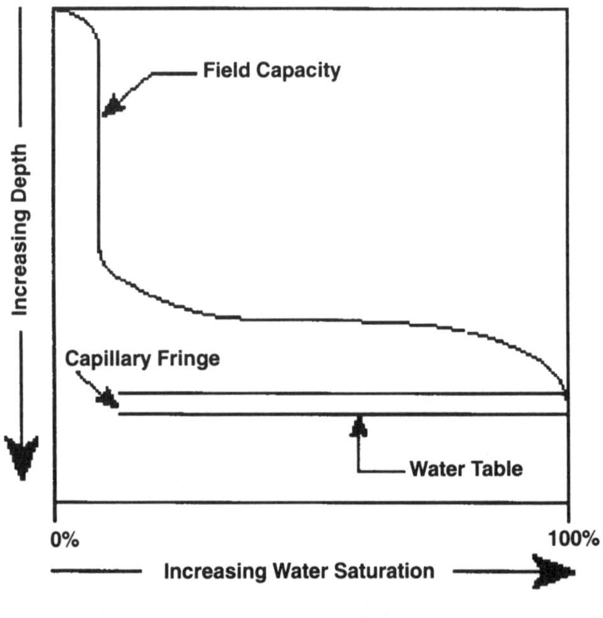


Fig. 5.16—Water-saturation profile of unsaturated zone.

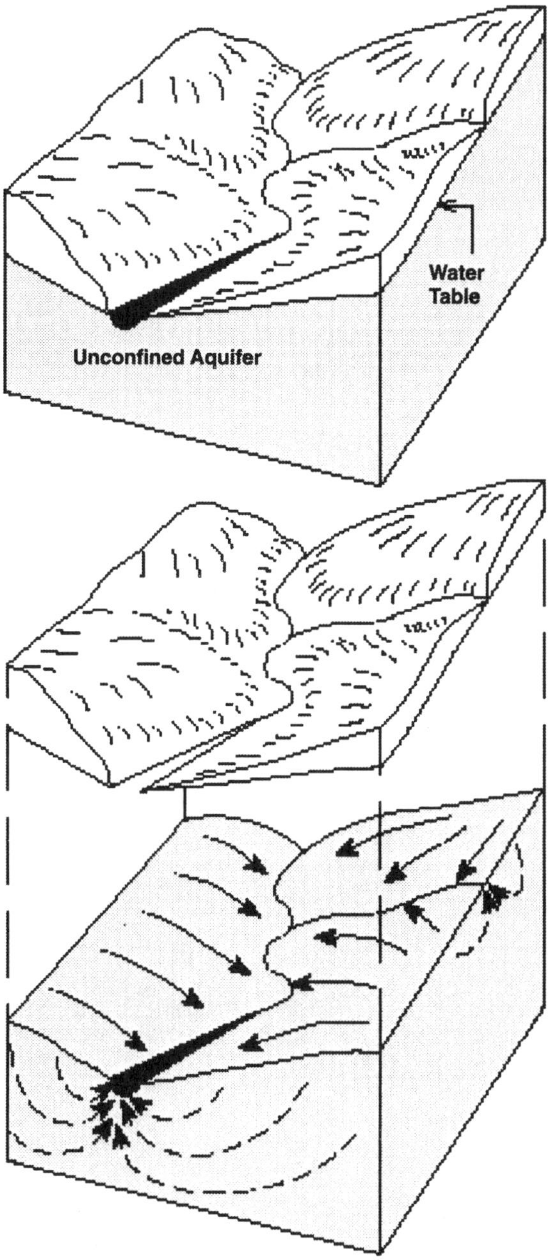


Fig. 5.17—Shaded area illustrates saturation zone below ground surface. The water table tends to follow topography in a subdued fashion. Solid arrows show direction of groundwater movement at the water table; dashed arrows show direction of groundwater flow within the aquifer (from Ref. 40).

etration. Under these conditions, a good rule of thumb is to wait 24 hours before measuring the water table.

If contaminants are released into a tight formation, the fact that the formation is not a recognized aquifer in a hydrogeologic sense does not alter the fact that remediation may be necessary. Regulations frequently refer to the water table as "the uppermost aquifer" and require that contaminated groundwater be addressed whether or not an aquifer in the hydrogeologic sense is actually present. In some states, cleanup standards vary with aquifer (water-table formation) classification and with site-specific conditions. To avoid confusion, water table and lithology at the water table should be used and the term aquifer should be restricted to formations that can yield usable quantities of water.

**5.3.3 Groundwater Movement.** Groundwater is dynamic. When dealing with contamination in the subsurface, it is advantageous to determine the direction and velocity of groundwater flow.

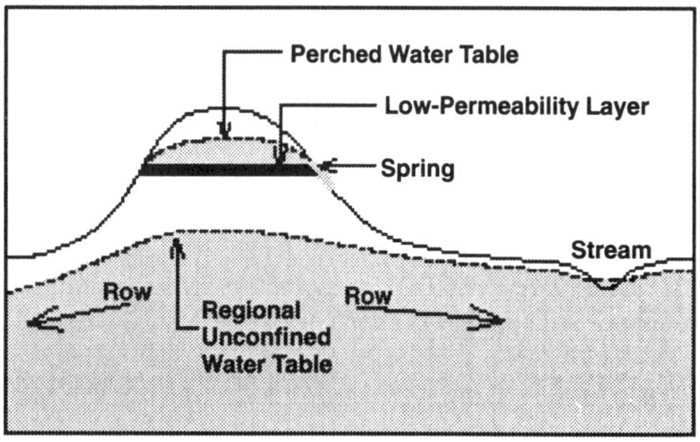


**Fig. 5.18—Spring where perched water table intersects hillside, with regional groundwater table shown below; regional flow is toward the stream.**

***Movement in the Unsaturated Zone.*** In the unsaturated zone, the primary component of flow is vertical as water moves from the surface to the water table. In high-permeability homogeneous zones, a contaminant slug or mass travels vertically with little lateral spreading until it reaches the capillary fringe or the water table. In low-permeability zones, lateral spreading tends to increase because of capillarity. In this case, the formation acts like a sponge. Lateral spreading also increases in nonhomogeneous zones that have stratigraphic variations in permeability.

***Movement in the Saturated Zone.*** In the saturated zone, water flows from areas of higher to lower water-table elevations, or, in a confined aquifer, it travels from areas with a relatively high potentiometric surface to areas with a lower potentiometric surface. If an aquifer is unconfined and relatively homogeneous, the water table tends to follow topography in a subdued fashion (**Fig. 5.17**). Groundwater therefore tends to flow "downhill" toward streams, rivers, and other bodies of water (discharge areas). If an impermeable layer is encountered as a lower boundary, water may flow laterally along the barrier until it emerges as a spring or seep. This is a perched water table (**Fig. 5.18**).

The general direction of groundwater flow in the saturated zone of relatively homogeneous and isotropic materials is easy to determine because it is approximately perpendicular to equipotential lines. Lines of equal water-table elevation are equipotential lines (the potentiometric surface contoured). For a single aquifer or water table, if three or more measurements of water-table elevation can be made (from wells, streams, seeps, and other such sources), equipotential lines can be drawn and direction of flow determined (at least three points are needed to define a plane). Of course, the more data points that are available, the better the direction of flow can be predicted. In heterogeneous or anisotropic media, the direction and velocity of flow are harder to predict. Higher-permeability zones caused by stratigraphic variations, fractures, solution changes, and other such factors become preferred paths of saturated flow. Preferred paths are hard to predict and may require detailed stratigraphic or historic analysis of a site.

***Rate of Flow in the Saturated Zone.*** The rate of of groundwater flow in the saturated zone is calculated with Darcy's law, which states that

$$q = KgA. \quad (5.1)$$

For a problem with $K = 30.48$ m/d, $g = 0.002$, and $A = 278.7$ m$^2$,

$$q = 30.48 \times 0.002 \times 278.7 = 16.99 \text{ m}^3/\text{d} \approx 17\,000 \text{ L/d.}$$

$q$ can be given in any volume/time measurement. The units of $q$ depend on the units used for $K$, $g$, and $A$ in the calculation. Consistency in units of measure within a given calculation is mandatory.

Hydraulic conductivity, $K$, is a function of the permeability of the medium and the density and viscosity of water. Its units are length/time. Regulations dealing with disposal sites often express $K$ in centimeter per second. This is not groundwater velocity. $K$ can

| TABLE 5.4—WATER CLASSIFICATION BY TDS | |
|---|---|
| Class | TDS (mg/L) |
| Fresh | <1000 |
| Brackish | 1000 to 10 000 |
| Saline | 10 000 to 100 000 |
| Brine | >100 000 |

be estimated for granular deposits on the basis of grain size and can be determined in the laboratory by use of samples of unconsolidated granular deposits or from rock cores. However, the most representative values of $K$ are obtained by field testing of wells with pumping or slug tests.

Hydraulic gradient, $g$, is measured in the direction of flow. It can be measured along any line perpendicular to lines of equipotential and consists of the drop in elevation between two points on this line divided by the distance between those points. Units are any form of length/length measurement.

$A$ is the cross-sectional area of the aquifer (or of the area of interest) perpendicular to the direction of flow. For an aquifer consisting of a buried Pleistocene stream channel 6.1 m wide by 3.05 m thick (saturated thickness), $A = 18.6$ m$^2$. To determine the rate of flow through the upper 9.1 m of an aquifer beneath a 30.48-m-wide disposal pit, $A = 278.7$ m$^2$.

**5.3.4 Chemistry of Natural Groundwater.** All groundwaters contain dissolved inorganic minerals. TDS is a broad and useful measure of water salinity. Reported TDS content of groundwater ranges from <25 to >300 000 mg/L. **Table 5.4** classifies water on the basis of TDS content.

Type and concentration of naturally occurring dissolved solids in groundwater depend on the geologic environment, age, and location of the groundwater.[41] Groundwater originates as rainfall or snow melt that infiltrates through the soil and vadose zone. During infiltration, the percolating water, which begins with almost no mineral content, dissolves minerals from the soil and rock through which it flows. At the same time, the oxygen content of the infiltrating water normally decreases and its $CO_2$ content and pH normally increase.

The major cations in groundwater are sodium, potassium, calcium, and magnesium. The major anions are bicarbonate, sulfate, and chloride. Shallow groundwaters tend to be low in TDS (<1000 mg/L) and most often are bicarbonate waters. Deeper and older groundwaters are usually higher in TDS and are more likely to be dominated by sulfates and chlorides. Oilfield reservoir waters, which the petroleum industry must handle and dispose of, are principally NaCl waters that are more saline than seawater and are often brines with TDS >100 000 mg/L. **Table 5.5** lists elements commonly present in oilfield waters and their concentrations.[42,43]

Shallow groundwaters normally contain little dissolved gas other than $CO_2$ and minor amounts of oxygen and nitrogen. However, shallow groundwaters can also occasionally contain small amounts of other gases, including methane and hydrogen sulfide.

**5.3.5 Potential E&P Effects on Groundwater.** Onshore E&P activities that can result in significant groundwater contamination include drilling, completions, and abandonments of oil and gas wells; field operations, including enhanced-oil-recovery (EOR) activities; and production, separation, and in-field transportation of oil, gas, and reservoir water. None of these activities needs to result in groundwater contamination, but all can do so.

The following are some of the more significant historic causes of E&P-related groundwater contamination.

1. Vertical downward percolation of reservoir water, drilling-fluid constituents, and petroleum and oilfield wastes from pits used during drilling and holding or disposal pits used during production.

2. Vertical upward movement and lateral escape of reservoir water, hydrocarbons, and hydrocarbon gases through wellbores of active and abandoned wells.

3. Vertical downward movement from spills or leaks of reservoir water, hydrocarbons, and oilfield chemicals and wastes.

| TABLE 5.5—CONCENTRATIONS OF ELEMENTS COMMONLY FOUND IN OILFIELD WATERS | |
|---|---|
| Element | Concentration |
| Sodium, chlorine, % | |
| Calcium, sulfates, %/ppm | |
| Potassium, strontium, ppm | >100 |
| Aluminum, boron, barium, iron, lithium, ppm | 1 to 100 |
| Chromium, copper, manganese, nickel, tin, titanium, zirconium, ppb (in most oilfield waters) | 1 to 1000 |
| Beryllium, cobalt, germanium, lead, vanadium, zinc, ppb (in some oilfield waters) | 1 to 700 |

Most known instances of groundwater contamination from these causes occurred in the past, before environmental implications of early oilfield practices were understood.

In our experience, of the existing impacts to groundwater, contamination from highly saline reservoir waters historically discharged into unlined pits is the most frequent and affects the largest volume of groundwater. Second in frequency of occurrence, but affecting a much smaller volume of groundwater, is the historic subsurface escape of reservoir waters from improperly constructed or improperly abandoned wells. Many well-documented instances exist of groundwater contamination by dissolved solids from naturally occurring salt deposits, brine seeps, and anthropogenic sources other than oilfield-related ones.[44] Therefore, occurrences of shallow salinized groundwater are not necessarily a result of oilfield activities. Various geochemical methods have been described to distinguish salination by oilfield brines from salination by other sources.[45]

Because of the very high TDS content of many petroleum reservoir waters, a small amount of such water can often contaminate a large volume of fresh water to higher-than-desirable TDS levels. With the previous example of a volumetric flow rate of 17,034 L/d under an unlined brine disposal pit, if the pit were leaking 0.16 m$^3$/d (159 L/d) of brine with 150 000 mg/L of TDS, it would have the capability of increasing the average TDS content of the groundwater flowing beneath the pit by approximately 1200 mg/L. If the natural groundwater had an original TDS content of 250 mg/L, a relatively fresh water, its level after contamination would be approximately 1670 mg/L, a brackish water that would be undesirable as a drinking water. In addition to TDS and chloride contamination from oilfield waters, contamination by accompanying minor and trace elements can also occur. However, the minor and trace elements seldom are a problem by themselves because high TDS and chloride concentrations prevent use of a brine-contaminated water until dilution and other effects have reduced the minor-/trace-element concentrations to below levels of concern.

Instances of contamination of groundwater by petroleum and natural gas from E&P activities do occur but are less common in the oilfield E&P environment than contamination from oilfield reservoir water. Crude petroleum can be released to the surface or shallow subsurface by spills, seepage from pits, tank or pipeline leaks, and leakage from wellheads or wellbores.

Crude petroleum varies in viscosity, and its viscosity frequently determines its mobility in the subsurface. A very close correlation exists between oil gravity and viscosity; the higher the oil gravity, the lower the specific gravity and the viscosity. If a sufficient volume of high-oil-gravity petroleum (>0.83 g/cm$^3$), such as condensate, is released, it can migrate through the unsaturated zone to the water table. Lower-oil-gravity petroleum tends to be more viscous and is frequently retained as residual in near-surface soils. The porosity and permeability of the near-surface medium also affect migration of crude petroleum. **Table 5.6** shows oil gravities of dif-

**TABLE 5.6—SOME TYPICAL GRAVITIES OF CONDENSATES, CRUDES, AND REFINED PRODUCTS**

| Product | Specific Gravity | Oil Gravity (°API) |
|---|---|---|
| Condensate | 0.73 to 0.80 | 62 to 45 |
| U.S. crude | | |
| Gulf coast | 0.78 to 1.01 | 50 to 9 |
| California | 0.78 to 1.02 | 50 to 7 |
| Midcontinent | 0.80 to 0.93 | 45 to 20 |
| Gasoline | 0.73 to 0.76 | 62 to 55 |
| Kerosene | 0.81 | 43 |
| Diesel fuel | 0.83 to 0.88 | 39 to 29 |
| Heavy fuels | 0.88 to 1.05 | 29 to 3 |

ferent types of crude petroleum and of refined products. Higher-oil-gravity petroleum is more likely to contain light aromatic compounds, which are soluble in groundwater.

Because of differences in the types and relative abundance of aromatic and polynuclear aromatic compounds in the crudes of different gravities, oil gravity of some crudes and condensates can be estimated in the field from the color of the cut fluorescence under ultraviolet light (**Table 5.7**). Very light condensates ( $>0.76$ g/cm$^3$) with only single-ring aromatic compounds fluoresce outside the visible spectrum and, therefore, may appear to the unaided human eye to have no cut fluorescence. This fluorescence can be detected with a spectrofluorometer.

**5.3.6 Contaminant Transport.** Average velocity, $\bar{v}$, of groundwater flow and of unretarded dissolved groundwater contaminants is calculated with a variation of Darcy's law which states

$$\bar{v} = Kg/\phi, \quad (5.2)$$

where $\phi$ = effective porosity and is a dimensionless number given as a decimal; i.e., $\phi = 20\% = 0.20$. For a problem with $K = 30.48$ m/d, $g = 0.002$, and $\phi = 0.2$,

$$\bar{v} = (30.48 \times 0.002)/0.2 = 0.3 \text{ m/d}.$$

Generally, groundwater velocities in unconfined formations average less than 0.3 m/d even in very permeable media. Exceptions include areas of very high permeability or steep gradients close to streams or other sources of discharge. Groundwater in deeper formations usually tends to travel more slowly than near-surface groundwater (**Fig. 5.19**).

The groundwater velocity calculated with Eq. 5.2 is an average value for molecules of groundwater or ions or molecules of a contaminant that is unretarded by chemical reaction or sorption. Chloride is such a contaminant. On the other hand, sodium is not because it can be subject to ion exchange for other common cations, particularly calcium or magnesium.

**TABLE 5.7—COLOR OF FLUORESCENCE VS. OIL GRAVITY**

| Oil Gravity (g/cm$^3$) | Color of Fluorescence |
|---|---|
| $>0.97$ | Brown |
| 0.97 to 0.90 | Orange |
| 0.90 to 0.85 | Yellow to cream |
| 0.85 to 0.80 | White |
| $<0.80$ | Blue-white to violet |

The tendency for a dissolved constituent to be retarded is characterized by its distribution coefficient, $K_d$, which, in turn, can be used to determine a relative velocity $v_c/\bar{v}$, where $v_c$ = velocity of a retarded contaminant and $\bar{v}$ = average unretarded velocity. Thus, if an aquifer were contaminated by an NaCl reservoir brine, the chloride ions would be expected to travel down the groundwater potentiometric gradient at the same velocity as water molecules, whereas the sodium ions might travel at a slower average velocity.

In addition to the possibility of retardation during travel in an aquifer, contaminants are subject to hydrodynamic dispersion, which is the combined processes of diffusion and mechanical mixing. Through dispersion, contaminants are diluted and spread out during travel. **Figs. 5.20 and 5.21** depict the effects of dispersion.

As a result of possible retardation and through dispersion, the concentration of a dissolved groundwater contaminant as it travels through an aquifer eventually is reduced to the point where it is not distinguishable as a contaminant.

**5.3.7 Condensate Migration and Transport in the Shallow Subsurface.** Condensate (or similar hydrocarbons) released to the subsurface can partition into four phases (**Fig. 5.22**): vapor phase, residual phase (adsorbed and trapped), free phase (mobile liquid hydrocarbons), and dissolved phase.

***Vapor Phase.*** The vapor phase exists only in the unsaturated zone. Most condensate constituents have relatively high vapor pressures, which means they evaporate rather quickly when exposed to air at 1 atm of pressure. In the unsaturated zone, air flow is restricted but condensate vapors can migrate through porous media by dispersion and by advection caused by changes in temperature or barometric pressure. These changes may be induced inadvertently in basements of buildings or on purpose at vapor-recovery wells (**Fig. 5.23**). The greatest movement tends to take place in the most permeable media. Frequently, these media are channel fills, fractures, or gravel-backfill areas along excavations or utility vaults.

***Residual Phase.*** *Unsaturated Zone (Adsorbed).* If condensate or petroleum products are released to the subsurface, a certain amount

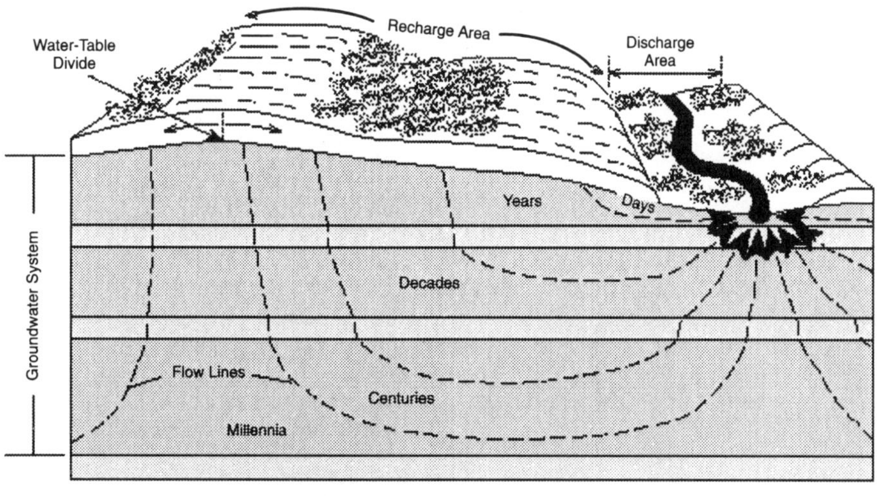


**Fig. 5.19—Residence times for various groundwater flow regimes; the deeper the system, the longer the period of flow from recharge to discharge (from Ref. 40).**

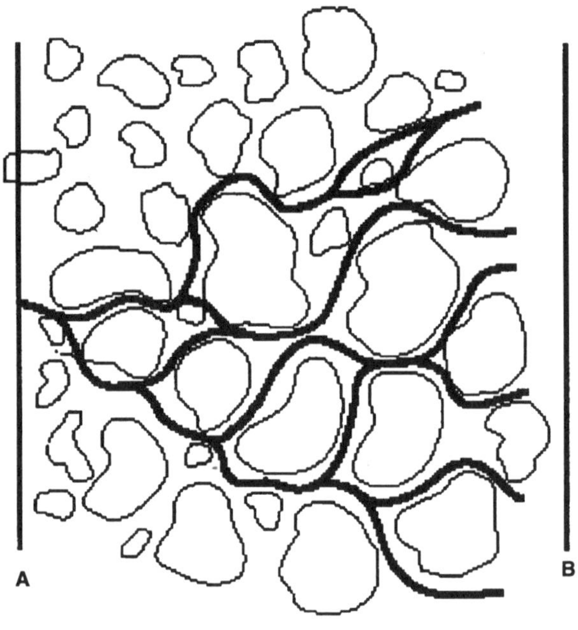


**Fig. 5.20—Flow paths in a porous medium that cause lateral hydrodynamic dispersion from A to B (from Ref. 46).**

is adsorbed as residual in the unsaturated zone in pore spaces or as a coating surrounding soil or formation particles (**Fig. 5.24**). This is similar in concept to formation field capacity or specific retention except that the fluid is hydrocarbon instead of water and we are now dealing with three phases in the pore system: air/oil/water. The percent volume of residual in the unsaturated zone varies according to grain size and other factors, but tends to be in the 0.5 to 4.0% range for porous granular materials (this is total volume, not percent saturation of porosity). This residual tends to continue to release contamination in the form of vapors and as dissolved hydrocarbons if recharge water percolates through it. If the unsaturated zone is very thick or if a petroleum release is small, the entire volume of a release may be consumed as residual. Generally, residual concentrations in the unsaturated zone do not migrate but act as a source, contributing to the vapor and dissolved phases of contamination.

*Saturated Zone (Trapped).* Residual also tends to be trapped in the saturated zone as small globules that are unable to pass from large pores through restricted water-wet pore throats. This occurs when free product passes through it or when the water table rises when free phase is present. Residual in the saturated zone tends to be present in higher percentages than in the unsaturated zone. This is because the third phase, air, is not present in the saturated zone and this volume is taken up by water and hydrocarbon. Obviously, residual hydrocarbon in the saturated zone is a continuing source of dissolved contamination.

Another problem is that residual in the saturated zone can change to free phase. This is a function of the oil or condensate saturation. If the water table drops in an area that contains residual in the saturated zone, the globules of residual may coalesce after water has evacuated most of the pore throats. This increases saturation locally and turns into a free-phase pancake in the capillary fringe,which disappears again with a seasonal rise in the water table. This process can create confusion.

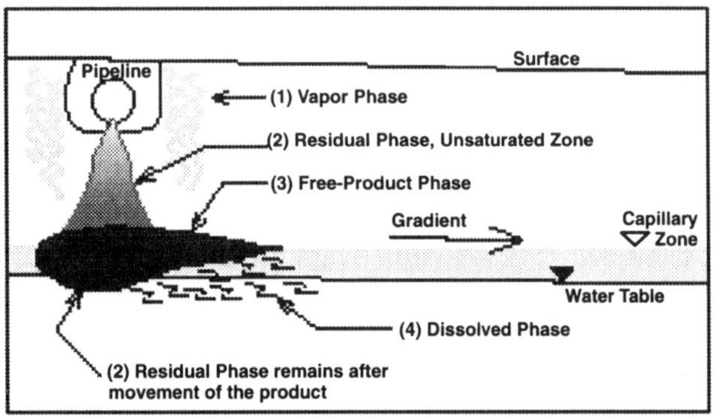


**Fig. 5.22—Condensate partitions into four phases when released into the shallow subsurface.**

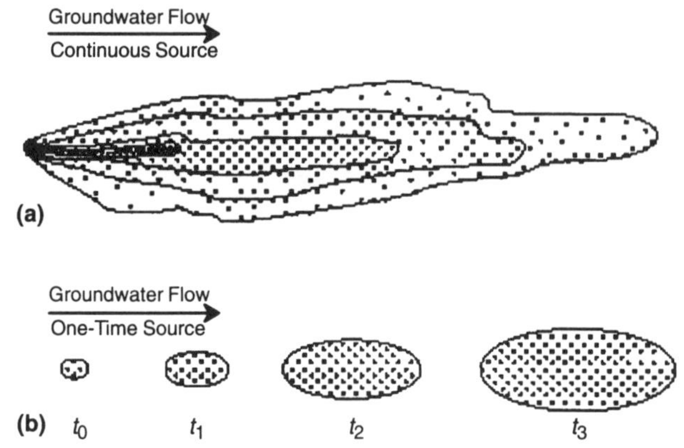


**Fig. 5.21—(a) Development of a contamination plume from a continuous point source and (b) travel of a contaminant slug from a one-time point source; density of the dots indicates solute concentration (from Ref. 46).**

***Free Phase.*** If the volume of release is sufficient to migrate to the capillary fringe, free phase (mobile liquid hydrocarbon) tends to "float" in this environment, somewhat above the actual water table. Free-phase hydrocarbon does not appear to float on the actual water table unless sufficient volume of product is present to depress the mobile-hydrocarbon/water contact down to the water table (Fig. 5.22) or unless the formation is so coarsely grained that little or no capillary fringe exists. As a result, static measurement of free-hydrocarbon thickness in an observation well usually is not indicative of the thickness of the mobile-hydrocarbon layer (**Fig. 5.25**). Product (or hydrocarbon) thickness measured in wells usually exceeds the actual thickness of sediments containing mobile free product. This phenomenon tends to be more pronounced in finer-grained formations because they have a thicker capillary fringe.

Thickness of product in an observation well is a function of hydrocarbon-saturation profiles and of where mobile rather than residual hydrocarbon exists (Fig. 5.25). This, in turn, is controlled by pore diameters; capillary pressures; hydrocarbon density, viscosity, and volume; and interfacial tension between water and hydrocarbon. Fortunately, much of this has been quantified and can be estimated by means of a computer program.[45]

Free phase (also known as nonaqueous phase) moves at a much slower velocity than groundwater because the pore system is water wet and relative permeability to hydrocarbon decreases as saturation decreases. At 100% saturation, 100% of the pore space is filled with a single liquid (100% water or 100% oil). Our system is never

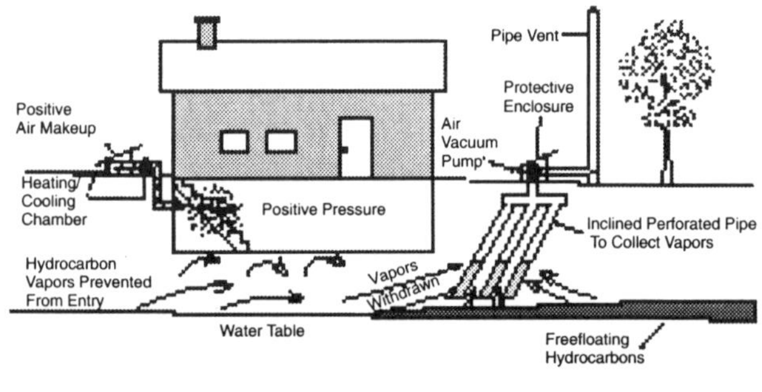


**Fig. 5.23—Generalized view of a soil-venting system installed to induce changes in vapor-phase migration. Vapor flows from high-pressure to low-pressure areas. Basements normally are low-pressure areas because of forced-air heating and cooling systems. In this case, positive pressure is inducted to keep vapors out. This should be done only if vapors are well below explosive limits in the basement; air should never be forced into an explosive environment.**

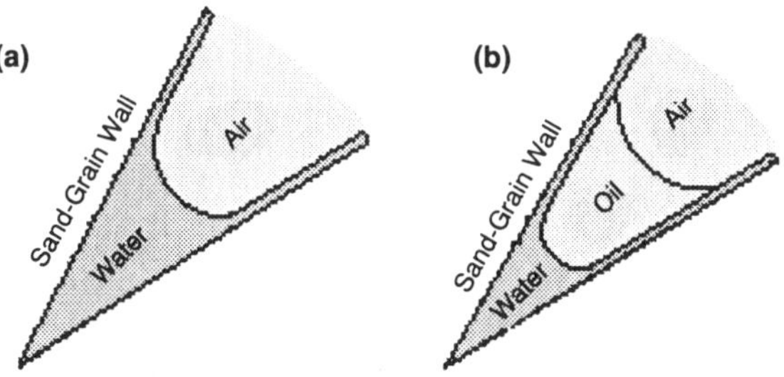


**Fig. 5.24—Residual systems: (a) two-phase system with water in a pore throat and coating a sand grain and (b) three-phase system with water, oil (condensate), and air.**

100% saturated with oil or condensate because the grains are coated with water and smaller pore throats contain water. Therefore, relative permeability to oil always is a fraction of the intrinsic permeability. This means that the "oil conductivity" is less than the hydraulic conductivity. In a practical sense, oil saturations of less than approximately 50% in the capillary fringe are essentially immobile and may be considered residual. Higher oil saturations are mobile but move at a fractional velocity compared with groundwater movement in the saturated zone.

***Dissolved Phase.*** Like crude oil, condensate is a mixture of different compounds. Crude-oil constituents are distilled into products (e.g., gasoline and diesel fractions). Condensate is very similar to refined gasoline. Some gasoline blends contain more than 270 different hydrocarbons.

Condensate generally is considered immiscible in water. In fact, many compounds in condensate dissolve on contact with water. The most commonly used marker compounds for dissolved condensate constituents are aromatic BTEX compounds, which are the constituents most commonly listed by state agencies in the regulation of releases of light hydrocarbons because they are the most water soluble of the hydrocarbon condensate compounds. Under U.S. EPA guidelines, the maximum concentration limit for benzene in public drinking-water supplies is 0.005 mg/L (or approximately 5 ppb). Allowable concentrations for toluene (1 mg/L), ethylbenzene (0.7 mg/L), and total xylenes (10 mg/L) are all substantially higher than the allowable concentration for benzene. As a result, the concentration of dissolved benzene tends to be the limiting factor for most groundwater-cleanup operations.

*Solubility and Equilibrium Saturation of Condensate Compounds.* Solubility alone does not determine the saturation concentration of an individual constituent when a light hydrocarbon is exposed to water. The various components of a hydrocarbon mixture reach an equilibrium saturation in water that is dependent on their water solubility and their relative abundance in the hydrocarbon mixture. For example, maximum solubility of pure benzene in water is approximately 1750 mg/L, but the equilibrium saturation of benzene in water resulting from a condensate/water mixture (similar to gasoline) seldom exceeds 65 mg/L. This is because benzene is soluble to some extent in both condensate (as an organic solvent) and water; therefore, benzene molecules tend to partition between the two solvents in some equilibrium saturation, depending on overall benzene concentration and relative solubilities.

In general, an estimate of the water/oil partitioning of BTEX compounds can be predicted with a derivation of Raoult's law.[47]

$$C_{iw} = R_i x_i \quad (5.3)$$

For example, if a condensate mixture contained 3.7% benzene and the water solubility for pure benzene is 1750 mg/L, the equilibrium saturation of benzene in fresh water that was in direct contact with that condensate would be estimated to be 65 mg/L. Under actual field conditions, dissolved concentrations of benzene (or other constituents of BTEX) rarely exceed 20% of the calculated equilibrium concentration unless free product is present as a sheen or as colloids. In this case, fresh groundwater in contact with a condensate containing 3.7% benzene would seldom contain more than approximately 13 mg/L dissolved benzene. This is because condensate and water are not as thoroughly mixed in the field as they are under laboratory conditions and because dissolved concentrations tend to decrease with distance from the source because of dilution, adsorption, volatilization, and biodegradation.

## 5.4 Oceans

The world's oceans are the most dynamic and difficult domains for man to work in and explore for minerals. This section briefly defines the characteristics of the work environment, the associated interactions between the oceans and offshore oil- and gas-related activities,

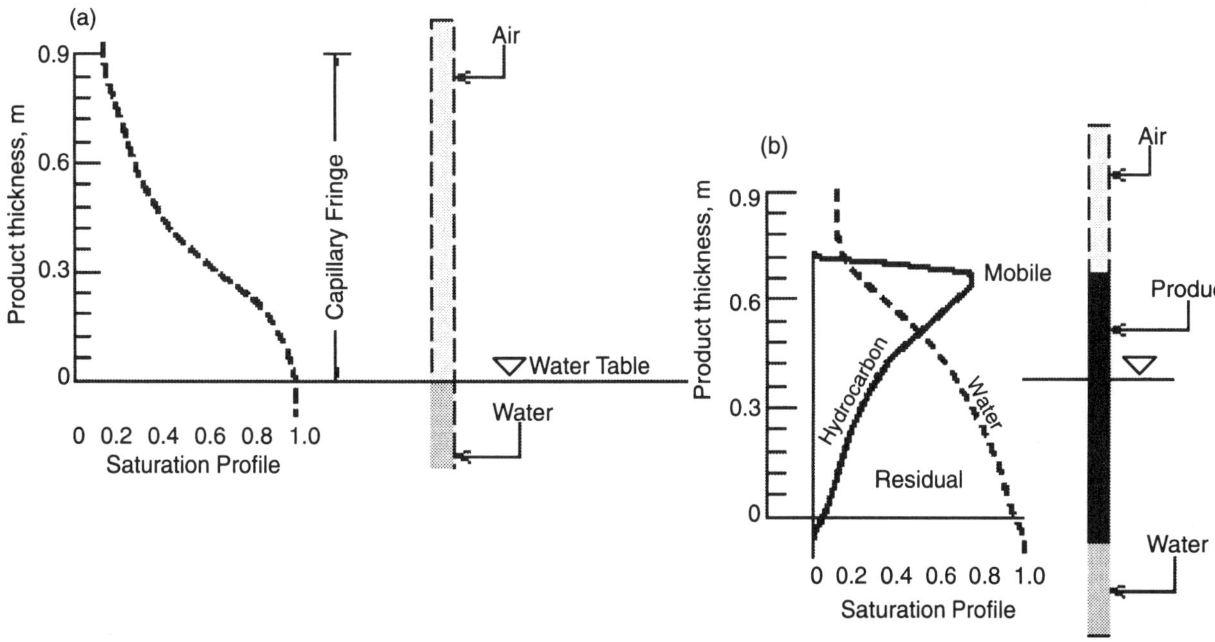


**Fig. 5.25—Illustration of how product thickness in a well does not represent thickness of the mobile product layer; lithology is a fine to medium sand. (a) Saturation profile of water from the water table through the capillary fringe with no product. (b) Saturation profile with product present as residual (immobile) phase in the saturated zone and as free (mobile) phase in the capillary fringe. Free phase was present when product (condensate) saturation exceeded 50% of total porosity.**

and the multiple government and private-sector initiatives to monitor environmental change associated with these activities.

**5.4.1 Introduction.** Since the beginning of the 20th Century, the oil and gas industry has ventured seaward from the coastal boundaries: from drilling piers extending from the beach, to floating barges, to fixed structures, to semisubmersible platforms, to today's modern drillships and tension-leg platforms. This evolution was slow at first because explorationists had to adapt their knowledge of oil and gas drilling onshore to an entirely new environment that posed many formidable operational challenges.

With entire volumes written to describe the intrinsic nature of the oceans, a brief description of the oceans and the industry's activities in them is impossible. This section covers some of the main characteristics of the offshore industry, its potential for environmental impacts, and the nature of the ocean environment as it relates to these important issues. Because of the nature of the limitations of this review, some of the discussions of "possible" environmental impacts may sound like environmental-impact statements. In reality, we have to discuss and understand what the possible impacts are to design studies properly, to write regulations realistically, and to monitor for unacceptable effects properly. The good news is that most of the possible effects have been found, in general, to be of minor consequence.

**5.4.2 The Oceans.** ***Zonation.*** Because the majority of the world's surface is covered by oceans and seas and because they are the original source for the deposition of sediment layers that, over geological time, generate petroleum reserves, the world's oceans are obviously a prime location for E&P. As a specific environment, that presented by the sea has great diversity in water depth, composition, temperature, ice, bottom characteristics, winds, waves, and currents. All these factors have a bearing on the engineering requirements for searching for oil and gas offshore. They also play crucial roles in determining the biological environments of the different regions and the fate and effects of introduced contaminants.

The world's beaches represent the land/water interface. They vary from flat to steep, from fine-grain clays to coarse-grain cobbles and solid rock. They are formed by a variety of geological processes, ranging from volcanic activity to erosion and terrestrial runoff. They are constantly changing in nature and being reworked by a variety of natural forces.

Shaped by a variety of forces, including wind, waves, and tectonic forces, the ocean basins (seafloor) have many different appearances. The shallow continental shelves, ranging from the coast to water depths of 110 to 150 m, can be broad, such as in the Gulf of Mexico or Bering Sea (e.g., $>150$ km), or quite narrow (e.g., 2 to 4 km off the coast of California). On a worldwide basis, the average is approximately 65 km. The shelves usually are typified by a slope gradient of 1:500 (or around 0.2%) and encompass most of the world's commercial fisheries.

The continental slope, the transition zone to the deep sea, is typified by a gradient of 1:20 and ranges from the lower shelf downward to the beginning of the deep sea at approximately 3000 m. Below this depth is the ocean abyss (or the deeps), which encompasses the zone downward to 6000 m and includes 76% of the ocean basins. Only 1% of the ocean is deeper. The mean depth of the seas is approximately 3800 m.

***Topography.*** The ocean bottom can vary considerably in topography, from large expanses of flat mud bottom to areas with considerable hills, valleys, canyons, outcrops, ridges, and other such structures. Detailed bathymetry over the past 50 years has considerably changed our view of the ocean bottoms.

Bottom sediments range from coarse sands to fine clays and muds. Most sediments common to the shelf and slope are terrigenous in origin (i.e., coming from the land through rivers and wind). In contrast, sediments of the deep sea are finer in texture, and mostly of oceanic origin (inorganic red clay and organic oozes, which are skeletal remains of microscopic plants and animals).

***Seawater.*** Seawater is a complex mixture of most known elements. Despite its complex nature, the more abundant dissolved salts are present in fairly constant ratios (known as conservative properties). These are predominantly chloride ion (55%), sodium ion (30.6%), magnesium ion (3.7%), and potassium ion (1.1%). Open ocean waters characteristically have salinities of approximately 35 parts per thousand (range of 33 to 37%), whereas waters closer to shore under the influence of freshwater mixing can range from 0.5 (fresh) to 30% (brackish). In rare instances, hypersalinity ($>40\%$) can occur.

Seawater physical/chemical properties have a variety of effects, including behavior in chemical reactions, freezing points, density, and transmissivity of light. These unique characteristics are important in understanding electrochemical processes (e.g., metal corrosion) and biological processes (e.g., osmotic regulation, photosynthesis).

Several factors influence seawater color, which relates in part to depth of light penetration. Light-penetration depth affects light-dependent biota (e.g., algae, corals). Although man can see to depths of 50 to 100 m, actual light can be detected in clear oceanic waters to depths of approximately 1000 m.

Seawater temperatures have a considerable range, and gradients are both horizontal (e.g., one latitude to the next) and vertical. Surface-water temperatures can range from $-2^\circ$C in the polar regions to 28°C in the tropics. Vertical distributions in the tropics would show surface-like temperatures from the surface down to 50 to 200 m. A rapid drop occurs from 500 to 1000 m, with a more gradual gradient in the deep sea. An example for the low latitudes would be 20°C at the surface, 8°C at 500 m, 5°C at 1000 m, and 2°C at 4000 m.[48] Depending on ocean currents and shelf/slope topography, vertical mixing of water masses can occur (e.g., upwelling currents), bringing colder, more-nutrient-rich waters to the surface.

Dissolved oxygen in seawater, which comes primarily from the atmosphere, can range from 0 to 8 mL/L (although some cases of supersaturation have been noted, primarily as a result of high photosynthetic activity). Most values range from 1 to 6 mL/L. Two of the controlling factors in oxygen-saturation values are water temperature and salinity. The solubility of oxygen in water decreases as temperature and salinity rise.

***Waves, Tides, and Currents.*** The most obvious dynamic expressions of the ocean's power are waves, currents, and tides. Their effects are most often seen along the coastlines, but they also are very evident in offshore regions. All the industry's activities offshore have to take into account wave, tide, and current behavior and effects on operations, both from an engineering as well as environmental standpoint (e.g., fate and effect of pollutants).

Several types of waves occur in the oceans. The most visible are surface waves, which are generated primarily by wind. Wave heights can vary upward to 30 m and can vary in period (time for two consecutive peaks to pass) from a few seconds to large oceanic swells of more than 15 seconds' duration. Waves are important to both surface and bottom structures as well as in the distribution of marine pollutants. Less common, tsunamis, or "tidal" (earthquake-generated) waves, can have significant coastal impacts.

The tides, a result of the gravitational effects of the sun and moon, range from one to two tidal cycles per day. Depending on the Earth's position and the geomorphology of the adjacent sea bottom and coastline, tidal excursions can range from almost imperceptible to as much as 15 m (e.g., Bay of Fundy).

Associated with this daily rise and fall of sea level are corresponding tidal currents. These current speeds can be quite high, exceeding 5 m/s in restricted locations. Currents can be driven by a variety of mechanisms, including wind, water-density and air-pressure differences, and tidal-forcing mechanisms. Current velocities can range from zero to a few meters per second. Subsurface currents affect offshore structures and play a significant role in operations in deep water. Bottom currents play a role in resuspension and transport of marine sediments. Although a complex issue, basic understanding is necessary in all facets of the industry's offshore activities.

**5.4.3 Potential E&P Effects.** As with any user of the ocean, the petroleum industry interacts with and affects the environment in myriad ways. In the following discussion, both the major physical, biological, and chemical effects are addressed. The intent is to provide an introduction to the primary literature related to the key topics, not a detailed treatise.

***E&P.*** The actual delineation between E&P activities is not always clear because companies define responsibilities differently for these two activities. Often, the initial drilling of exploratory wells is done by the production department under the supervision of the exploration organization. These definitional differences are ignored in the following discussion of general activities.

Exploration generally encompasses activities related to conducting seismic surveys, drilling a small number of single wells to test subsurface structures, and delineating the size of a prospective find. Development and production activities are more intense locally and can include installation of numerous platforms in a small geographic region and drilling a number of wells from a single platform. These activities result in a longer duration of drilling discharges and higher overall volumes of all over-the-side discharges. In addition, the overall life of development and production operations (and associated discharges) is longer.

***Acoustics.*** One of the first exploration activities that occurs offshore is geophysical surveys (i.e, seismic). This acoustic profiling of subsurface strata usually is conducted from specially equipped ships that use a variety of sound-energy-generating techniques. Old practices used underwater explosives to generate the high-energy sound waves required. More modern techniques use air guns (sleeve exploders and gas guns are also used) that generate sounds downward and laterally. The high energy from these sound waves can have a variety of possible environmental impacts, including causing behavioral changes in marine mammals and interfering with their communications.

Selected studies have addressed the potential effect on marine fish and birds. The issues include physical damage and avoidance behaviors that might affect fishing success. Avoidance in localized areas has been seen in tests conducted in California[49] and Norway.[50] Tests have shown that direct damage to marine species can occur only very close (within several feet) to the noise-generating source.

Other acoustical issues include the effects offshore of industrially generated sounds, including those from platform and ship machinery, boats and propellers, aircraft, dredging, ice breaking, and other such equipment and procedures. Greene[51] reviewed the variety of sources and their levels. These sounds have potential effects on fish populations (although studies have suggested that the displacement is temporary), marine mammals, and birds. Low-level aircraft flying in marine nesting areas can cause considerable disturbance and possibly impact nesting birds and marine mammals (e.g., seal rookeries).

***Physical Bottom Disturbance.*** *Structural Impact.* A wide variety of industry activities can impact submerged hard bottoms (e.g., banks, reefs, outcrops). Bottom-founded structures, such as barges, jackup platforms, and subsea wellheads, can have localized bottom impacts. In addition, structures and vessels that require anchoring can cause physical impact to the bottom. The large anchors associated with drillships and semisubmersibles can cause damage to the bottom (e.g., hard bottom outcrops and reefs) and to biota. This is because of both the physical presence (e.g., 16 000-kg anchors) and the scouring effects from dragging and chain movement.

*Burial.* Covering and smothering of benthic (bottom) habitat can result from a variety of industrial activities. In those regions where pipeline burial is required, trenching activities displace sediment in the pipeline corridor. This causes a narrow path of disturbance where local fauna may be directly impacted or buried/smothered by the displaced and resuspended sediments.

*Debris.* Although a small issue, a variety of debris, including scrap metal equipment, tools, sand-blast waste, and spent welding rods, accidentally falls to the bottom beneath offshore platforms over the course of years. This bottom impact is very localized beneath the platform.

***Offshore Discharges.*** In the normal conduct of business, the industry is allowed to discharge a variety of effluents to the ocean (under permit from the U.S. EPA). The primary discharges, drill muds and cuttings and produced water, have been the subject of extensive study and debate over the past 2 decades. After exhaustive study, the industry and the U.S. government have concluded that, under current regulatory schemes, the impact of these discharges on the environment is minimal and very localized. Most studies cannot detect significant detrimental impacts to the biota beyond 200 m from the discharges.

*Drilling Muds.* One of the most hotly debated effluents released offshore has been drilling fluids. This issue reached its peak in the late 1980's. As a result of the concerns of U.S. government agencies and environmental groups, a very-large-scale research program, encompassing both laboratory and field work, was conducted to answer the environmental questions. A loose calculation places the value of this joint industry/government research at well over U.S. $100 million.

The issues related to the discharge were both physical and chemical. The first concerned spatial impacts of the solids settling directly on the ocean floor [estimated volumes of 795 to 4770 $m^3$ of bulk mud and 477 to 954 $m^3$ of drilling solids (cuttings)]. Additional concerns were based on the possibility of water-column effects and distant impacts to sensitive habitats (such as coral reefs and sea-grass areas) from transport of suspended solids in the drilling-mud plume.

Typical drilling scenarios include relatively-high-volume bulk discharges early in the life of a well where large hole diameters and occasional mud-type changes require bulk discharges. These can be at rates up to approximately 80 $m^3/h$. In addition, continuous mud and cuttings discharges occur as a result of solids-removal equipment (shakers, desanders, desilters) that is removing the larger particulates for disposal over the side and recycling most of the mud back to the mud system.

Second are the potential environmental impacts to biota caused by the compositional makeup of drilling fluids. Refs. 52 and 53 are the two main publications that summarize the broad literature that addresses these many issues. Through evolution of regulation, mud-formula modification, and toxicity testing, offshore mud programs evolved to less toxic systems. Although early systems had little environmental impact, the newer, practically nontoxic systems have even less impact. An ongoing evaluation of mud composition related to biological toxicity resulted in the elimination of those components that contributed the most to toxicity. Use of such materials as diesel oil for lubricity, KCl in shale control, and chromates as extenders has, in some cases, been eliminated or severely reduced.

Numerous studies conducted over many years have shown that the impact from drill mud and cuttings discharges is primarily restricted to the deposition zone immediately beneath the discharge point. Beyond 100 to 200 m, no significant effects can be attributed to drilling-mud discharges.[52,53] This is in part because of the relatively nontoxic nature of the muds in use and in part because of the very high natural dispersion that occurs on discharge to the ocean.

***Produced Water.*** Produced water is separated from oil, gas, and condensate when it is brought to the surface. As a result of contact in the formations, it can incorporate a mixture of various salts, metals, radionuclides, and organic materials. In addition, different chemical additives are introduced into the produced-water-treatment system to control a variety of treatment problems, including corrosion, microbial activity, and foaming. Although present in low concentrations, these chemicals have to be accounted for in assessing the environmental fate and effects of produced water.

Discharge characteristics are extremely variable and very platform specific. More than 50% of the discharges in the Gulf of Mexico are above the water surface and freefall from the platform. Others range from a discharge just below the water surface to near the ocean floor. Recent regulatory limits on the maximum volume per discharge (3975 $m^3/d$) and the imposition of toxicity limits based on a dispersion model have forced some operators to split their high-volume discharge streams into multiple outlets.

Typical discharges in the Gulf of Mexico average approximately 636 $m^3/d$. Only a few percent of the platforms in the Gulf of Mexico produce more than 3975 $m^3/d$. Often, multiple producing platforms pipe their production to a central production facility for treatment and disposal of produced water. In some locales, the produced hydrocarbons are shipped to shore for treatment and the separated produced water is piped back offshore for disposal.

Offshore produced waters are characterized by high salinity (ranging from seawater salinities of 30 to 300%) and an oil and grease content that must meet a regulatory limit of 42 mg/L (daily maximum) and 29 mg/L (average). They also contain a variety of dissolved organics and inorganics at low concentrations. In addition, Gulf of Mexico produced waters contain radium 226 and 228, which average approximately 175 pCi/L.

***Platform Removal.*** Over the past decade, as more aging offshore infrastructure completed its useful life, the rate of platform removal increased. From 1984 to 1994, 1,007 platforms were removed (while 1,719 new ones were installed).[54] Standard removal practices use bulk explosive charges lowered inside jacket legs (to a depth below the mud line). The shock waves that emanate from the underwater explosions can damage or kill fish, turtles, and mammals. Per U.S. government regulation, structures must be severed 4.8 m below the mud line. Shaped charges and manual cutting by commercial divers have also been used.

In recent years, with the recognition that threatened or endangered sea turtles have an affinity for offshore platforms as do some marine mammals (e.g., porpoises), special measures have been put in place to ensure that the area is clear of animals before explosive charges are used. More recently, the U.S. Minerals Management Service (MMS) conducted studies to evaluate the mortalities to near-platform fish during removal operations.

***Spills.*** Oil spills are the highest-visibility environmental incidents associated with the offshore oil and gas industry. This is rather ironic because this is the type of impact that is least likely to occur according to the spill statistics kept by the U.S. Coast Guard and the U.S. MMS.[54] Although many small spills occur in the process of normal business, large spills are very infrequent. The two major reasons for spills are loss of well control (i.e., blowouts) and ruptured pipelines. The U.S. still has a record of not having had a blowout from exploratory drilling in more than 2 decades. Pipeline spills can occur because of pipeline failure or because of accidental breakage (e.g., boat collisions, anchor snags). Transfer spillage and the possibility of spillage from offshore storage facilities also exist. In some regions, in addition to the transit of imported oil and products by tanker and barge into U.S. territorial seas, there is infrequent transport of oil from offshore production sites to shore locations by tankers (e.g., California). This last mode is unusual and primarily a result of local regulations that do not allow pipelines to come ashore.

**5.4.4. Fate and Effects.** Of primary interest, and particularly unique because of the ocean medium, are the fate and effect of materials spilled in or discharged to the water. In the U.S. alone, more than U.S. $0.5 billion (our estimate) has been expended in the past 2 decades to study the fate and effects of pollutants in the marine environment. The literature resulting from these extensive research programs is quite voluminous. This chapter references major, summary publications that contain most of the literature. To compound the scientific issues further, much of the research has been reported in the form of reports to companies and agencies and has not been distributed internationally by the standard peer-reviewed publication process. For this discussion, we assume that most of the industry's potential impacts offshore relate to the release of materials to the surface, the water column, or the bottom and address the effects these discharges have on the ocean, and the effects the ocean has on the materials released.

***Physical Factors.*** As discussed in Sec. 5.4.3, a wide variety of physical/chemical characteristics come into play when considering the oceans and their effects. Winds, waves, and currents not only affect the physical offshore structures (e.g., pipelines, platforms), but also the materials released.

Waves are instrumental in the movement of bottom materials in the near-shore environment. The larger the wave, the deeper the surge effects are felt. Major bottom reworking and scouring can occur at depths of more than 30 m in large storm events. Near-bottom water velocities generated from this type of storm surge are strong enough to resuspend and transport bottom sediments. This phenomenon spreads bottom contamination over larger areas, thus reducing its concentration.

Currents are also significant in the transport of marine pollutants. In contrast to bottom-sediment transport, current velocity is less important in dispersion of water-column-borne contaminants. Its primary influence is on the mixing time required for dilution to occur. Currents can carry pollutants over large distances and therefore are important in understanding ocean impacts. Often, surface currents are driven primarily by surface winds, while deeper moving-water layers are unaffected by surface winds. In events like oil spills, both surface and subsurface currents are important because they both affect the fate of the oil.

***Chemical Factors.*** The salinity of seawater can affect the materials discharged chemically and their distribution in the water column because of density differences caused by dissolved solids. Seawater chemistry can have significant effects on pollutants that are mixed with it. A variety of organics and metals can form complexes with discharged materials. In some cases, these reactions render the pollutants less toxic and more degradable. In other cases, it might have the opposite effect. The oxidative and buffer capacities of seawater also affect the chemical behavior of materials discharged (e.g., neutralizing of acidic materials or reduction of toxic $Cr^{+6}$ to the less toxic $Cr^{+3}$).

Near-shore water-mass differences also affect the transport of contaminants. In many locales, denser, high-salinity marine waters flow under the lighter, low-salinity surface waters resulting from land runoff. In offshore waters, horizontal density layers (pycnoclines) can affect vertical transport of near-neutral buoyancy materials.

***Biological Factors.*** Biological factors can play a significant role in the fate and effects of pollutants in the marine environment. At the lowest levels, marine bacteria, fungi, and yeasts all are involved in the degradation and transformation of inorganics and organics in the water column and in the bottom sediments. In some instances, this degradation can detoxify chemical compounds. In other cases, the process might actually create intermediary products that might be more detrimental (e.g., methylation of elemental mercury). Biological agents also can act as pathways for assimilation of toxic materials and their byproducts to higher trophic levels.

Moving to a slightly larger size, marine plankton, small drifting plants (e.g., algae) and animals in the sea can be vectors for the uptake and transport of contaminants in the sea. In some cases, marine zooplankton can ingest materials from the water column and incorporate them into fecal material, thus accelerating their transport to the ocean bottom.

Benthic (near-bottom dwellers) organisms and infauna (animals living in the sediments) can affect the transport of marine contaminants as well as contribute to effects on higher trophic levels. Living in intimate contact with the sediments, they have the opportunity to ingest contaminated materials that fall to the ocean bottom. In addition, animal activities, both on the surface and in the sediments, tend to mix and disperse bottom materials. Small marine worms (polychaetes) actively burrow in the upper 10 to 20 cm of the sediment, which helps transport surface-deposited materials to deeper layers of the sediment while reoxygenating the sediments.

An issue always of concern is the bioaccumulation of toxics from the sediments and waters by marine organisms, which then have the potential to be transferred up to higher trophic levels in the food chain. This transfer also can occur in land animals such as bears along the shore and birds that feed on marine fish. The ultimate consumer of marine products is man, which is a primary issue when addressing marine pollution issues.

It is reassuring to note from numerous offshore studies that there is almost no detectable bioaccumulation of operations-related contaminants in near-field marine biota. This is true for inorganics, organics, and naturally occurring radioactive materials, such as $^{226/228}Ra$.

***Effects Monitoring.*** Over the past 2 decades, more than U.S. $750 billion has been spent on environmental studies related to offshore oil and gas. A large portion of this has been spent by the U.S. government through various programs funded and managed by the MMS, the Natl. Oceanic and Atmospheric Admin. (NOAA), and the EPA. In addition, numerous coastal states in the U.S. have expended monies to conduct environmentally oriented oceanographic research. The petroleum industry has also conducted a wide variety of environmental studies, both individually and through cooperative programs managed through trade associations, such as the API and the Offshore Operators Committee. Refs. 52 through 58 give the results of these endeavors.

***Industry.*** A strong demand has been placed on the petroleum industry to provide environmental data regarding the fate and effects of their activities in the marine environment. In part because of limited government funding and lack of specific focus and timeliness, the industry has had to conduct much of its own environmental research and monitoring. While some of the work has been done inhouse by industry experts, most of it has been done externally by scientists from academia, consulting companies, and sometimes the government. Studies have been conducted from the high arctic to the coral reefs of the tropics and from the shallow coastal waters to the deep sea off the continental slopes.

Much of the industry research has focused on obtaining real-world data on the fate and effects of the discharges from the vicinity of actual offshore sources. These investigations have included detailed studies on the fate of drilling muds and cuttings in the water column and on the ocean floor. They have also included laboratory assessments of drilling-fluid toxicities as well as field assessments of actual effects on resident benthic communities.

Similarly, very detailed studies have been conducted on produced waters. A detailed dispersion model was developed that predicts downstream behavior and concentrations of produced waters in the water column. This model also predicts the fate of drilling-mud plumes and estimates their solids-deposition loadings on adjacent sea bottoms. Field validation studies have been conducted for both the produced-water and drilling-fluid dispersion models. Recent studies have also addressed such issues as the discharge of produced sands.

Through a variety of venues, the industry has been involved in oil fate and effects studies since the early 1970's. The data from these studies have provided important insight into the behavior of spilled oil in the sea. More importantly, the data have expanded our understanding of the impacts and recoveries that occur and have been important in setting priorities for environmental protection as well as in selecting appropriate cleanup technologies to be used for different habitats. Most of the research conducted on actual use of oil spill dispersant chemicals was conducted by the petroleum industry.

***U.S. Government.*** The primary focus for offshore environmental research related to the oil and gas industry has been the MMS. Under authority of the Outer Continental Shelf Lands Acts of 1953 and 1978, the MMS was given the authority to lease submerged marine lands for minerals development.[59] As part of this responsibility, the MMS was to develop relevant environmental information for lease decision making and was mandated to monitor the effects of the leasing program (i.e., impacts of activities that resulted from the leases). Since inception, the MMS has funded/conducted/contracted more than 1,300 projects at a cost of approximately U.S. $609 million. Summaries of more than 700 of these projects are now available on the Internet at *http://www.mms.gov/espis*. These studies range from laboratory toxicity studies to deep-ocean physical oceanography. Biological, physical, chemical, and geological aspects of the ocean as they relate to oil and gas development have been studied. Short-term effects as well as longer-term, chronic impacts resulting from development and production have been studied.

Other U.S. government agencies have also been involved, some independently, others in consort with MMS activities. These include the NOAA, the Natl. Marine Fisheries Service, the EPA, and the U.S. Geological Survey (USGS). On occasion, the agencies have cofunded projects with the industry. In addition, many academic institutions and faculty have conducted oil-industry-related environmental research over the years. Many of these studies are contained in the cited literature.

### 5.4.5 Conclusions

This synopsis makes it evident that the oceans are very complex and can have diverse influences on the petroleum industry's offshore activities. The basic oceanographic sciences are the key to understanding the oceans and to conducting work there safely. Understanding "processes" is of the utmost importance in answering questions related to fate and effects, from the transport and behavior of myriad chemicals to the intricate web of biological interrelationships.

An extensive literature base exists on environmental studies related to offshore impacts. Although much still remains to be learned, we are by no means operating in a vacuum of limited knowledge.

In most aspects of the E&P business that relate to environmental impact, environmental studies have provided vital information on which reasonable (usually) regulations have been based. This is made evident by the guidance and restrictions placed on most of the industry's activities to mitigate potential impacts, such as minimum flight restrictions for aircraft in Alaska to protect bird colonies or the prelease surveys conducted on the outer continental slope to determine the presence of deep-dwelling chemosynthetic communities in the proposed drillsites. The industry operates with toxicity limits and specific component restrictions on drilling fluids as well as with dispersion (and toxicity) guidelines on produced-water discharges. The list is long but very representative of how the environmental sciences have been used to mitigate impacts of the industry's offshore activities.

The preceding discussions touched on a broad variety of processes and effects that hypothetically can occur and, in some cases, do occur on a small scale in very localized areas. Risking criticism for broad generalization, much of this information as it relates to offshore impacts can be, and is, integrated and summarized in the following.

Effluent discharges in the U.S. OCS have minimal impact partly because of their relatively nontoxic nature and partly because of rapid natural dispersion. Biological impacts have been predominately limited to within 100 to 200 m of most platforms, with chemical contamination primarily limited to the same zone. Beyond these distances, it is virtually impossible to document significant biological effects and the chemical signal that is still detectable is very low and not related to toxic components (or levels). Most impacts appear to be related more to direct burial beneath the platforms. Impacts caused by seismic surveys appear to be minimal (if at all), and marine mammal response to U.S. OCS activity appears to be small and primarily that of local avoidance.

## Nomenclature

$A$ = cross-sectional area, $L^2$
$C_{iw}$ = equilibrium concentration of Component $i$ in water phase, m/$L^3$, mg/L
$g$ = hydraulic gradient, L/L
$K$ = hydraulic conductivity, L/t
$K_d$ = distribution coefficient
$K_{oc}$ = partitioning coefficients that are normalized to the organic-matter content of the soil
$q$ = flow rate, $L^3$/t
$R_i$ = solubility of Pure Component $i$ in water phase, m/$L^3$, mg/L
$t_0$ = intial time, t
$t_1, t_2, t_3,$ = timesteps, t
$\bar{v}$ = average velocity of groundwater flow, L/t
$v_c$ = velocity of a retarded contaminant, L/t
$x_i$ = molar fraction of Component $i$ in oil
$\phi$ = effective porosity, dimensionless

## References


1. Dunn, I.S., Anderson, L.R., and Kiefer, F.W.: *Fundamentals of Geotechnical Analysis*, John Wiley & Sons, New York City (1980).
2. *Soil Taxonomy: A Basic System of Soil Classification for Making and Interpreting Soil Surveys*, USDA Handbook 436, U.S. Government Printing Office, Washington, DC (1975).
3. Brady, N.C.: *The Nature and Properties of Soils*, eighth edition, Macmillan Publishing Co. Inc., New York City (1974).
4. Buol, S.W., Hole, F.D., and McCracken, R.J.: *Soil Genesis and Classification*, Iowa State U. Press, Ames, Iowa (1973).
5. Fanning, D.S. and Fanning, M.C.B.: *Soil Morphology, Genesis, and Classification*, John Wiley & Sons, New York City (1989).
6. *Western Fertilizer Handbook*, Potash and Phosphate Inst., Norcross, Georgia (1985).
7. Troeh, E.R. and Thompson, L.M.: *Soils and Soil Fertility*, fifth edition, Oxford U. Press, New York City (1993).
8. Tisdale, S.L., Werner, W.L., and Beaton, J.D.: *Soil Fertility and Fertilizers*, Macmillan Publishing Co. Inc., New York City (1985).
9. Stevenson, F.J.: *Cycles of Soil*, John Wiley & Sons, New York City (1986).
10. Bohn, H.L., McNeal, B.L., and O'Connor, G.A.: *Soil Chemistry*, Wiley-Interscience, John Wiley & Sons, New York City (1979).
11. Gupta, R.K. and Abrol, I.P.: "Salt Affected Soils: Their Reclamation and Management for Crop Production," *Advances in Soil Science*, B.A. Stewart (ed.), (1990) Chap. 11, 224–88.

12. Levy, R.: *Chemistry of Irrigated Soils*, Van Nostrand Reinhold Co., New York City (1984).
13. Deuel, L.E. Jr. and Freeman, B.D.: "Amendment to Louisiana Statewide Order 29-B, Suggested Modifications for Barium Criteria," paper SPE 18673 presented at the 1989 SPE/IADC Drilling Conference, New Orleans, 28 February–3 March.
14. *Chemistry of Clays and Clay Minerals*, A.C.D. Newman (ed.), John Wiley & Sons, New York City (1987).
15. McBride, M.B.: *Environmental Chemistry of Soils*, Oxford U. Press, New York City (1994).
16. *Minerals in Soil Environments*, J.B. Dixon and S.B. Weed (eds.), second edition, Soil Science Soc. of America, Madison, Wisconsin (1989).
17. Paul, E.A. and Clark, F.E.: *Soil Microbiology and Biochemistry*, Academic Press, New York City (1989).
18. Tate, R.L.: *Soil Microbiology*, John Wiley & Sons, New York City (1995).
19. Truitt, M.M.: *Soil Mechanics Technology*, Prentice-Hall Inc., Englewood Cliffs, New Jersey (1983).
20. McCarthy, D.F.: *Essentials of Soil Mechanics and Foundations: Basic Geotechnics*, fourth edition, Regents/Prentice-Hall, Englewood Cliffs, New Jersey (1993).
21. *Diagnosis and Improvement of Saline and Sodic Soils*, USDA Handbook 60, U.S. Government Printing Office, Washington, DC (1954).
22. *Standard D-2487-98, Standard Practice for Classification of Soils for Engineering Purposes*, American Soc. for Testing Materials, West Conshohocken, Pennsylvania (1998).
23. Alexander, M.: *Biodegradation and Bioremediation*, Academic Press, New York City (1994).
24. "Metals Criteria for Land Management of Exploration and Production Wastes," Technical Support Document, API Publication No. 4600, API, Washington, DC (1995).
25. Keren, R. and Bingham, F.T.: "Boron in Water, Soils, and Plants," *Advances in Soil Science*, B.A. Stewart (ed.), Springer-Verlag, New York City (1985) **1**, 229–76.
26. "Amendment to Statewide Order No. 29-B," Office of Conservation, Louisiana Dept. of Natural Resources, Baton Rouge, Louisiana (1990).
27. "Testing Methods for Evaluating Solid Waste," third edition, SW-846, U.S. EPA, Washington, DC (1986).
28. *Atlas of Gas-Related Produced Water*, GRI, Chicago (1995).
29. Richter, B.C., Dutton, A.R., and Kreitler, C.W.: Identification of Sources and Mechanisms of Salt-Water Pollution Affecting Ground-Water Quality: A Case Study, West Texas," Report No. 191, Texas Bureau of Economic Geology, Austin, Texas (1990).
30. Morton, R.B.: "Effects of Brine on the Chemical Quality of Waters in Parts of Creek, Lincoln, Okfuskee, Payne, Pottawatonie, and Seminole Counties, Oklahoma," Circular 89, Oklahoma Geological Survey, Norman, Oklahoma (1986).
31. "Identification of Sources of Ground-Water Salinization Using Geochemical Techniques," EPA/600/2-91/064, U.S. EPA, Washington, DC (December 1991).
32. "VADSAT: A Vadose and Saturated Zone Transport Model for Assessing the Effects on Groundwater Quality from Subsurface Hydrocarbon Releases and Petroleum Production Waste Management Practices," API, Washington, DC (1995).
33. "A Guide to the Assessment and Remediation of Underground Petroleum Releases," Publication 1628, second edition, API, Washington, DC (April 1989).
34. Bossert, I. and Bartha, R.: *The Fate of Petroleum in Soil Ecosystems: Petroleum Microbiology*, R.M. Atlas (ed.), Macmillan Co. Inc., New York City (1984).
35. Testa, S.M. and Winegardner, D.L.: *Restoration of Petroleum-Contaminated Aquifers*, Lewis Publishers, Chelsea, Michigan (1991).
36. Howard, P.H.: *Handbook of Environmental Fate and Exposure Data for Organic Chemicals*, Lewis Publishers, Chelsea, Michigan, (1990) **1 through 4.**
37. "Technical Support Document for Land Application of Sewage Sludge," EPA822/R-93-001, U.S. EPA, Washington, DC (1992) **1 and 2.**
38. *40 CFR, Standards for the Use or Disposal of Sewage Sludge*, U.S. EPA, Washington, DC (1993) Parts 257, 403, and 503.
39. Hem, J.D.: "Study and Interpretation of the Chemical Characteristics of Natural Water," Water-Supply Paper 269, USGS, Reston, Virginia (1959).
40. Heath, R.C. "Basic Groundwater Hydrology," USGS Water Supply Paper 2220, U.S. Government Printing Office, Washington, DC (April 1983).
41. *Waste Disposal Effects on Groundwater*, D.W. Miller (ed.), Premier Press, Berkeley, California (1980) 512.
42. Richter, B.C. and Kreitler, C.W.: *Geochemical Techniques for Identifying Sources of Groundwater Salinization*, CRC Press, Boca Raton, Florida (1993).
43. Whittemore, D.O.: "Geochemical Differentiation of Oil and Gas Brine from Other Saltwater Sources Contaminating Water Resources: Case Studies from Kansas and Oklahoma," *Environmental Geosciences* (1995) **2,** No. 1, 15.
44. Parker, J.C.: "Multiphase Flow and Transport in Porous Media," *Review of Geophysics* (1989) **27,** No. 3, 311.
45. "Laboratory Study on Solubilities of Petroleum Hydrocarbons in Groundwater," Publication 4395, API, Washington, DC (1985).
46. Fetter, C.W.: *Applied Hydrogeology*, third edition, Prentice-Hall Inc., Upper Saddle River, New Jersey (1994) 128–55.
47. Bennett, C.O. and Myers, J.E.: *Momentum, Heat, and Mass Transfer*, third edition, McGraw-Hill Inc., New York City (1962) 532–35.
48. Pickard, G.L.: *Descriptive Physical Oceanography: An Introduction*, Pergamon Press, New York City (1996) 200.
49. Greene, G.D. Jr.: "A Pilot Study of Possible Effects of Marine Seismic Airgun Array on Rockfish Plumes," report prepared for Seismic Steering Committee, Greenridge Sciences Inc., Santa Barbara, California (15 May 1985).
50. Dalen, J. and Raknes, A.: "Scaring Effects in Fish and Harmful Effects on Eggs, Larvae and Fry by Offshore Seismic Explorations," paper presented at the 1986 Intl. Congress on Acoustics Associated Symposium on Underwater Acoustics, Halifax, Canada, 16–18 July.
51. Greene, C.R. Jr.: "Man-Made Noise," Report NTIS No. PB91-168914, LGL Ecological Research Assocs. Inc., Bryan, Texas, U.S. MMS, Washington, DC (1991) 462.
52. "Drilling Discharges in the Marine Environment," Natl. Research Council Panel on Assessment of Fates and Effects of Drilling Fluids and Cuttings in the Marine Environment, Natl. Academy Press, Washington, DC (1983) 169.
53. Engelhardt, F.R., Ray, J.P., and Gillam, A.H.: *Drilling Wastes*, Elsevier Science Publishers Ltd., Oxford, U.K. (1989) 867.
54. "Offshore Statistics, Third Quarter," U.S. MMS, Washington, DC (1995).
55. "Petroleum in the Marine Environment," Natl. Academy of Science Workshop on Inputs, Fates and the Effects of Petroleum in the Marine Environment, Ocean Affairs Board, Natl. Research Council, Washington, DC (1975) 107.
56. "Oil in the Sea: Inputs, Fate, and Effects," NAS, Natl. Academy Press, Washington, DC (1985) 601.
57. Boesch, D.F. and Rabalais, N.N.: *Long-Term Environmental Effects of Offshore Oil and Gas Development*, Elsevier Applied Science Publishers, New York City (1987) 708.
58. Tetley, M. and Wells, K.: "Bibliography: Scientific Journal Articles Based on MMS Environmental Research," OCS Statistical Report, MMS 93-0069, U.S. MMS, Washington, DC (1993) 312.
59. *Code of Laws of the United States of America*, Title 43, Secs. 1801 through 1866.

## SI Metric Conversion Factors

| | | | |
|---|---|---|---|
| °API | 141.5/(131.5 + °API) | | = g/cm$^3$ |
| atm | × 1.013 250* | E + 05 | = Pa |
| bbl | × 1.589 873 | E − 01 | = m$^3$ |
| cycles/sec | × 1.0* | E + 00 | = Hz |
| ft | × 3.048* | E − 01 | = m |
| ft$^2$ | × 9.290 304* | E − 02 | = m$^2$ |
| °F | (°F − 32)/1.8 | | = °C |
| gal | × 3.785 412 | E − 03 | = m$^3$ |
| in. | × 2.54* | E + 01 | = mm |
| mile | × 1.609 344* | E + 00 | = km |
| psi | × 6.894 757 | E + 00 | = kPa |
| yd$^3$ | × 7.645 549 | E − 01 | = m$^3$ |

*Conversion factor is exact.

# Chapter 6
# Pollution Evaluation

**William A. Brommelsiek,** Chevron U.S.A. Production Co.
**James W. Collins,** James W. Collins and Assocs.
**Daniel K. Jernigan,** Chevron U.S.A. Production Co.
**Frederick V. Jones,** Marathon Oil Co.
**Vick Newsome,** BP-Amoco Corp.
**Ricky E. Clements,** East Texas Salt Water Disposal Co.
**Bart Sims,** Railroad Commission of Texas
**F. Douglas Skinner,** Radian Intl.
**John D. Admire,** Phillips Petroleum Co.
**Andrea Osborne,** U.S. Environmental Protection Agency
**Kevin J. Grice,** Texaco Group Inc.
**Donald W. Butler,** Hunt Oil Co.
**Steven H. Souders,** U.S. Environmental Protection Agency

## 6.1 Introduction

E&P operations vary worldwide, and those that are cost-effective and protect the environment are indicative of proper management. This chapter discusses the practice of pollution evaluation, which recognizes the potential sources of releases and potential pollutants related to E&P operations. E&P operations typically include exploration (e.g., seismic-data acquisition), drilling and completion, production operations, and support activities. An understanding of the sources and pollutants associated with these operations can help prevent adverse environmental impacts and increase overall cost-efficiency.

## 6.2 Exploration Operations

A variety of survey techniques have been developed to assist in the exploration for hydrocarbons. These include seismic, gravity, magnetic, and geochemical surveys. **Table 6.1** lists common potential sources of release and associated pollutants from exploration. Although most field operations related to exploratory surveys have minimal environmental impact, seismic surveys deserve special consideration because they are more widely used and a greater number of potential sources of releases and potential pollutants are associated with them.

Seismic surveying requires generation of acoustic or shock waves at specific points along a relatively straight survey line. Onshore, these waves usually are created with explosives or by a technique known as Vibroseis. When explosives are used, detonation occurs in 3.5- to 4.7-in.-diameter holes drilled below the weathered (or low-velocity) layer, sometimes as deep as 164 ft. The Vibroseis technique uses three to five massive vehicles to vibrate the ground at specific points along the survey line. Both methods generate acoustic waves that travel through the earth. Density and velocity variations in the underlying strata cause parts of the wave energy to be reflected to the surface, where their strengths and arrival times are detected by geophones. As the survey progresses, the geophones are repositioned along the survey line by crews traveling by truck or helicopter.

The degree to which seismic operations impact the environment is related to the characteristics of the region where they are conducted. In some areas, vegetation may have to be cleared, roads built, and rivers crossed. In other areas, desert or tundra may have to be traversed. Some areas require the use of mud-assisted drilling rigs rather than hydraulic augers to reach below the low-velocity layer. In all cases, vehicular traffic and land-use requirements are of primary concern. Erosion caused by heavy vehicles and clear-cutting of vegetation can cause sedimentation problems on land and in surface waters. Spills and leaks from vehicles and support equipment may contaminate soils and pollute surface waters. Improperly plugged shotholes may induce surface and groundwater contamination.

Potential pollutants from onshore seismic operations include oil and grease, hydraulic fluids, fuels, engine coolant, total dissoved and total suspended solids (TDS and TSS, respectively), chlorides, and solid waste (i.e., scrap materials, such as flags and marking tape). Other potential pollutants that may be of concern are explosives and cable residues (e.g., organic compounds, nitrates, and heavy metals).

Offshore seismic operations use compressed-air and -water guns to generate acoustic energy that travels through the water and into the seafloor. The reflected energy is detected at sea level by hydrophones towed by a survey vessel. A variety of shipboard operations use equipment that can serve as potential sources of pollutant release. Sec. 6.7 addresses these sources and the respective pollutants.

An area of concern common to both onshore and offshore seismic operations is the potential impact of acoustic energy on terrestrial and aquatic organisms. The degree to which these organisms are affected and the steps necessary to lessen the severity of the impact are subjects of continuing research.

## 6.3 Drilling and Completion Operations

**6.3.1 Introduction.** A potential field may be identified after the data gathered during exploration activities are processed and evaluated. Then, drilling sites are selected and prepared for installation

**TABLE 6.1—EXPLORATION OPERATIONS**

| System | Pollutant Sources | Potential Pollutants |
|---|---|---|
| Seismic operations | Air and water guns | Acoustic energy |
| | Soil exposed by survey line and road construction | Contaminated debris |
| | Shot holes | Hydrocarbons (oil/grease, hydraulic fluids, fuels) |
| | Vehicles | Nitrates and other compounds from explosives |
| | Marine vessels | Total dissolved solids (TDS) |
| | Explosives and cable | Total suspended solids (TSS) |

of the drilling rig, mud-circulating systems, and other associated equipment. The size of the site and volume of waste generated depend on a number of factors, including anticipated well depth, drilling-rig size, drilling-fluid type and volume, cuttings that must be stored on the site, and drilling-fluid-circulating system. Once the site is prepared, the drilling rig and -fluid-circulating system are installed and drilling starts. Drilling operations also involve several support activities, such as delivery or on-site formulation of drilling fluids, setting surface and intermediate casing in the hole, cementing the casing in place, well logging and testing, and a variety of maintenance operations. If hydrocarbons are discovered in sufficient quantities to be considered economic, the well is completed (i.e., prepared for production).

Like drilling operations, completion operations require a host of support activities that typically are performed by service companies. These activities can include installation of production casing, cementing, perforating the production zone(s), fracturing and stimulating the producing formation(s), well logging and testing, and installation of production tubing and pumping or lifting equipment. Unfortunately, many wells are not economically viable and must be properly plugged and the site reclaimed. All these processes involve operations and activities that may create potential sources of contaminant releases to the environment. These activities and potential sources of releases and pollutants associated with drilling and completion activities are discussed next; **Table 6.2** provides a complete list.

**6.3.2 Site Preparation.** Once a prospective site has been selected, the area needs to be prepared for drilling operations. If adequate oil and/or gas reserves are found, the area needs further development to prepare for production operations. In any case, construction of roads and pads includes leveling and grading the base area to ensure that the slope of the location is correct to collect storm-water runoff. Additionally, pits and cellars for the drill rig are constructed. Earth-moving equipment and fuel-storage facilities are the main potential sources of pollutant release during road and pad construction.

Erosion is a major concern during and after construction of roads and pads. In some areas of the world, poor-quality construction materials must be used. In other cases, high rainfall and undulating terrain can compound erosion problems if the construction operation is not performed properly. Sediment runoff to surface waters is the main potential pollutant (sediment runoff rates from construction sites are typically 10 to 20 times greater than those from agricultural land). Other potential pollutants are phosphorus, nitrogen ($N_2$), and nutrients from fertilizers; pesticides; construction; chemicals; and solid wastes. Vehicles are another source of potential pollutant release. Potential pollutants associated with vehicles include hydraulic fluids, motor oils, and fuels.

After the site has been prepared, the drilling rig is moved into place and constructed or organized for drilling operations. Rig-up typically is conducted in stages, with each operation moving in and out as needed, removing equipment and scrap material as it leaves. In some operations, much of the equipment and supplies are on site from the first day and crowding problems may create a potential for releases. Rig-up activities generate a great deal of scrap metal, packing material, paper, cardboard, and wood. Packing; strapping; welding waste; oil and grease; and broken tools, bits, and pieces of rubber hose are typical solid wastes that must be stored and subsequently managed according to applicable regulations. During rig-up activities, the major potential pollutants are scrap metal, used and discarded parts, oils and lubricants, scrap wood, paper, plastics, and normal domestic wastes.

**6.3.3 Drilling-Fluid Systems.** Drilling-fluid systems result in the second-largest E&P waste stream and offer the largest potential source of releases to the environment from drilling operations. In general, drilling fluids have been found to have low toxicity and minimal impact on the environment. With respect to drilling fluids as a potential pollutant, attention should be focused on additives and fluid components that may increase the toxicity of the fluid. This section describes the various types of drilling fluids and drilling-fluid-circulating systems.

Drilling fluids (muds) consist of a continuous phase (either water, oil, or air) and additives, such as clays or chemicals. Most muds are water-based, with either fresh or salt water as the continuous phase. Oil-based muds are used to drill under extreme conditions, such as high downhole temperatures or when drilling hydratable shales, and are formulated with either diesel oil, mineral oil, or synthetic oils as the continuous phase. The least common fluid system consists of air as the continuous phase and is used to drill low-pressure impermeable formations, such as limestones and dolomites.

A variety of additives are used in water- and oil-based muds to control density, corrosion, pH, viscosity, and formation hydration and to reduce friction between the formation and the drill bit. Additives used in air-fluid systems generally are limited to water and surfactants to produce foam in the air and enhance the ability of the fluid to lift cuttings out of the hole. Drilling-fluid containment is achieved with in-ground reserve pits or steel tanks.

Depending on the economics, type and volume of fluid used, and environmental considerations, either earthen pits or steel tanks are used to contain drilling fluids and cuttings. Many land drilling operations use steel tanks for the active mud system and earthen pits to contain excess mud, cuttings, and waste water. Closed-loop systems use steel tanks for the active mud system and to contain cuttings and other wastes. When earthen pits are used, releases of water- or oil-based muds can occur if fluids leach into the ground or if the pits are overtopped. Releases from tanks can occur if the tanks leak or are overtopped. Other sources of potential releases from the drilling-fluid-circulating systems include flowlines, mud pumps, the mud-makeup area, and discarded or damaged chemical and materials containers (i.e., sacks and drums).

Air-fluid-circulating systems are similar to liquid-mud systems and consist of compressors, water- and chemical-injection pumps, and flowlines. Earthen pits are used to collect foam, water, and cuttings that are circulated out of the hole. These pits can be a potential source of releases from fluids leaching into the ground and from spills resulting from overtopping of the pit. As with liquid-fluid systems, other potential sources of releases include flowlines, pumps and compressors, chemical-storage areas, and discarded or damaged containers.

Solids control is an important component of water- and oil-based-fluid systems. Solids control involves removal of formation cuttings and fine solids to maintain desired drilling-fluid characteristics. Shale shakers, desanders, desilters, and centrifuges are examples of solids-control equipment. These units can be potential sources of releases (e.g., leaks). Discarded equipment, such as worn shale-shaker screens, are also potential sources of releases associated with solids control.

**TABLE 6.2—DRILLING AND COMPLETION OPERATIONS**

| System | Pollutant Sources | Potential Pollutants |
|---|---|---|
| Site preparation<br>Roads and pads | Contaminated soil<br>Soil exposed by site preparation<br>Vehicles | Chlorides<br>Heavy metals<br>Hydraulic fluids<br>Lubricating oils and grease<br>TDS and TSS |
| Rig-up activities | Equipment used to assemble rig<br>Fuel tanks<br>Lubricating equipment<br>Welding equipment | Fuel<br>Hydraulic oil<br>Lubricating oils and grease<br>Packing materials<br>Paper<br>Scrap metal and wood<br>Solvents<br>Welding solder |
| Drilling and completion<br>Fluid systems | Blowout preventers (BOP's)<br>Compressors<br>Empty containers<br>Flowlines<br>Fluid tanks<br>Fuel-storage tanks<br>Generators<br>Material-storage areas<br>Mixing tanks<br>Pumps<br>Reserve pits<br>Rig floor and sump<br>Solids-control equipment<br>Trucks<br>Used filters | Cement<br>Cement additives<br>Chlorides<br>Drill cuttings<br>Formation test fluids<br>Hydraulic oil<br>Lubricating oils and grease<br>Mud additives<br>Metals<br>Organic chemicals<br>TSS and TDS<br>Radioactive-source material |
| Rig operations | Casing equipment<br>Cellar<br>Cementing equipment<br>Debris<br>Drill line<br>Fishing tools<br>Flowlines<br>Milling tools<br>Painting equipment<br>Pipe-dope buckets<br>Pulled drillpipe<br>Rathole/mousehole<br>Reserve pits and tanks<br>Rig wash/cleaners<br>Well-logging equipment | Caustic cleaners<br>Chlorides<br>Diatomaceous-earth filter media<br>Fuels<br>Heat-transfer fluids<br>Hydraulic fluids<br>Lubricating oils and grease<br>Metals<br>Pipe dope<br>Produced oil<br>Radioactive waste<br>Solvents<br>Well-completion and -treatment fluids (chemicals) |

Potential pollutants from drilling-fluid systems include the drilling fluids and a variety of chemical additives, such as caustics, salts and metals, formation cuttings and fluids (including hydrocarbons), metal cuttings, cement and cement additives, and lubricants.

Drilling-rig operations involve a number of activities that present potential risk of releases to the environment. For example, when making or breaking drillpipe connections, drilling mud in the drillpipe and kelly spills on the rig floor. Excess pipe dope brushed on drillpipe threads to lubricate and protect pipe threads also may drop on the rig floor. Additionally, pipe dope squeezed inside the drillpipe during a connection is circulated into the mud system. The rig floor is washed down regularly with water and occasionally with rig-wash chemicals or detergents. Generally, the rig floor is washed after making or breaking a series of drillpipe connections. Rig-wash water, which gathers potential pollutants from the mud, pipe dope, and other materials spilled on the rig floor, accumulates in the rig cellar (a sump located under the rig) from which the water is diverted to the reserve pit, a tank, or other containment vessel. The rathole and mousehole can also serve as conduits for releases of drilling muds and other fluids to the environment.

As drilling progresses, casing must be run into the hole to protect freshwater aquifers and to prevent caving of the hole. During casing operations, drilling fluids and excess pipe dope may be spilled on the rig floor and subsequently washed into the rig cellar and reserve pit, tank, or other containment vessel. Casing is set in place by circulating cement down the hole and up the back side of the casing. Oilfield cements typically are Portland cements with additives to control cement density, fluid loss, and setting rates. During cementing operations, cement can spill on the rig floor or the ground or can accumulate in the drilling mud. Additionally, cement placed in the open hole below the casing shoe is drilled out when drilling operations begin. Cement and metal cuttings from drilling out the shoe and cement circulate out of the hole and accumulate in the drilling fluid.

Other potential sources of releases include generators, blowout preventers (BOP's), used filters, standpipe or flowlines, painting equipment, drill-line equipment, well-logging trucks and equipment (e.g., wireline, mud logging), fishing and milling tools, worn or damaged equipment and drillpipe, drillpipe-thread protectors, debris (e.g., oily rags), fuel-storage vessels, discarded pipe-dope buckets and brushes, and storm water.

Potential pollutants from drilling-rig operations include drilling fluids and drilling-fluid additives (chemicals and metals), grease and metals (lead, copper, zinc) from pipe dope, cement and cement additives (chemicals and metals), metal cuttings, scrap metal, hydraulic fluids and lubricants (used oil and grease), fuels, painting waste, radioactive materials from well-logging operations, laboratory wastes (chemicals), solvents, domestic waste, and sewage.

**6.3.4 Completion Operations.** Completion fluids are used to condition the wellbore, to pretreat and fracture the producing formation, to stimulate and clean the producing formation, to control downhole pressure, and to circulate solids out of the hole.

Fracturing fluids generally consist of either fresh or salt water, diesel, crude oil, alcohols, carbon dioxide ($CO_2$) or $N_2$, and a proppant (such as fine sand). Other chemicals may be added to control fluid loss, pH, corrosion, and bacterial growth. Stimulation fluids

**TABLE 6.3—PRODUCTION OPERATIONS**

| System | Pollutant Sources | Potential Pollutants |
|---|---|---|
| Facility construction-site preparation<br>Roads and pads | Contaminated soil<br>Soil exposed by site preparation<br>Vehicles | Chlorides<br>Fuels<br>Heavy metals<br>Hydraulic fluids<br>Lubricating oils and grease<br>TDS and TSS |
| Equipment and construction | Equipment used for assembly<br>Fuel tanks<br>Lubricating equipment<br>Welding equipment | Lubricating oils and grease<br>Packing materials<br>Contaminates<br>Scrap metal and wood<br>Solvents<br>Welding solder |
| Field production systems | Casing and tubing<br>Chemical-injection systems<br>Compressors<br>Crude-oil storage<br>Desalination systems<br>Drums and containers<br>Electrical equipment (e.g., transformers and capacitors)<br>Electrostatic coalescers<br>Engines<br>Field/portable laboratories<br>Flowlines (heaters, connections, valves)<br>Gas-lift systems<br>Heat exchangers<br>Hydraulic systems<br>Leak-, drip-, and spill-containment systems<br>Load-line connections<br>Produced sand<br>Produced-water storage<br>Rod pumps<br>Separation equipment<br>Sorbent pads<br>Submersible pumps | Chlorides<br>Crude oil<br>Condensate<br>Dielectric fluids [including polychlorinate biphenyls (PCB's)]<br>Excelsior<br>Fuels<br>$H_2S$<br>Hydraulic fluid<br>Inhibitors<br>Laboratory waste<br>Lubricating oils and grease<br>Metals<br>Naturally occurring radioactive materials (NORM)<br>Paraffin<br>Scale<br>Surfactants<br>Sweetening/dehydration chemicals<br>Tank bottoms (basic sediment and water)<br>Treatment chemicals<br>TDS and TSS |
| Workovers and recompletions | Cementing trucks and equipment<br>Coiled-tubing units<br>Drums and containers<br>Fracture trucks and tanks<br>Logging units<br>Produced sand<br>Produced water<br>Sorbent pads<br>Tubing and casing<br>Vacuum trucks<br>Wellheads<br>Workover pits/tanks<br>Workover rigs | Acids/corrosives<br>Cement and additives<br>Chlorides<br>Condensate<br>Crude oil<br>Drum rinsate<br>Explosives<br>Fracture gels and additives<br>Fuels<br>$H_2S$<br>Hydraulic fluids<br>Lubricating oils and grease<br>Paraffin<br>Treatment chemicals<br>Well blowdown |

typically are a mix of water with 15 to 20% hydrochloric acid (HCl), hydrofluoric acid (HF), or a combination of both.

In many cases, completion fluids are similar to drilling fluids; however, some completion fluids can be as simple as fresh or salt water with additives. Completion-fluid additives include materials and chemicals to control density, fluid loss, pH, corrosion, and bacterial growth and to clean or condition the wellbore.

Completion-fluid systems and equipment may be similar to those used for drilling-fluid systems and can include earthen pits (reserve pits and emergency pits), steel tanks, solids-control equipment (desanders and desilters), pumps, and flowlines. Tanker trucks and fracture tanks frequently are used to contain completion and stimulation fluids. Pits; tanks; equipment; mixing and materials-storage areas; damaged or discarded containers; and flowback of treatment, stimulation, or fracturing fluids all are potential sources of contaminant releases.

Potential pollutants from completion-fluid systems include fluids and additives; grease and metals (lead, copper, zinc) from pipe dope; chemicals spilled or leaked in mixing and storage areas; formation fluids, including crude oil; produced sand; acids and caustics; chlorides; hydraulic fluids and lubricants; metal cuttings; cement and cement additives; solvents; domestic waste; and sewage.

Rig operations during completions are much like those during drilling. However, drilling activities during completion operations usually are limited to reaming and drilling out cement after running casing into the hole. Also, because well-completion operations involve a large number of activities that are outsourced to service companies, movement of equipment on and off the site increases. This additional activity and movement of equipment around the rig increases the number of potential sources of releases that would normally be present during drilling operations. Potential releases can occur from fuel-storage tanks; chemical-storage or -mixing areas; reserve or emergency pits; tanks; the rig floor when making or breaking connections; the rig cellar; pipe-dope buckets and brushes; equipment; equipment repair and replacement; used filters; materials-delivery or vacuum trucks; well-logging, casing, and cementing operations; perforating; well treating and stimulation; and well testing.

Potential pollutants from completion operations include completion fluids and additives; produced sand; cement cuttings and returns; metal cuttings; formation fluids, including crude oil and produced water; chlorides; acids and caustics; grease and metals in pipe

**TABLE 6.3—PRODUCTION OPERATIONS (continued)**

| System | Pollutant Sources | Potential Pollutants |
|---|---|---|
| Gas processing | | |
| Dehydration | Glycol filters<br>Media-handling systems<br>Vapor-control systems | Chemicals<br>Condensate (hydrocarbon)<br>Fuels<br>Glycols<br>Lubricating oils and grease<br>Naturally occurring metals<br>Paraffin |
| Sweetening | Amine filters<br>Liquid-absorption systems<br>Solid-adsorption systems | Amines<br>Chemicals<br>Fuels<br>$H_2S$<br>Lubricating oils and grease<br>Mercury (and other naturally occurring metals)<br>Paraffin |
| Ancillary operations | Chemical-injection equipment<br>Cooling towers<br>Electrical equipment (e.g., transformers, capacitors)<br>Engines<br>Flowlines<br>Hydrocarbon-loading racks<br>Line pigging<br>Metering equipment<br>Oil filters<br>Pumps<br>Separators<br>Storage tanks | Brine<br>Chemicals<br>Crude oil<br>Dielectric fluids (includes PCB's)<br>Engine coolant<br>Fuels<br>$H_2S$<br>Hydraulic fluid<br>Lubricating oils and grease<br>Metals<br>Natural gas liquids (NGL's)<br>NORM<br>Scale<br>Solvent |
| Produced-water management | Casing/tubing leaks<br>Injection/disposal units (e.g., pumps, engines)<br>Injection/disposal wells<br>Filters<br>Flowlines<br>Produced water<br>Saltwater pits<br>Shut-in and abandoned wells<br>Storage tanks<br>Transformers<br>Treatment equipment<br>Vacuum trucks<br>Wellheads | Chlorides<br>Coolants<br>Crude oil<br>$H_2S$<br>Heat-transfer fluids<br>Lubricating oils and grease<br>Metals<br>NORM<br>PCB's<br>Salts<br>Scale<br>Solvents<br>Tank bottoms/sludges (basic sediment and water)<br>Treatment chemicals<br>Vacuum-truck rinsate |

dope; lubricants (used oil and grease); solvents; scrap metal; debris; trash; and sewage.

## 6.4 Production Operations

**6.4.1 Introduction.** Production of oil and gas requires a series of operations that range from facility construction to transferring treated oil or gas to a transportation system. Crude oil is extracted from subsurface formations by a variety of methods and typically requires subsequent treatment. Natural gas requires more complex treatment and processing to remove impurities. An important aspect of production operations is management of produced water, which is, by far, the largest-volume waste stream in oil and gas E&P operations. This section describes the major aspects of oil and gas production and treatment operations, with each section addressing the most common potential sources of pollutants and associated pollutants. **Table 6.3** lists these sources and pollutants.

**6.4.2 Site Preparation.** Drillsites of successfully completed wells usually are expanded to accommodate production facilities. This expansion can include addition of new roads and pads, tank batteries, production facilities, flowlines, pipelines, and support facilities. Concerns dealt with in construction of drilling pads and roads are also of concern during production construction operations.

Construction of production equipment can take place in one area in a field or at multiple locations throughout the field. The construction operations include such potential pollutants as scrap metal and packing and shipping wastes (e.g., paper, wood, and strapping). Other potential pollutants are those discussed in Sec. 6.3.2.

**6.4.3 Field Production Systems.** Production of crude-oil wells may require artificial lift, such as gas-lift systems or rod or submersible pumps. Additional activity at the wellsite may include separation of produced water and other impurities in the oil or gas. Potential pollutant sources at the wellsite include downhole tubing and casing, wellhead assemblies, surface pumping equipment, and chemical-treatment injection systems.

Produced oil and gas can be treated at the wellsite or sent to a treatment and storage facility, such as a central tank battery or gas plant. Sec. 6.4.5 discusses gas-processing operations.

Before storage in stock tanks, crude oil may undergo treatment at the tank battery in a variety of separation devices (e.g., heater treaters, two- or three-phase separators) to remove any associated gas or water. Produced water also is stored in tanks at the battery. Releases may result from equipment failures caused by corrosion, excessive wear, abnormal temperature and/or pressure, and human error. Other sources of releases include flowlines, tanks, tank-battery equipment, pipe connections, pump seals, valves, and load-line connections.

Produced oil and water are the most common potential pollutants associated with field production systems. Potential pollutants in production operations are treatment chemicals used in field production systems, scale/sludge from flowlines and equipment, basic sediment and water from separation vessels and storage tanks, and naturally occurring radioactive material (NORM). Table 6.3 provides a more extensive list of potential pollutants from field production systems.

**6.4.4 Workovers and Recompletions.** Producing oil and gas wells may require workovers and recompletions to correct downhole equipment problems or to increase production rate. Workovers and recompletions usually require use of a mobile workover rig. A workover rig presents many of the same potential sources of releases as drilling. Workovers may require displacement of wellbore fluids with specially prepared fluids. The removed well fluid (i.e., blowdown) may be directed either to an earthen workover pit or a portable tank, each of which is a potential source of release. In some instances, a workover rig uses a circulating-fluid system similar to that used on a drilling rig, which presents potential pollutant sources similar to those discussed in Sec. 6.3. Workovers may also require removal of downhole tubing and equipment, which may be a significant potential source of release.

Workovers intended to re-establish or increase well production may require mobile service and support equipment, such as fracturing-fluid pump trucks and tanks, cement trucks, and vacuum trucks. These services typically bring treatment chemicals to the wellsite and are potential sources of release.

Potential pollutants from workover and recompletion operations include hydraulic fluids, lubricating oils, fuels, weighting agents, surfactants, muds, crude oil, produced water, acids, inhibitors, gels, solvents, formation solids, scale, and NORM.

**6.4.5 Gas Processing.** Natural-gas processing can be divided into three categories: dehydration, sweetening, and ancillary operations.

***Dehydration.*** Natural-gas dehydration makes gas marketable by removing water vapor from the produced-gas stream. Dehydration can occur at field facilities in close proximity to the wellhead or in centrally located gas-processing plants. Two methods used for gas dehydration are absorption of the water vapor into liquid desiccants or adsorption of the water vapor onto solid desiccants. Solid desiccants include alumina, silica gel, calcium chloride, and molecular sieve. Liquid desiccants include ethylene, diethylene, and triethylene glycols. Both solid and liquid desiccants have an affinity not only for the water in the gas stream, but also for hydrocarbon constituents, such as benzene, toluene, ethylbenzene, and xylenes (BTEX) and some naturally occurring metals like mercury. Commonly, vents from the dehydration system are routed through a vapor-control system, such as a condenser, vapor-recovery system, or thermal oxidizer, to collect the potential pollutants.

Dehydration and vapor-control systems are potential sources of releases. Releases from the dehydration system can result from improper handling of the desiccant during transfers and storage; leaking piping, tubing, connections, and pumps; and improper disposal of spent dehydration desiccant. Also, glycol filters in the desiccant systems are a potential source of release of glycol if improperly handled, recycled, or disposed of. The potential for releases of condensed fluids to the ground and into water can exist as a result of leaky piping, tubing, leaky connections, leaky pumps, and improper handling of the condensed liquids.

Potential pollutants from dehydration systems include glycols, hydrocarbons, mercury, and natural-gas or wellbore-treatment chemicals. Hydrocarbons are associated with the glycol filters, the desiccant system, and condensed fluids from the vapor-control system. In some producing regions, the gas stream contains relatively high concentrations of mercury. This naturally occurring metal can accumulate in the glycol filters, desiccant system, and condensed fluids. In addition, natural gas and wellbore-treatment chemicals injected upstream of the dehydration system may accumulate in the desiccant, glycol filters, and condensed liquids.

***Sweetening.*** Natural-gas-sweetening processes reduce the hydrogen sulfide ($H_2S$) and $CO_2$ acid-gas concentrations in gas to marketable levels. Like dehydration, gas sweetening can take place at field facilities or gas plants. Acid gases can be removed by absorption into liquid desiccants or adsorption onto solid media. Gas-sweetening media include amine, iron sponge, caustic, and molecular sieve. Also, gases with high concentrations of $H_2S$ may be treated by catalytic processes that result in production of elemental sulfur. In addition to acid gas, media used in the sweetening process may accumulate the gas-stream hydrocarbon constituents and naturally occurring metals (e.g., BTEX and mercury, respectively).

Releases from the sweetening system can arise from improper handling of the media during storage and transfer; leaking piping, pumps, and connections; and improper disposal. In addition, filters in amine-circulation systems represent a potential source of release if they are improperly handled and disposed of. Primary pollutants of concern from sweetening systems include amines, hydrocarbons, naturally occurring metals, and chemicals. Hydrocarbons, naturally occurring metals, and chemicals are associated with accumulations of these pollutants in the sweetening media and may appear in the amine filters.

***Ancillary Operations.*** Gas processing includes several ancillary operations in addition to dehydration and sweetening that have potential for releases to the environment. These include gas metering and compression; inlet separation; natural-gas-liquid (NGL) recovery, storage, and loading; corrosion protection; process cooling; and line pigging. Leaks and spills from systems handling hydrocarbon liquids present the highest potential for releases. Additional release potential exists from compressor engines, oil filters, gas-metering equipment, chemicals used in the corrosion-protection system, contact cooling water, and transfer and storage of chemicals and lubrication oils. In addition, scale formation throughout the process can result in accumulation of NORM.

The primary pollutants of concern in ancillary systems are condensates; crude, lubricating, and hydraulic oils; fuels; and solvents. Additional pollutants of concern are the chemicals used throughout the operation, produced water, naturally occurring metals, NORM, and engine coolant.

Releases of condensate, crude oil, and produced water are associated with inlet separation, fuel scrubbing for fired and compression equipment, natural-gas recovery, condensate storage and loading, and line pigging. Releases of lubricating and hydraulic oils, fuels, coolants, and solvents are associated with compression equipment, filters, and oil-transfer and -storage operations.

Chemical-injection applications present other potential sources of release. Such applications include those related to corrosion inhibition, hydrate reduction, and cooling-water treatment. Chemical releases occur as a result of leaks, spills, and drips in the transfer, storage, and use of the chemicals. In some operations, cooling water may be discharged to the ground or surface water. As discussed under gas sweetening and dehydration, chemicals introduced into the gas stream can accumulate in the sweetening and dehydration media. Treating chemicals may also accumulate in inlet-gas separators and pipeline-pigging liquids and wastes.

In the past, mercury commonly was used in manometers and gas-metering equipment. Potential releases of mercury are associated with meter runs and meter-maintenance operations. In addition, as discussed previously, naturally occurring metals tend to condense and accumulate in the natural-gas-handling processes. These metals generally are associated with inlet separation, fuel scrubbing for fired and compression equipment, line pigging, and natural-gas handling.

NORM-contaminated scales can be deposited throughout the gas-processing train. In gas-processing facilities, the long-lived radon daughters, lead 210 and polonium 210, may be found in pumps, tanks, and product lines associated with ethane/propane ($C_2/C_3$) processing and distribution. In this equipment, NORM typically plates out as a thin film on equipment walls such as T valves and elbows, with little or no loose accumulations. These films adhere to the metallic walls of the equipment and may not be readily visible.

**6.4.6 Produced-Water Management.** Water produced with oil and gas includes formation water and water injected into the reservoir to increase productivity. Produced water constitutes the major waste-stream source by volume from the entire E&P process. The constituents of concern in produced water are chlorides, oil, heavy metals, radionuclides, treating chemicals, and dissolved oxygen ($O_2$).

After passing through an oil/water separator, produced water usually is injected back into another subsurface formation for disposal, into a producing reservoir for pressure maintenance (waterflooding), or as steam for thermally enhanced oil recovery or for discharge into the sea. However, where chlorides and other constituents in produced water are low enough, regulations may allow the

**TABLE 6.4—SUPPORT OPERATIONS**

| System | Pollutant Sources | Potential Pollutants |
|---|---|---|
| Maintenance | Vehicle repair and inspection | Acids |
| | Batteries | Asbestos |
| | Contaminated soil | Biocides |
| | Contaminated wash water | Caustics |
| | Engines and brakes | Chlorides |
| | Filters | Contaminated debris |
| | Scrap material | Dielectric fluids (includes PCB's) |
| | Sorbent pads | Heavy metals |
| | Spill-cleanup equipment | Hydraulic oil |
| | Process equipment | Lubricating oils and grease |
| | Batteries | NORM |
| | Blowdown equipment | Organic chemicals |
| | Boilers | Organic salts |
| | Cleaning equipment | Painting waste (spent sandblast media, |
| | Compressors | paint, and thinners) |
| | Contaminated soil | Scale |
| | Electrical equipment (e.g., transformers and capacitors) | Scrap metal |
| | Filters | Solvents |
| | Generators | Treating chemicals |
| | Pumps | |
| | Scrap material | |
| | Washdown | |
| | Wire and cable | |
| | Painting | |
| | Painting equipment | |
| | Sand blasting | |
| | Spills | |
| Waste and Chemical Storage | | |
| | Batteries | Acids |
| | Contaminated soil | Alumina |
| | Drums and containers | Amines |
| | Filters | Chemicals |
| | Motors | Chlorides |
| | Pumps | Contaminated debris |
| | | Crude oil |
| | | Heavy metals |
| | | Fuels |
| | | Glycols |
| | | Hydraulic oil |
| | | Lubricating oils and grease |
| | | Metals |
| | | Scrap metal |
| | | Silica gel |
| Domestic/Office | | Solvents |
| | Canteen | Ammonia |
| | Filters | Biochemical and chemical oxygen demand |
| | Living quarters | Bacteria |
| | Medical | Contaminated debris |
| | | Detergents and soaps |
| | | Medicines |
| | | Oil/grease |
| | | Medical waste |
| | | Scrap food |
| | | Sewage liquids and sludge |
| | | Solids |
| | | Treating chemicals |

water to be discharged for beneficial purposes, such as crop irrigation, livestock or wildlife watering, and groundwater recharging.

Potential sources of release of produced water include injection/disposal wells, point-source discharges, flowlines and gathering lines, tanks, pits, vacuum trucks, and treatment-related equipment.

Releases from injection and disposal systems can occur both above and below ground. At the surface, potential sources of releases include pumps, engines, motors, piping, metering equipment, and power sources. Subsurface sources of releases typically are tubing, packers, and casing. Produced water also may impact the quality of surface and underground sources of drinking water if injection/disposal wells are completed improperly or plugged and abandoned improperly. Potential sources of releases at produced-water storage facilities include pit and tank liners, piping, metering equipment, pumps, fuel sources, chemicals and treatment equipment, and vacuum trucks.

Potential pollutants from produced-water-injection and -disposal operations include chlorides, treatment chemicals, NORM, hydrocarbons, heavy metals, and used oils. NORM most commonly occurs in scale that forms or accumulates in above-ground flowlines and processing equipment; equipment associated with saltwater disposal/injection wells; tanks, casing, and tubing; and soils contaminated as a result of well workovers, saltwater releases, and pipe-cleaning operations.

## 6.5 Support Operations

E&P operations often require support functions that include maintenance and storage operations, living quarters, and office facilities. Activities include vehicle/equipment inspection, maintenance, and repair; chemical and waste storage; power generation; painting; and support of domestic facilities and offices. Potential sources of pollutants unique to these activities include vehicles, transformers, capacitors, batteries, domestic-waste systems, and sandblasting equipment. Associated potential pollutants include battery acids, lead, polychlorinate biphenyls (PCB's), refrigerants, solvents, sanitary waste, and office and medical wastes. **Table 6.4** lists common potential sources of release and associated pollutants.

### 6.6. Closure and Abandonment

E&P sites and facilities ultimately require closure and abandonment. Proper closure ensures that no contaminants or sources of contaminants remain on site and that the location is restored. Closure may be addressed in three areas: disassembly and removal of equipment, buildings, and other surface structures; closing or plugging of shotholes, pits, and wells; and site reclamation.

Disassembly and removal of equipment, buildings, and other structures involve potential sources for release of pollutants. For example, tanks, treating and storage vessels, meters and gauges, power-generation equipment, field laboratories, and chemical-storage warehouses may release pollutants if containment, capture, and proper disposal of their contents are not performed.

Potential sources of releases (particularly to groundwater) associated with E&P operations should be mitigated when a site is abandoned. These include seismic shotholes, ratholes, production and injection wells, and pits. Shotholes, ratholes, and wells should be plugged properly to eliminate avenues for releases to the environment. Pits should be closed in a manner that prevents leaching of contained substances into the adjacent environment.

Finally, reclamation of the site should be addressed when the site has been cleared and all potential sources have been mitigated. Reclamation may require planting or seeding vegetation to return the site to its original condition. In some instances, reclamation begins with remediation of contaminated soils and/or groundwater. Management of wastes from remediation operations may present potential sources for release (e.g., leakage or spillage from drums that contain recovered contaminants).

Potential pollutants associated with closure and abandonment operations include all pollutants discussed in preceding sections.

### 6.7 Over-Water E&P Operations—Additional Pollution-Evaluation Considerations

**6.7.1 Introduction.** Most aspects of over-water E&P operations are identical to those already discussed. However, various types of equipment and activities are unique in this operating environment. This section highlights the additional pollution-evaluation considerations related to over-water E&P operations in onshore (e.g., lacustrine), offshore, and coastal areas.

**6.7.2 Drilling and Completion Operations.** Operations conducted in lacustrine and coastal environments often require dredging of bottom sediments and placement of pilings. Typically, barge-mounted dredging equipment is mobilized and dredged material is transported to the shoreline or other open-water areas. Piles are driven to secure the placement of drilling barges and support equipment. The equipment used to conduct these types of operations provides a potential source of release.

Jackup rigs, floating rigs with dynamic positioning or anchoring, or drilling platforms placed on the seafloor may be used in offshore drilling and completion operations. Most potential sources of release for this type of equipment are identical to those in onshore operations. Some unique potential sources of releases associated with drilling and completion operations in over-water environments include containment/discharge systems for drilling mud and drill cuttings, deck-drainage systems, fire-control systems, drip- and leak-containment systems, and seafloor BOP's. Potential pollutants from over-water drilling and completion operations include those presented in Sec. 6.3.

**6.7.3 Production Operations.** Although similar to most onshore operations, over-water production operations can include unique potential release sources. Subsea equipment, drip- and leak-containment systems for decks, and fire-control systems are examples. Potential pollutants are those discussed in Sec. 6.4.

Produced water merits additional mention with respect to offshore production operations. More so than in onshore operations, produced water may be discharged to the environment as permitted or regulated by international conventions or local guidelines. Therefore, discharged produced water may be a potential source of release of pollutants. Consideration should be given to such potential pollutants in discharged produced water as dispersed and soluble oil, heavy metals, radionuclides, and treating chemicals.

**6.7.4 Facility Support Equipment and Vessels.** Over-water operations often require use of support equipment that is normally not required for onshore operations. Such items as cranes, helicopters, marine vessels, engines, pumps, batteries, drums and tanks, boilers, bilges and decks are potential sources of release. Potential pollutants include fuels, lubricating oils, hydraulic fluids, filters, heavy metals, rinsate, boiler blowdown, and bilge and ballast water.

**6.7.5 Domestic- and Sanitary-Waste Systems.** Because of the remoteness of over-water operations, provisions sometimes are made for housing the crew and gang labor. General work areas, living quarters, lavatories, showers, laundry rooms, galleys, and recreation rooms are sources of potential releases of contaminants. Sec. 6.5 discusses potential pollutants associated with domestic- and sanitary-waste systems.

**6.7.6 Facility Closure and Abandonment.** Closure and abandonment of over-water operations present unique potential sources of release. These include marine risers, subsea production and test equipment, and platforms and spars. Potential pollutants include crude oil, test fluids, muds, cuttings, and cement.

### 6.8 Air Emissions

**6.8.1 Introduction.** Air emissions are the most ubiquitous of all environmental releases from E&P operations and the most difficult to identify because they usually are undetectable without the aid of special monitoring equipment. Air emissions are of concern because of their physical and sensory impacts on people (e.g., physiological impacts, visibility, and odor) and their effects on the environment (e.g., plant stress, acid precipitation, and global climate change among others). For the purposes of this section, air pollutants addressed include those (criteria pollutants) for which the U.S. Environmental Protection Agency (U.S. EPA) has established ambient-air quality standards.[1] These pollutants include the following.

1. Reactive organic compounds (ROC's), also known as nonmethane hydrocarbons or volatile organic compounds (VOC's).
2. Oxides of $N_2$ ($NO_x$).
3. Sulfur dioxide ($SO_2$).
4. Particulate matter smaller than 10 $\mu$m in diameter ($PM_{10}$).
5. Carbon monoxide (CO).
6. Lead (Pb).

Also addressed are methane ($CH_4$) and $CO_2$ because of their link to suspected global climate change. $H_2S$, BTEX, *n*-hexane (*n*-$C_6$), chromate, ammonia, chlorine, and formaldehyde are considered because of their identification as hazardous air pollutants (HAP's) by several regulatory agencies. Depending on the nature of the hydrocarbons and other chemicals present in individual operations and on national or local regulations, other compounds may also need to be evaluated as HAP's.

A source test is the most accurate method to quantify and speciate air emissions. However, such tests usually require modifications to the equipment to ensure representative sampling, require special equipment and training, and are expensive. Where possible, published methods to calculate or estimate emissions are identified.

The precision and accuracy required in emissions estimates are determined by the following.

1. Precision defined in the regulations that require preparation of an inventory (e.g., the emission estimate for each source of a specif-

| TABLE 6.5—INFORMATION FOR ESTIMATION OF STORAGE-TANK EMISSIONS |
|---|
| Material stored |
| Crude analysis (inputs for flash calculation) |
| Vapor pressure (taken from literature or computed) |
| Vapor molecular weight (taken from literature or computed) |
| HAP fractions in liquid (if speciated HAP emissions are required) |
| HAP fractions in vapor phase (may be computed from liquid composition) |
| Tank description |
| Capacity and dimensions |
| Color (for fixed roof only) |
| Liquid-storage temperature |
| Liquid-storage pressure |
| Type of rim seal(s) (for floating roof only) |
| Number and type of fittings (for floating roof only; may be estimated) |
| Type of deck (bolted or welded) |
| Condition of seals and tank walls (for floating roof only) |
| Annual throughput |
| Environmental data |
| Annual mean wind speed (for external floating roof only) |
| Diurnal temperature change (for fixed roof only; may be approximated) |
| Optional headspace analysis (source test of representative tank) |

ic HAP might be required to be accurate to within 10% of the facilitywide emissions of that substance).

2. Magnitude of the emissions and the estimated environmental or public-health impact of the emissions.

3. Cost of source testing or other measurements required to refine the emissions estimates.

4. Source-test method specified by the reviewing agency. The precision and accuracy of source tests often are specified in the applicable published source-test method. If the pollutants are highly toxic, care should be taken to ensure that method detection limits are low enough to quantify emissions accurately.

To use source-test data to estimate emissions, the process must be evaluated to determine whether a functional relationship exists between the emissions rates and some process variable. The process variable chosen typically is related to fuel use or production rate. For steady-state operations, emissions may depend only on hours of operation.

If possible, target detection limits should be established before the technical approach is finalized. Detection limits are well-established for the majority of the criteria pollutants [e.g., particulate matter, $NO_x$, sulfur oxides ($SO_x$)] by the source-test methods and generally are not an issue. However, detection limits for many HAP's and atypical pollutants may be a particularly important component of the test program and directly affect the usefulness of the data.

**6.8.2 Tanks/Sumps/Pits.** ***Equipment.*** The equipment covered in this section includes those processing and storage devices that operate at or near atmospheric pressure and release volatile components through open vents or directly from the liquid's surface. In production operations, the most common sources are fixed-roof tanks, followed by open-top tanks, floating-roof tanks, sumps, and pits.

***Emission Types.*** Significant VOC emissions can be emitted from this source, depending on tank size and working, breathing, and flashing losses. The emission rates normally are estimated with tank calculations found in U.S. EPA's *AP-42*,[2] which does not address losses from flashing. Emissions of concern from tanks, sumps, and pits are $C_3$ through heptane ($C_{7+}$), BTEX, $n$-$C_6$ and $H_2S$. Some individuals may be interested in $CH_4$ emissions as well.

***Emission Estimation/Measurement.*** Emissions from fixed-roof tanks result from flash, standing storage (breathing) losses, and working losses. Flash occurs when the pressure on the crude drops below the vapor pressure of the light organic species (e.g., $CH_4$, $C_2$) contained in the crude oil, resulting in boiling off of these light species. Standing-storage losses are vapors lost from the tank vapor space as a result of thermal expansion as the tank is heated by solar radiation and daytime-temperature fluctuations. Working losses are caused by displacement of vapors by incoming product. Tank vents normally are equipped with pressure/vacuum devices that allow the tank to operate at a slight internal pressure or vacuum to prevent release of vapors during very slight changes of temperature or tank-fluid level.

Emissions from floating-roof tanks result from withdrawal loss (clingage of liquid to sliding surfaces) and vapor losses through rim seals and roof fittings. Withdrawal loss occurs as the roof moves downward during product removal. Internal-floating-roof tanks have both a permanent fixed roof and an internal floating deck. The fixed roof protects the deck and tank walls from the weather. The internal floating roof eliminates the VOC-rich "vapor space" between the liquid surface and the roof of a fixed-roof tank.

**Table 6.5** provides information necessary to estimate storage-tank emissions. The general procedure for estimating HAP emissions from tanks involves three steps.

1. Gather data on the tank and the product stored.

2. Estimate total VOC emissions with tank equations in U.S. EPA *AP-42*[2] or from flashing estimates.

3. Estimate HAP fraction of emitted VOC's with Raoult's law,[3] headspace-sampling data, or other method.

*Flashing Losses.* Standard tank-emission estimating methods published in U.S. EPA *AP-42* do not address flash emissions. Tank flashing losses are based on evaluation of (1) pressure drop from the last separator to the tank, (2) production throughput, (3) oil gravity of production (°API), and (4) production temperature. Flash emissions then are added to the estimated working and breathing losses. Because flash components generally dominate the emissions, the flash composition is likely the best estimate of the composition of working and breathing losses unless headspace-analysis data are available.

American Petroleum Inst. (API) and Gas Research Inst. (GRI) have developed flash-emission equations that require the following inputs to determine flashing losses: (1) pressure drop from last separator before tank (psig), (2) production throughput (B/D), (3) product temperature at the last separator (°F), and (4) oil gravity at the last separator (°API).[4]

If the API emissions equations are unsatisfactory, direct-measurement techniques are available that involve sealing all tank openings and setting up a flowmeter on the tank vent. This can be done with an orifice-plate flowmeter attached to a strip chart; however, the flow may reverse, especially at night. A sample of the tank emissions should be taken during the test to determine the composition of the

**TABLE 6.6—VAPOR-LOSS CALCULATIONS**

Flashing losses—fixed-roof tanks

Equations currently being developed by API.

Working and breathing losses—fixed-roof tanks[2]

Total losses from a fixed-roof tank that no longer is experiencing flashing losses are described by

$$L_t = L_s + L_w,$$

where $L_s = 365 V_V \rho_v K_e K_s$ (where 365 = D/yr) and $L_w = 0.0010 M_v p_v Q K_n K_p$ (where for turnovers > 36, $K_n = (180 + n)/6n$, and for turnovers < 36, $K_n = 1$).

Working and breathing losses—floating-roof tanks[2]

Total losses from a floating-roof tank that no longer is experiencing flashing losses are described by

$$L_t = L_r + L_{wd} + L_f.$$

Methods for low-vapor-pressure crude oils

Standard methods and assumptions used to estimate tank emissions can give unrealistic estimates of emissions in some cases, particularly in heavy-crude and water tanks. The methods presented in Ref. 2 assume that the headspace of a product tank is in equilibrium with the tank contents; for crude tanks, these methods assume that the RVP of the crude is known. These assumptions are not applicable for many crude-oil tanks. Alternative approaches are needed for the following reasons.

1. The vapor space in crude-oil tanks often is made up of flash gas or of gas from the tank vapor-balance system, not of species evaporated from the liquid product.

2. The default crude-oil vapor molecular weight listed in Ref. 2 of $M_v = 50$ is intended to be applicable to crude oils with RVP = 5 stored at 60°F; it may not be accurate for crudes stored at elevated temperatures.

3. The standard Reid vapor test is an appropriate tool for estimating the storage-temperature true vapor pressure, $p_v$, of products with an RVP $\geq$ 10. It is a very poor measure of properties of crude oils [particularly heavy crudes (RVP = 1)] where $CO_2$ and other inerts evolving off the crude make up most of the measured RVP.

The RVP problem is compounded when the produced crude is in the form of alternating slugs of heavy crude and water. Drawing a representative sample of such a stream is not possible. In contrast, tank headspace is relatively homogeneous and can be sampled. The disadvantage of headspace sampling is the difficulty in gathering a representative sample from tanks where headspace composition may change over time.

Analytical data can be used in Ref. 2 tank equations in place of default assumptions to generate refined, site-specific, tank-emission estimates. Two of the key parameters used in Ref. 2 tank equations are $M_v$ and $p_v$. The discussion in Ref. 2 assumes that $M_v$ for most tanks can be approximated from Raoult's law. Because Raoult's law is not very useful in the case of crudes or crude/water mixtures, Ref. 2 offers a single default $M_v$ of 50 (Ref. 2, Table 12.3-2) for use with crude oil.

The discussion in Ref. 2 assumes that $p_v$ can be approximated from the RVP of the product; however, it offers no suggestions on how to deal with heavy crudes and crude/water mixtures where the RVP is not a meaningful parameter. When actual measured $M_v$ and $p_v$ are used in the fixed-roof-tank equation, two of the other equation variables, the vapor-space-expansion factor, $K_e$, and the saturation factor, $K_S$, must also be defined so as to be consistent with the measured data.

Methods for high-water-vapor fractions in the headspace

If the product is stored at elevated temperatures and it contains water, water vapor may occupy a significant fraction of the tank headspace. Correcting the $p_v$ estimate to account for water vapor can reduce tank-emission estimates by up to 50%. The following are methods for determining the amount of water vapor present in tank headspace, listed in order of preference.

1. Ask the source-test technician who does the headspace analysis to measure and report water vapor.

2. Measure the headspace-water-vapor content yourself.

3. Estimate headspace-water-vapor content from the vapor pressure of water at tank temperature.

Choice 3 works well in water tanks where water is exposed to the headspace; it is not reliable where the water is effectively capped by oil or where vapor-balance gases move among tanks.

emitted hydrocarbons. Because most of this flow will be flash (rather than working and breathing losses), the flow should be proportional to the production rate during the test period.

*Working and Breathing Losses.* Tank working and breathing losses normally are used for crudes that are stored at atmospheric pressures and that have small amounts of gas separating them from the fluid. The most current and universally accepted method for estimating these losses is presented in Chap. 12, Supplement E, of U.S. EPA *AP-42*,[2] which provides separate estimating methods for fixed and floating-roof tanks. (Associated with this is a menu-driven computer program called TANKS that computes tank emissions.) U.S. EPA *AP-42* tank equations are not accurate for low-specific-gravity crudes (>20°API) and generally overpredict emissions.

Key factors that affect the release of VOC's from floating-roof tanks include type and integrity of the floating-roof rim seal(s), condition of tank walls, and number and type of fittings. For external-floating-roof tanks, atmospheric conditions (specifically, average wind speeds) are critical. For internal-floating-roof tanks, effectiveness of vapor recovery (if any) also must be considered.

Speciation of HAP emissions from fixed-roof tanks normally is done with Raoult's law. This method predicts the composition of the vapors in the tank on the basis of composition of the liquid and physical properties of those components. Speciation of HAP emissions from internal- and external-floating-roof tanks involves speciating both the liquid and the vapor space above the liquid. Losses from deck fittings and rim seals are assumed to be equilibrium vapor lost from cavities below the roof. Raoult's law or source testing can be used to determine the composition of this vapor. Clingage-loss emissions (i.e., liquid clinging to the side of a floating-roof tank as the floating roof lowers) have the same composition as the liquid in the tank. Clingage typically is a very small fraction of total emissions.

In some cases, where Raoult's law and Reid vapor pressure (RVP)[5] are poor predictors of headspace composition and vapor pressures, headspace sampling can be undertaken. Sampling is most useful for crude oil or oil/water mixtures that exhibit little or no flashing emissions. Tank headspace normally is stratified with $CH_4$ and air at the top and heavier organics near the liquid surface. The gases near the top dominate emissions; however, a common practice is to pull a vapor sample from just above the liquid surface. This produces a conservative sample and reduces the possibility of contaminating the sample with outside air. **Table 6.6** provides further insights into vapor-loss calculations. A copy of U.S. EPA's "Tanks 2" equation should be obtained to make the calculations for working or breathing (standing) losses in the table.[6]

Methods for Constant-Level and Surge Tanks. To compute emissions from a tank that operates at constant or controlled level, use Ref. 2 (Chap. 12) to handle the turnover factor, $K_n$, in the working-loss equation and the product throughput factor, $Q$, in the working-loss and floating-roof withdrawal-loss equations. Unfortunately, the text in Ref. 2, Chap. 12, is a little unclear. The issue is total change in liquid height over time. This change defines the volume of vapors pushed out of the vent (working-loss equation) or the actual linear motion of the deck (withdrawal-loss equation). It does not matter whether the total change is the sum of many small changes or a few large changes. What matters is whether the tank is simultaneously filled and emptied so that total-liquid-height change is less than would be indicated by the volume of product throughput. If inlet/outlet pumping rates and pumping schedules are known, it is possible to estimate total change in liquid height over the year and, hence, total vapor displacement. (If specific data are not available, U.S. EPA *AP-42* suggests that a typical total annual displacement value, $Q$, for constant-level tanks is four times the tank volume. For constant level tanks, $K_n$ in the fixed-roof-tank working-loss equation should be 1.0.)

If U.S. EPA *AP-42* tank equations were based solely on ideal gas theory, estimated emissions would be linear functions of the vapor pressure at daily average liquid surface temperature, $p_v$, and vapor molecular weight, $M_v$. If that were the case, $p_v$ and $M_v$ could be computed for some subset of total organics (such as nonmethane hydrocarbons or benzene) and the desired emission computed directly from the tank equation. However, the adjustments used to fit tank equations to measured tank-emission data are nonlinear. Therefore, partial pressure and molecular weight of a single species or a set of species cannot be substituted into the tank equations to get the same answer that can be obtained by estimating VOC's and then speciating the total by weight fractions. The standard approach is to use VOC's (including $CH_4$) in the tank equations, then speciate the estimated VOC annual emissions (lbm/yr) into annual emissions of the various organic species.

*Sumps/Pits/Separators.* Emissions from well cellars, oil/water separators, pits, and sumps may be estimated with one of the three following methods.

1. Default or "average" emission factors (lbm-yr/unit).

2. Computations based on the content of the unit. U.S. EPA's WATER8[7]/CHEMDAT8[8] computer models may be used for various types of sumps and impoundments.

3. Direct-measurement or mass-balance methods; however, these are generally not applicable to E&P facilities.

Selection of an emission-estimation method depends on the accuracy required and the availability of data about the composition of the material being handled. Average emission factors are easy to use but tend to give relatively high emission estimates. Another shortcoming of emission factors is the shortage of published factors that are applicable to all source types.

Guidance on estimating emissions from sumps and surface impoundments is presented in U.S. EPA *AP-42*,[2] which recommends use of mass-transfer equations. Because the recommended mass-transfer equations are tedious to use, the U.S. EPA developed the PC-based software. Use of such models as WATER8/CHEMDAT8 to estimate VOC emissions requires a modest amount of site-specific data, but the extra effort provides more accurate emission estimates. (Use of these models to estimate VOC emissions from wastewater-treatment facilities is accepted by U.S. EPA and other U.S. regulatory agencies and generally is preferable to use of emission factors found in U.S. EPA *AP-42* or elsewhere.)

If these models are used to calculate emissions, the result is reported by chemical species. However, if emission factors are used to estimate total emissions, they must be speciated if HAP-emission estimates are required. The basic techniques for speciating emissions are Henry's law,[3] Raoult's law,[3] and source testing. Henry's law is usable as long as concentrations of organic species in the water are below their solubility limit. If an oil layer is present on the water surface, Raoult's law should be used. Source testing of sumps and oil/water separators is very difficult because there is no defined air-exhaust stream with a measurable flow rate.

| TABLE 6.7—EMISSION-REDUCTION TECHNIQUES |
|---|
| Fixed-roof tanks |
| Paint with light-colored paint to reduce solar absorptivity. |
| Install vacuum/pressure vent controls. |
| Convert to an internal floating roof (an air sweep or inert blanketing may be required to avoid generating an explosive mixture in the vapor space). |
| Add a vapor-recovery system (vapors may be recovered by means of condensers, captured on activated carbon for subsequent recovery or destruction, or destroyed in a flare). |
| External-floating-roof tanks |
| Install double-rim seals or upgrade existing dual-rim seals. |
| Use a welded-seam rather than a bolted roof. |
| Add gaskets or seals to all fittings and hatches. |
| Convert to an internal floating roof (an air sweep or inert blanketing may be required to avoid generating an explosive mixture in the vapor space). |
| Internal-floating-roof tanks |
| Use self-supporting (e.g., geodesic) roofs to avoid support legs that penetrate the floating roof. |

*Air-Flotation Cells/Deperators.* Emissions from air-flotation devices or deperators are a concern only if $H_2S$ is present in the produced water or if hydrocarbons are present in the flotation gas. Emissions of VOC's can be calculated on the basis of manufacturers' specifications for flotation gas and mass percentages of VOC's in a gas.

***Control Techniques.*** A variety of techniques may be used to reduce tank emissions. The technique selected depends on the material being stored and the configuration of the existing tank (if any). **Table 6.7** lists emission-reduction techniques. Control of emissions from tanks normally is accomplished by capturing the emissions in a vapor-recovery unit or incinerating the emissions in a flare. Emissions from tank flashing can be substantial and may provide a positive cash flow when attached to a vapor-recovery unit.

**6.8.3 Fugitive Emissions.** ***Equipment Description.*** Fugitive emissions are routine emissions of process gases or liquids from small leaks in equipment, usually at a joint, seal, or rotating surface. Sources that contribute to fugitive emissions include such components as valves, flanges, sampling connections, pumps, compressors, and pressure-relief valves as well as sources like storage tanks, loading operations, process drains, agitators, cooling towers, and separators. Typically, most of a facility's fugitive emissions result from valves, flanges, and pumps. Pneumatic systems include gas-actuated chemical pumps and actuator valves. These may be sources of $CH_4$ or VOC's if field gas is used because these devices usually vent to the atmosphere.

***Emissions Types.*** Pollutants of concern are VOC's, such as $C_3$ through $C_{7+}$, BTEX, $n$-$C_6$, and $H_2S$.

***Emissions Estimation/Measurement.*** The various methods for estimating fugitive emissions include use of average, leak/no-leak, and stratified emission factors; screening-value to leak-rate correlations; and screening and bagging studies.

*Average-Emission-Factor Methods.* The basic procedure for estimating fugitive emissions from equipment can be summarized by the following three steps.

1. Count the number of components in a given service category (e.g., light liquid, heavy liquid, gas).

2. Multiply the component count by the average emission factor to estimate total VOC emissions.

3. If HAP emissions are to be calculated, speciate the total VOC-emission estimate by multiplying the calculated total VOC rate by the fraction of the HAP.

API developed emission factors for E&P fugitive emissions by monitoring the emissions of large numbers of components in liquid and gas service.[9] Emission factors are presented in units like pounds mass per hour per source and are used to estimate emissions from a specific component type (e.g., valve, flange, or other such component). Emission factors are also available in U.S. EPA *AP-42*.[2]

These factors are for total hydrocarbons, including $CH_4$ and $C_3$. Therefore, to determine VOC emissions, the calculated percentage of $C_3$ through $C_{7+}$ hydrocarbons needs to be introduced to determine actual VOC's. If the process stream served is a mixture, it is assumed that the fugitive emissions have the same composition as the material flowing through the component. For example, if the material flowing through a valve is a liquid, it is assumed to leak as a liquid and evaporate into the air.

The service category is related to the physical state of the material in the line (gas or liquid), the volatility range (light or heavy liquid), and the composition in the line. Each valve should be classified into one of these four service categories. Pumps should be distinguished by either light- or heavy-liquid service. Emissions from flanges and all other components are assumed to be independent of service category. Emission factors for each component type have been developed for applicable categories.

Although this is the quickest method to obtain a fugitive-emission estimate, it may overestimate emissions from facilities with an inspection and maintenance program, leakless equipment, or modern sealing technology. The U.S. EPA and most U.S. states allow facilities to reduce their "uncontrolled" emissions estimates by 50 to 90% when a formal inspection and maintenance program is adopted. The exact percentage is determined by local regulations and is a function of the leak-detection and -repair (LDAR) frequency and the leak definition used.

*Emission-Factor Methods.* With the exception of average emission factors, emission-factor methods (leak/no-leak, stratified, and correlation equations) require facility screening data. Screening, a relatively quick measurement method, uses a portable detector that indicates whether a source is leaking and gives a rough measurement of the magnitude of the emissions. Concentrations measured near potential leak points are called screening values. A screening survey is simply a compilation of component screening values obtained for a specific process unit or facility. (Note that if screening values are obtained as part of an LDAR or inspection and maintenance program, screening values obtained both *before and after* the repair must be recorded.)

U.S. EPA Method 21[10] outlines proper screening procedures for process equipment. Before making a screening measurement, the instrument response to the ambient air near the component (about 3 ft upwind) is determined. This background reading is subtracted from the component screening values to correct for the ambient level of hydrocarbons.

*Leak/No-Leak Emission Factors.* With screening data, the frequency of leak occurrence for a specific unit can be accounted for. Leak/no-leak emission factors are used as a means of taking equipment-leak occurrence into account when estimating emissions. The U.S. EPA currently defines a leak as a total-VOC screening value of 10,000 ppm or greater.[11] Separate emission factors are used for those components screening < 10,000 ppm (nonleakers) and > 10,000 ppm (leakers).

Recording actual concentrations is recommended rather than qualitative indications of leaks or nonleaks. This allows use of better emission factors, which can lead to lower fugitive-emission estimates. Note that this and all subsequent estimation methods reflect the actual fraction of leaking components at a facility. Therefore, it is not appropriate to reduce leak/no-leak or stratified factors further to account for an LDAR program. LDAR credits apply only when average emission factors are used.

*Stratified Emission Factors.* As an additional refinement, stratified emission factors can be used to estimate emissions. The process is almost identical to the one described previously for leak/no-leak, except that emission factors are divided into sources screening (by volume) at 0 to 1,000 ppm, 1,001 to 10,000 ppm, and > 10,000 ppm. Using stratified emission factors requires a more rigorous screening value measurement and record-keeping program because the concentration reading must be classified into three categories instead of two as in the case of leak/no-leak estimates.

*Screening-Value to Leak-Rate Correlation Equations.* The most exact method of fugitive-emission estimation, short of field testing, is use of screening-value to leak-rate correlations. These equations and their confidence intervals represent a statistical analysis of many pairs of screening-value/measured-leak-rate data sets for specific source types and service categories. To use these correlations, recording of actual screening values is absolutely necessary.

*Screening and Bagging Studies.* If truly unit-specific emission factors are desired, quantitative emission-rate data can be obtained from screening and bagging studies. Bagging, or tenting, consists of enclosing a source in an airtight plastic or aluminum-foil shroud and analyzing the steady-state air stream (of known rate) withdrawn from the enclosure.

***Control Technologies.*** There are usually no fugitive-emission control requirements on E&P operations, with the possible exception of gas plants. If required, controls for fugitive emissions normally consist of an inspection and maintenance program. This ensures that components are inspected on a periodic basis, with repairs normally occurring during equipment shutdowns.

Emissions from a facility can be reduced in several ways. The U.S. EPA and most U.S. states allow a facility to reduce uncontrolled emission estimates by 50 to 90% when a formal inspection and maintenance program is adopted. Equipment modifications that essentially eliminate emissions from a source also can be used. No reduction currently is given for the decreased leak frequency of high-tech seals and gaskets (such as flexible graphite) that are not inherently leakfree.

Implementing or expanding a facility's LDAR program often can be the most cost-effective way to reduce fugitive emissions. An LDAR program consists of tagging components in a facility for identification purposes, screening these components with U.S. EPA Method 21, recording screening values (ideally, an actual reading and not a qualitative leak/no-leak estimate), and repairing those components with unacceptable screening values.

Some types of components have been recognized to be inherently leakfree, primarily because no physical path exists for emissions. Regulators generally allow assumption of a zero emission rate for such components.

**6.8.4 Internal-Combustion Engines.** ***Equipment.*** This section addresses internal-combustion engines including two- and four-cycle rich-burn ( $< 4\%$ $O_2$ in the stack) or lean-burn ( $> 4\%$ $O_2$ in the stack) fossil-fuel engines and turbines.

***Emissions Types.*** Emissions from internal-combustion engines and turbines vary with engine type, combustion-chamber design, type of fuel, operating conditions, and abatement devices used. $NO_x$ are the primary criteria pollutants of interest from internal-combustion engines and turbines. CO and ROC emissions may be a concern in the case of spark-ignited engines. $SO_2$ emissions are proportional to the fuel sulfur content. Benzene and formaldehyde are the HAP's of primary concern, although carcinogenic metals and polyaromatic hydrocarbons may be emitted from units that fire diesel fuel, production-field gas, or pipeline-quality gas. If a selective catalytic reduction (SCR) $NO_x$ control unit is used, ammonia emissions are also of concern.

***Emissions Estimation/Measurement.*** Emission rates are normally determined with emission factors based on data obtained from (1) EPA's U.S. EPA *AP-42* manual,[2] (2) manufacturers' data, and (3) stack testing. A factor that affects emission rate is the fuel. Fuel parameters, such as Btu content, sulfur content ($H_2S$ or free sulfur), and excess $N_2$, need to be evaluated for their potential impact.

*Stationary Engines.* Emission factors for internal-combustion engines generally are expressed as mass of pollutant per unit of fuel used or mass of pollutant per unit of work output (i.e., g/kW-hr or g/hp-hr, respectively). Regulatory emission limits, however, are often given in terms of ppm (dry) by volume at some standard $O_2$ or $CO_2$ concentration. [Generally, stack testing of natural-gas-fired internal-combustion engines is required to determine only criteria-pollutant emissions from new units (for permitting reasons).]

Emission estimates for internal-combustion engines are calculated as follows.

1. Identify source type (e.g., 2,000-hp four-cycle lean-burn internal-combustion engine firing natural gas).

2. Determine annual fuel use or work output (may be estimated from fuel-use data or from engine rated power and an assumed or known load factor).

3. Locate appropriate emission factors.

4. Compute emissions, $E$, with

$$E = mF, \quad (1)$$

where $m$ = fuel use and $F$ = emission factor. U.S. EPA *AP-42*[2] emission factors are recommended for estimates of criteria-pollutant emissions from internal-combustion engines unless the engine is equipped with emission controls [e.g., low-$NO_x$ kit, SCR, selective noncatalytic reduction (SNCR)]; specific vendor guarantee or source-test data are available; and/or permit conditions include specific emissions levels.

Ref. 2 includes tables listing typical effectiveness of emission-control technologies, such as engine derating, timing retardation, and SCR for selected engine designs. Ref. 2 does not include emission factors for internal-combustion engines burning liquid $C_3$ gas.

For engines subject to cyclical- or variable-loading operations (e.g., drilling-rig and rod-pump engines), an average engine-load factor must be determined. This average load factor typically is much lower than the peak load. For cyclically loaded engines, an engineering evaluation of load demand can be used; for pumping-unit engines, an evaluation of rod-load demands can be used. To develop average engine load for drilling-rig engines, an evaluation must be made of normal loading patterns and their temporal relationship to the complete operation cycle being used.

If an internal-combustion engine is equipped with an $NO_x$-abatement system, such as a low-$NO_x$ kit, SCR, or SNCR, $NO_x$ emissions should be estimated with source-test data from a similar unit, with continuous-emissions-monitoring-system data, or from manufacturer performance-guarantee data. However, caution should be exercised in applying performance-guarantee data because a performance level achievable during an initial source test may not be maintained over the life of the equipment.

For engines that fire diesel or heavier oils, regulatory agencies may require stack testing to quantify metals and toxic organic levels in the stack gas. Accurate estimation of toxic-metal emissions from diesel-fuel combustion may be determined by stack testing; however, it normally is generated from analysis of the fuel. Metals present in the fuel are assumed to be 100% emitted. A data set that may be used to approximate metals emissions from diesel internal-combustion engines is the distillate oil-fired turbine-emission factors presented in U.S. EPA *AP-42*.[2]

Emission factors for HAP's may be presented either as standard emission factors (i.e., g emitted/hp-hr) or as speciation profiles (i.e., a breakdown of total VOC's into specific constituents). In addition, the U.S. EPA has published default speciation profiles[12] that may be used where source-specific data are not available. Other sources of HAP species data for internal-combustion engines include industry source-test studies, California AB-2588 source-test data,[13] and U.S. EPA *AP-42*.[2] (We recommend that existing petroleum industry source tests be checked first, U.S. EPA *AP-42* HAP-emission factors second, and then the U.S. EPA default speciation profiles to select data that correspond best to internal-combustion-engine type and fuel being used. Note that the $CH_4$ fraction of the stack gas presented in default speciation profiles rarely agrees with the $CH_4$-emission factor presented in U.S. EPA *AP-42*. The method selected to deal with this inconsistency depends on the data being used and how the values compare. Some work on engine emissions indicates that BTEX emissions correlate to fuel BTEX where fuel BTEX is > 1 lbm/MMcf. The data indicate no apparent correlation of BTEX emissions.)

*Turbine Engines.* U.S. EPA *AP-42* emission factors can be used to estimate criteria-pollutant emissions from turbines unless the turbine is equipped with emission controls (water injection, SCR, CO catalyst). These emission factors correspond to simple cycle turbines. For systems equipped with supplementary-fired heat-recovery steam generators, the emissions contribution from the supplementary heater (duct burner) should be addressed.

Stack testing of turbines generally is required only to determine $NO_x$ emissions (for permit compliance). Testing for CO emissions and ammonia "slip" through SCR systems may also be required. For turbines that fire on oil or refinery-fuel gas, stack testing may be required by regulatory agencies to quantify HAP levels in the stack gas.

U.S. EPA *AP-42* includes some HAP-emissions data plus data on metals emissions from distillate-oil combustion in turbines. Other than source-test data gathered for California AB-2588 toxics inventories, public domain data on HAP emissions from turbines are limited. HAP-species data for turbines are available from three sources: California AB-2588 source-test data, U.S. EPA *AP-42*, and the U.S. EPA default speciation profile for internal-combustion engines. (The gas-turbine-species profile included in Ref. 12 indicates that VOC's emitted from turbines are 70% $CH_4$ and 30% formaldehyde. This is not considered to be credible data. Therefore, this species profile is not used by regulators or by industry. The U.S. EPA toxic-species profile for gas-fired internal-combustion engines often is used as a surrogate for the gas-turbine-species profile.)

***Control Technologies.*** Control of VOC's, $NO_x$, and CO can be achieved by use of a catalytic converter. A smaller $NO_x$ reduction can be achieved by tuning the engine to run fuel rich; however, this technique increases CO emissions. Control of $NO_x$ and CO also can be achieved by use of low-$NO_x$ technology and SCR, water injection, or steam injection.

*Stationary-Piston-Engine Control Technology.* Control technologies available for stationary internal-combustion engines vary by engine type. Engine types are two-cycle engines, rich-burn four-cycle engines, lean-burn four-cycle engines, and diesel engines.

Several general control methods are used: minor tuning changes, extensive combustion modifications, and tailpipe controls. Minor changes include retarding ignition timing, installing air/fuel ratio controllers, ignition modifications, turbocharging, aftercooling, power deration, and others. Combustion modifications, such as pre-combustion chambers, are more extensive. Turbochargers or other devices may have to be installed to maintain rated power. Tailpipe controls include SCR and SNCR control units. Note that SCR systems require ammonia injection. Tailpipe controls require a detailed evaluation of catalyst-added backpressure, minimum exhaust-gas temperatures, and other such factors.

Control efficiencies can be compared with U.S. EPA *AP-42* or obtained from manufacturer's data. **Table 6.8** lists the ranges of $NO_x$-emission reductions that may be expected.

*Turbine Engines.* Several technologies reduce $NO_x$ emissions from turbines. **Table 6.9** lists these technologies with the ranges of $NO_x$-emission reductions that may be expected.

Gas-fired turbines may also be equipped with CO-oxidation catalysts. Oxidation catalysts typically reduce CO emissions by approximately 80%. ROC emissions also are reduced (typically by approximately 50%) when a CO-oxidation catalyst is used. Oxidation catalysts are not used with fuel-oil-fired turbines because the catalyst would oxidize fuel sulfur to $SO_x$.

**6.8.5 Fired Equipment.** ***Equipment.*** This broad category includes equipment that is fired with a natural- or forced-draft burner and that typically is characterized as an external-combustion device. This category also addresses continuous, intermittent, and emergency flares and incinerators used for destruction of hydrocarbons. Flares and burners used in heaters, boilers, or steam generators often are used as control devices for waste-gas streams and have an estimated efficiency of 98% for destruction of total hydrocarbons. Incinerators, relying on long residence times and elevated temperatures can achieve destruction efficiencies as high as 99.9999% for total hydrocarbons.

***Emissions Types.*** Emissions resulting from an external-combustion unit vary with type of unit, heat duty, fuel composition, and operating conditions. $NO_x$ are the primary criteria pollutants of interest from boilers, heaters, flares, and steam generators. $SO_x$ may be a concern as well, depending on fuel sulfur content. Benzene and formaldehyde are the primary HAP's of concern, although hazardous metals and hydrocarbons may be emitted from units that fire crude oil, fuel

**TABLE 6.8—RANGES OF $NO_x$ REDUCTIONS**

| Engine Type | Technology | Nominal $NO_x$-Emission Reduction (%) |
|---|---|---|
| Lean burn—two cycle | Combustion modification | 71 to 77 |
| | SCR | 80 to 95 |
| Rich burn—four cycle | Conversion to lean burn | 80 to 92 |
| | SNCR | 80 to 96 |
| Lean burn—four cycle | Site-specific enhancements | 80 to 92 |
| | SCR | 80 to 95 |
| Diesel | Derate | 5 to 23 |
| | 4° timing retard | <40 |
| | Air-to-fuel ratio | 7 to 8 |
| | SCR | 80 to 95 |

Percentage removals are only approximate, actual emission reduction is facility specific.

**TABLE 6.9—TURBINE-ENGINE RANGES OF $NO_x$ REDUCTIONS**

| Technology | Nominal $NO_x$-Emission Reduction (%) |
|---|---|
| Water or steam injection combined with SCR | 90 to 98 |
| SCR | 80 to 90 |
| Water or steam injection | 50 to 80 |
| Low-$NO_x$ combustors (dry control) | 50 to 80 |
| Low-$N_2$ fuels | Depends on fuel $N_2$ content |

Values are approximate. Ranges are shown because of variations in performance that occur as a result of such factors as turbine design, fuel type, cycle configuration, and operating schedule. Actual reductions must always be determined on a case-by-case basis.

oil, or refinery-fuel gas. To avoid high particulate emissions and opacity problems, the burner must be operated without creating smoke except for short periods during initial ignition. Also, if fired equipment is not maintained in good operating order or an incinerator is not operated at its design temperature, the potential for creating formaldehyde or other HAP's increases significantly.

***Emissions Estimation/Measurement.*** Emission rates from external-combustion units normally are determined by use of emission factors based on data obtained from (1) U.S. EPA *AP-42*,[2] (2) manufacturers' data, or (3) stack testing. The ability to use published emission factors strongly depends on the similarity of the equipment configuration and the fuel analysis. Such issues as Btu content, fuel sulfur ($H_2S$ or free sulfur) and $N_2$ content, burner-heat-release rate, and excess-air ratios need to be evaluated for their potential impact on emission-factor use.

*Heaters/Boilers/Steam Generators/Heater Treaters.* Emission factors for combustion sources generally are expressed as mass of pollutant per unit volume of fuel (i.e., lbm/1,000 gal or lbm/Mcf); mass of pollutant per unit of fuel energy (i.e., lbm/MMBtu or ng/J); mass of pollutant per unit time (i.e., ton/yr or lbm/hr); or concentration of pollutant in stack gas (i.e., ppm by volume).

Emission estimates for combustion sources can be calculated with the following steps.

1. Identify source type (e.g., 15-MMBtu/yr-capacity package boiler firing on natural gas).
2. Determine annual fuel use.
3. Locate appropriate emission factors.
4. Evaluate compatibility of the emissions factors.
5. Compute emissions with Eq. 1.

For a system equipped with an $NO_x$-abatement system, such as SCR or SNCR, $NO_x$-emission estimates should be based on source-test results from similar equipment or performance-guarantee data. Operators of SCR and SNCR often are required to estimate emissions of ammonia slip through the system. Ammonia emissions are computed from the maximum ppm by volume ammonia measured or expected in the stack gas (usually 5 to 15 ppm by volume). $SO_x$ emissions can be estimated on the basis of fuel sulfur content ($H_2S$, mercaptans, elemental sulfur), assuming that all fuel sulfur is converted to $SO_2$.

Hazardous emissions from combustion can be broadly classified as hazardous because of metals or organics. Estimates of hazardous-metal emissions from liquid- and solid-fuel combustion may be based on stack testing but normally are generated from analysis of the fuel. Metals are assumed to be 100% emitted, subject to particulate-control efficiency (if any) and minus the metals discharged in ash (if any). If the metal content of the fuel is known, emissions can be estimated on the basis of mass of fuel burned. Alternatively, emission factors can be generated from metal content, density, and heat content of the fuel.

Emission factors for hazardous emissions may be presented in two formats: as standard emission factors (i.e., lbm emitted/MMBtu fired) or as speciation profiles (i.e., a breakdown of total VOC's or total $PM_{10}$ emissions into specific constituents).

Hazardous organic emissions from combustion sources are a function of fuel, burner type, size and configuration of the combustion source, and operating conditions. For external-combustion units that burn natural gas and light fuel oils, stack testing generally is required to determine only criteria-pollutant emissions from new units (for permitting reasons). For units that burn heavy oils or waste solvents, stack testing may be required by regulatory agencies to quantify hazardous pollutant levels in the stack gas.

*Flares.* Flares present a special challenge to emission estimators for two reasons: they cannot be source tested by conventional stack source-testing methods and the flow of gas to a flare may vary by several orders of magnitude over time. Flare emissions would not be expected to vary linearly with gas flow, although a linear relationship is generally assumed because of a lack of detailed performance data.

In applying emission factors to flares, the fact that emission factors are estimates of the conversion of $N_2$ in the air to $NO_x$ must be remembered. If the flared stream contains chemically bound $N_2$, the bulk of this $N_2$ should be added to the $NO_x$ predicted by the emission factor. All sulfur in the flared stream should be assumed to be converted to $SO_2$ and all chlorine to HCl.

Flare emissions normally consist of two components: (1) pilot- and purge-gas-flame emissions and (2) flare-gas emissions. To estimate flare emissions, the following data are required.

1. Firing rate of pilot burner(s).
2. Pilot-burner fuel type (e.g., natural gas, $C_3$).

3. Purge-gas flow rate.
4. Purge-gas type.
5. Flared-gas flow.
6. Flared-gas composition.
7. Supplemental-gas flow (if any).

Estimates of criteria-pollutant emissions from flares may be generated with emission factors published in U.S. EPA *AP-42*.[2] These factors are based on tests of a flare burning a propylene/$C_3$ mixture. Vendor data specific to individual flare design may also be available. Criteria-pollutant emission factors from other external-combustion sources may be selected if the other source type is deemed more representative of a given flare because of the nature of the fuel stream. Pilot-flame emissions are straightforward to estimate because both firing rate and gas composition are well-defined. Emissions from natural-gas-fired pilots should be estimated with emission factors for natural-gas-fired boilers unless more applicable data are available from the vendor. Flare emissions of $SO_2$ may be significant if sulfur (e.g., $H_2S$) is present in the flared-gas stream. Sulfur should be assumed to be 100% converted to $SO_2$ in a flare.

Flare-gas-combustion emissions consist of unburned flare gas as well as products of combustion. The quantity of unburned flare gas emitted is a function of the destruction and removal efficiency of the flare. U.S. EPA tests of flares (Ref. 2) have indicated that 98% destruction and removal efficiency is achievable with air- or steam-assisted flares that fire gas streams of 300 Btu/ft$^3$ or greater. Most regulatory agencies accept the assumption that 98% of each of the combustible species in the flared-gas stream are destroyed in a flare.

Laboratory analyses of the flared-gas stream may be required to determine hazardous-species fractions in the flared gas. However, if the composition of the flash gas has been estimated, the flash-gas-species profile may be acceptable as an approximation of flared-gas composition.

If information is not available on the composition of the flared-gas stream, a published default flare-emissions-species profile may be used. Two flare-emissions profiles have been published by the U.S. EPA[12]; these profiles are also included in Appendix T of U.S. EPA's SPECIATE database.[14] Note that these profiles are composites, published for the use of regulators who compile regionwide emissions inventories, and are not particularly applicable for estimating HAP emissions from a single flare.

***Control Technologies.*** Control technologies for emissions from external-combustion units typically focus on $NO_x$ and $SO_x$. $NO_x$ controls can be categorized as those that change the combustion process to limit the formation mechanisms of $NO_x$ ($O_2$ controllers, flue-gas recirculation, staged combustion, and SNCR) and those that work to scrub $NO_x$ out of the exhaust stream (SCR).

Three factors that contribute to $NO_x$ formation are high temperatures; sufficient residence time in the combustion environment; and, in the case of liquid fuel, fuel $N_2$ content. $O_2$ controllers, flue-gas recirculation, and staged-combustion technologies tend to lower the combustion temperature and therefore the ability to form $NO_x$. SCR and SNCR technologies work to convert $NO_x$ formed back into water and $N_2$.

For boilers, heaters, and steam generators, $SO_x$ emissions can be reduced by use of flue-gas desulfurization units (also known as stack-gas scrubbers). Reductions of $SO_x$ emissions by more than 95% are common. Stack-gas scrubbers can reduce particulates by up to 50%. Further particulate removal can be achieved with electrostatic precipitators.

Techniques for preventing flares from smoking include advanced tip design or air or steam assist. Correct design of a flare is critical to its efficient operation, and the designer should consider the maximum anticipated load in building a smokeless flare.

**6.8.6 Gas-Processing Vents/Stacks.** Sources of air emissions discussed in this section are point sources that typically are emitted directly into the atmosphere. Emissions from safety vents normally are associated with emergency conditions and usually are not included in emission calculations for normal operations.

***Equipment Description.*** *Glycol Dehydration.* Glycol-dehydration units are used widely to remove water from produced-natural-gas streams to prevent formation of hydrates and corrosion in the pipeline. During the absorption step, the glycol, which has a high affinity for aromatic compounds, removes a fraction of BTEX and, to a lesser extent, other VOC's from the natural gas. The rich glycol leaves the bottom of the absorber and enters a flash tank (normally present only on larger units), which separates much of the absorbed natural gas from the glycol. Then, the rich glycol enters a regenerator where water is distilled and stripped from the glycol, after which the lean glycol is pumped back to the top of the absorber. Most of the BTEX and VOC's typically are released along with water vapor to the atmosphere.

*Gas Sweetening.* Gas-sweetening units remove acid gases, primarily $H_2S$ and $CO_2$, from sour natural gas. The most widely used natural-gas-sweetening processes use various ethanolamines (such as di- and monoethanolamine and methyl diethanolamine) or various proprietary mixtures. The rich amine, containing the acid-gas species as well as absorbed natural-gas components, is routed through a flash tank, which removes most of the more volatile natural-gas components. The chemically absorbed acid-gas species are removed from the rich amine in the regenerator, then a condenser on the gas outlet returns most of the vaporized amine and water to the top of the tower. The overhead gas stream consists primarily of acid gases with a smaller amount of water vapor. Most of the absorbed organics present in the rich amine after flashing, including BTEX and other VOC's, are also present in the acid-gas stream. In cases where $H_2S$ concentration is low and $CO_2$ is the major acid-gas species being removed, the regenerator overhead gas stream may be vented directly to the atmosphere. Other options for dealing with the acid-gas stream include routing it to a sulfur-recovery unit, an incinerator, or a flare (where the $H_2S$ is converted to $SO_2$).

*Sulfur-Recovery Units.* Sulfur-recovery units are used to convert $H_2S$ in acid-gas streams to elemental sulfur or some other sulfur form. The most commonly used sulfur-recovery method continues to be the Claus process where a portion of the $H_2S$ is converted to $SO_2$, which subsequently reacts to form elemental sulfur and water vapor.[15] A typical Claus plant consists of a thermal stage and two or three catalytic stages in series. The sulfur formed in each stage is condensed and flows to a storage pit before being loaded into tank cars or tank trucks. The major source of air emissions in a Claus sulfur-recovery unit is the tail-gas stream exiting the final conversion stage. In addition, several small vent streams are associated with the sulfur-storage and -handling operations.

*Cooling Towers.* Cooling towers are used to remove heat from various process streams by contacting the cooling water with air. As the water evaporates, a portion of the chemicals used in treating the cooling water (e.g., biocides, which may include chromates) may be emitted to the atmosphere. If chromates have been used in the cooling water, caution should be exercised during dismantling and disposal of the towers.

***Emissions Types.*** Species emitted from most process vent and stack sources discussed include $CH_4$, BTEX, and other VOC's. In some special instances, sulfur species, such as $H_2S$, carbonyl sulfide (COS), carbon disulfide ($CS_2$), and $SO_2$, may be emitted.

Emissions from uncontrolled glycol-regeneration vents and directly vented amine-regenerator vents may represent a significant source of BTEX and other ROC's, depending on gas throughput, BTEX concentrations in the gas feed, and glycol-recirculation rate. In addition to BTEX and ROC's, directly vented amine-regenerator-vent gas may contain $H_2S$ and other sulfur species (e.g., $CS_2$ or COS). Tail gas from sulfur-recovery units contains $SO_2$ and unconverted $H_2S$ and COS and may also contain unburned hydrocarbons, including BTEX and other ROC's. Sulfur-storage vents consist primarily of $H_2S$.

Pollutants of concern from cooling towers are $PM_{10}$ from solids in the cooling water; VOC's that may leak into cooling water from the process; hexavalent chromium ($Cr^{+6}$), nickel, or other toxics that may be present in cooling-tower additives; and chloroform, which may be formed as a byproduct of the reaction of chlorine with organics in the system. Most cooling-tower additives have been reformulated to eliminate $Cr^{+6}$. If the additives used at a facility contain $Cr^{+6}$ or other toxic metals, the metal concentration in the "drift" (aerosol liquid-water droplets) must be determined (usually by analysis).

***Emissions Estimation/Measurement.*** Direct measurement of emissions may be possible for some of the vents and stacks discussed here. However, obtaining a representative sample may be difficult in some cases, especially when the stream in question contains high concentrations of water vapor. Glycol-dehydrator and amine-sweetening-regenerator-vent emissions may be measured indirectly by sampling and analyzing the regenerator rich and lean streams. In the case of glycol-dehydrator-emission measurement, a GRI study[16] found that the atmospheric rich/lean (ARL) method is accurate, cost-effective, and easy to use. The ARL method gives low results for VOC's from units that do not have a flash tank or that use stripping gas to enhance glycol-regeneration efficiency. For these cases, factors developed as part of Ref. 16 can be used to correct the ARL measurements.

GRI-GLYCalc has gained wide acceptance by both U.S. industry and regulatory communities as a conservative regenerator-vent-emission estimation technique.[17] Use of GRI-GLYCalc requires the user to provide an accurate extended natural-gas analysis, including speciation of the BTEX species, which typically are found in concentrations of less than a few hundred ppm in natural gas. Additional user-supplied input data include natural-gas throughput, lean-glycol concentration and recirculation rate, flash temperature, and pressure.

Sweetening-unit-regenerator overhead-gas emissions (or composition of the sulfur-recovery-unit feed gas) can be measured by direct sampling of the acid-gas stream together with a gas-stream flow rate. Evacuated stainless-steel cylinders should be used to provide good emission measurements. Sampling of regenerator rich- and lean-amine streams, coupled with the amine-recirculation rate, can provide an alternative to direct stack measurement. Measurements have shown that concentrations of BTEX and other VOC's in the lean amine typically are negligible. Therefore, sampling only the rich amine may provide sufficiently accurate measurements.

GRI's computer program GRI-HAPCalc can be used to obtain an estimate of the $n$-$C_6$ and BTEX content of sweetening-unit overhead gas.[18] The estimation is based on a limited number of measurements and requires the user to provide information on the BTEX content of the sour natural gas. The program uses emission factors based on three BTEX concentration ranges and assumes that 5% of each of the BTEX species and 0.01% of the $n$-$C_6$ in the natural gas are present in the regenerator overhead-gas stream.

Tail-gas emissions from a Claus sulfur-recovery unit may be measured directly or estimated with emission factors. U.S. EPA *AP-42*[2] factors are available for $SO_2$ emissions from several typical process configurations. These factors assume typical sulfur-recovery efficiencies and that all the sulfur species in the tail gas are converted to $SO_2$ in an incinerator or flare.

Molten sulfur from Claus unit condensers contains an equilibrium amount of dissolved $H_2S$. Available measurements suggest that the maximum $H_2S$ content is approximately 300 ppm by weight. Maximum uncontrolled emissions can be estimated with this concentration and the sulfur throughput rate. The sulfur pit or sulfur-storage tank probably is the principal source of $H_2S$ emissions. Additional emissions result during sulfur-tank-car or -truck loading; however, this is the smallest emission source in terms of mass of $H_2S$ emitted.

The mechanisms of emissions from cooling towers are volatilization or air stripping of contaminants from cooling-tower water and loss from the tower in cooling-tower drift. The method used to estimate emissions depends on the physical characteristics of the pollutant of interest. VOC's are assumed to volatilize from the water, while metals and other particulate matter are assumed to be lost as aerosols in cooling-tower drift. The quantity of VOC's lost in drift generally is insignificant compared with the quantity volatilized.

The following should be determined to estimate particulate ($PM_{10}$) emissions.

1. TDS and TSS in the feedwater.
2. Cycles of concentration of the tower.
3. Tower-circulation rate (gallons per minute).
4. Tower-drift fraction (gallons per gallons circulated).

Estimates of particulate emissions are based on the concentration of solids in the drift that is lost from the cooling tower. Drift fraction varies widely among different cooling towers. Modern cooling towers may be equipped with drift eliminators to minimize particulate emissions. Emission factors for particulate emissions are based on local water quality, additives used in the tower, and tower characteristics. These factors should be developed on a site-specific basis. Typically, all emitted particulate is assumed to be $PM_{10}$.

U.S. EPA *AP-42* proposes a default $PM_{10}$ emission factor. This emission factor may be used if site-specific data are not available; however, it assumes very high cooling-tower drift and water-solids content. The use of the U.S. EPA *AP-42* $PM_{10}$ emission factor is likely to result in a relatively high $PM_{10}$ estimate. Obtaining site-specific data for drift fraction and water TDS is recommended to avoid overestimating $PM_{10}$ emissions.

Tower circulation and drift rates are used to estimate gallons per year of water lost as drift. Once the concentration of solids in the drift water is known, determination of pounds mass per year of particulate emissions is straightforward.

To estimate emissions of specific HAP's, such as $Cr^{+6}$ or manganese, data are needed on the specific concentrations of these compounds. Laboratory analyses of tower-water samples generally are required to determine these concentrations.

Cooling-tower-water VOC's are difficult to quantify because they result from leaks. Emission factors for VOC emissions from cooling towers are generated by measuring the contaminant concentrations in the risers and basin. The VOC concentration is unlikely to remain constant, and changes cannot be predicted. Because the VOC concentration in cooling water may vary constantly, "typical" refinery cooling-tower VOC emission factors from U.S. EPA *AP-42*[2] normally are used. Ref. 2 lists controlled and uncontrolled VOC emission factors for refinery cooling towers. Control is defined as measures that minimize leaks of hydrocarbons into the cooling water. These factors may be used if site-specific data are not available. Of course, if no possibility of VOC leakage into the cooling water exists, VOC emissions are zero. HAP species present in the VOC emissions are assumed to be present in proportion to their concentrations in the cooled process stream.

Chlorine is a reportable HAP under several U.S. state and federal regulations. Whether chlorine is emitted from cooling towers is frequently questioned. In fact, unless chlorine gas is lost directly from the chlorine-injection system, chlorine as $Cl^{-2}$ is not likely to be emitted from cooling towers. Chloroform may be formed when chlorine reacts with organic species in the tower. These contaminants, if present, are at low parts-per-billion concentrations.

***Control Techniques.*** The two most common methods of controlling emissions from glycol dehydrators are combustion (i.e., flaring or incineration) and condensation. A properly operating combustion device can provide good destruction of BTEX and other VOC's but also results in loss of potentially valuable hydrocarbons. In addition, combustion of the vent stream is difficult because of the possibility of condensation of hydrocarbons and water in the flare or incinerator, which may result in smoking and/or incomplete combustion of the vapors. Supplemental fuel is often required to obtain an adequate flame temperature.

Condensation has become the most widely used method of controlling dehydrator emissions. Several types of condensers are currently in use, including air cooled, glycol cooled, and direct quench. Control efficiencies for condensers are typically lower (±85%) than for combustion devices (±98%) because they are limited to temperatures at or above ambient levels. For this reason, it may be difficult to achieve desired control efficiencies in warmer locations or during the summer months.

Gas-sweetening-unit regenerator emissions may be controlled by routing the regenerator overhead-gas stream to a flare or incinerator or by adding a sulfur-recovery unit. Options to control sulfur-recovery-unit emissions include flaring or incineration of the tail gas or addition of one of a number of available tail-gas-treatment processes (e.g., the Claus offgas-treatment process[15]). Tail-gas-treatment processes may be needed to increase overall sulfur recovery because the Claus process has inherent efficiency constraints as a result of equilibrium considerations.

Sulfur-storage pits and tanks often are purged with air to prevent the accumulation of explosive concentrations of $H_2S$. The vent streams from the sulfur pit and tank may be directed to the atmosphere at a safe location or to the sulfur-recovery unit or tail-gas-treatment process thermal oxidizer.

**6.8.7 Roads/Site Activities.** ***Equipment.*** Emission sources are vehicles, off-road dirt-moving equipment, and storage piles.

***Emissions Types.*** Fine particulates ($PM_{10}$) and ROC's from piles containing hydrocarbon contaminated solids (soils and tank bottoms).

***Emissions Estimation/Measurement.*** Road emissions can be estimated with U.S. EPA *AP-42*[2] calculations. The emissions are a function of the vehicle miles traveled and soil type. Evaporative hydrocarbons emissions from piles can be estimated from bench-pan volatilization measurements. (This technique usually overpredicts emissions because of natural biodegradation of hydrocarbons that may occur in the piles).

***Control Technologies.*** Dust emissions from roads can be reduced by controlling vehicle speeds, by applying water, and by oiling or surfacing the roads. Emissions caused by dirt moving can be reduced by application of water. Emissions from storage piles also can be reduced by spraying water, applying solidifying chemicals, or covering.

**6.8.8 Maintenance/Construction/Other Activities.** ***Equipment.*** Welding power sources are typically oil-powered engines; Sec. 6.8.4 discusses emissions from these sources.

*Painting/Coating/Sandblasting.* Painting/coating operations are required to protect metal or other surfaces from damage that might occur as a result of oxidation, corrosion, or erosion. Most paint and coatings are applied by spraying or brushing. The method used most commonly to prepare a surface to receive paint or other coating has been abrasive blasting, with sandblasting being the most widely used medium. Most of the concern with sandblasting has been associated with health impacts caused by particulate inhalation, such as silicosis or exposure to lead.

*Solvent Cleaning.* Solvent-cleaning operations may include cleaning of boilers and other fired heaters (e.g., acid cleaning, followed by neutralization and rinsing) as well as chemical stripping as an alternative to sandblasting for surface preparation. Boiler-cleaning operations can generate quantities of liquid waste but typically do not produce air emissions. Solvent cleaning or chemical stripping usually involves spraying or brushing a paste over the desired area; the stripper breaks down the paint, and the surface is subsequently scraped. This technique does not produce respirable dust but may result in the emission of VOC's, depending on the chemical makeup of the stripping chemical used.

*Maintenance Depressurization.* Maintenance depressurization is performed to provide worker access to the equipment to be maintained. It involves blocking in a section of the unit where maintenance activities are to take place and removing the process liquids or gases by pressure reduction and/or pumping.

*Well-Casing Vents.* Well casing vents may be a source of emissions if the casing valve is left open (pumping wells with very low bottomhole pressures) or when wells are unloaded to a pit, sump, or tank. These activities may result in release of $CH_4$, VOC's, BTEX or other HAP's, and $H_2S$.

*Loading Operations.* Vessel-, rail-car-, and truck-loading operations can cause release of VOC's and BTEX and other HAP's by displacement of vapors from the receiving container. Depending on the blanketing gas used on a vessel, displaced gases may also contain $NO_2$ and $CO_2$.

***Emission Types.*** Pollutants of concern from painting and coating operations consist of VOC's; specific compounds emitted depend on the organics used in the paint or coating formula. Solvent-cleaning emissions consist of VOC's and possibly HAP's, with the specific compounds emitted depending, again, on the specific solvent used in a given application. Emissions from maintenance depressurizing include whatever components are present in the specific section of pipe or equipment item being depressurized. Species of concern include HAP's and ROC's and may also include sulfur species (e.g., $H_2S$ and COS).

***Emission Estimation/Measurement.*** Emissions from painting and coating operations can be estimated on the basis of knowledge of the composition of the paint or coating in use, quantity applied in a given period of time, and time required for the surface to dry or cure. Solvent-cleaning-emissions estimation is similar to that for painting and coating operations. It can be based on composition of the solvent, amount used, and fraction of the solvent evaporated.

Maintenance-depressurizing emissions can be estimated on the basis of knowledge of the quantity of material present in the section of the process of interest. To calculate the mass of potential emissions in the case of a gas, pressure, temperature, and, possibly, compressibility factors must be known in addition to the equipment volume.

***Control Technologies.*** Materials substitution (i.e., use of a paint or coating with a smaller concentration of the pollutants of concern) is one technique that can be used to control air emissions from painting and coating. In some instances, depending on the size of the area being coated, tenting and vapor collection and recovery can be used as another alternative. Solvent-cleaning emissions can be controlled by materials substitution (i.e., use of a less volatile solvent) or, possibly, use of a vapor-collection system. Maintenance-depressurizing emissions can be controlled by routing the line or equipment contents to a tank (in the case of liquids) or flaring (in the case of a gas).

**6.8.9 Emissions Impact Analysis.** When gases are released to the atmosphere, they mix with the air and become dispersed, resulting in reduced concentrations as they move away from the source. Actual measurement of concentrations with air-sampling devices is ideal; however, it is very difficult and costly. For this reason, dispersion modeling, which can be used to evaluate multiple sources and locations relatively inexpensively, is used. A good, useful dispersion model requires a work plan, model selection, emission information, source description, description of the surrounding area, and weather data.

As a general rule, the simpler the model, the less data and time needed. However, these simple models use many worst-case assumptions and usually overpredict concentrations. The more complex models require more data and time but generally provide more realistic predictions. Sources of usable emissions data include source tests, engineering estimates, design data, and data from other sources. Collection of site-specific emissions data can be time-consuming and costly; however, the benefits of a more accurate model must be considered.

The better the knowledge about an emission source, the better it can be modeled. The information required to define an emission source includes its shape (point, area, volume, line), size (length and width), location, height, operating conditions (including temperature), and emission rate. Some models require more information than others because of their complexity.

Rugged terrain, vegetation, and physical structures (such as buildings or other equipment) can cause added atmospheric turbulence and faster mixing. These effects may result in lower concentrations than would be expected from a source without these features. Hills or a nearby building just downwind from the source can cause elevated concentrations. Weather data should be gathered as close to the emission site as possible. Hourly average data over several years are preferred. The nearest airport is often the best source of these data. Where data are lacking, assumptions can be made. However, assumptions tend to be worst case and result in overprediction of concentrations. The following model equation shows the relationship between the various factors.

$$C = q/(\pi\sigma_y\sigma_z v), \quad \text{(2)}$$

where $C$ = concentration at plume centerline, $q$ = substance emission rate, $\sigma_y$ = dispersion coefficients in $y$ cross-plume direction, $\sigma_z$ = dispersion coefficients in $z$ vertical cross-plume direction, and $v$ = wind speed. Generally, higher wind speeds result in lower concentrations, lower stability (little turbulence or mixing) results in higher concentrations, and higher emission rates lead to higher concentrations.

**6.8.10 Additional Air-Emissions Information Sources.** ***U.S. EPA Sources.*** The U.S. EPA has published a variety of documents to assist in preparation of emission inventories. While, in general, these documents were prepared to assist regulatory agencies in the preparation of U.S. state and local emission inventories, these data are also commonly used to estimate emissions from individual emission sources. Many documents are available in addition to those listed here. Current information can be obtained from U.S. EPA's CHIEF bulletin board by computer modem (telephone: 1-919-541-5742 1200/2400/9600 bps). This bulletin board provides access to a wide array of documents and other information sources.

*U.S. EPA Office of Air Quality Planning and Standards (OAQPS).* OAQPS coordinates preparation of most air-emission-related publications and also maintains computerized databases and references and offers training courses through the U.S. EPA's Air Pollution Training Inst. The best way to obtain current information from OAQPS is to access the CHIEF bulletin board.

*Natl. Technical Information Service (NTIS).* U.S. EPA documents available for sale can be purchased from NTIS (voice telephone 1-703-487-4650). The U.S. EPA Publications Bibliography lists U.S. EPA publications available through NTIS. This document is available free from NTIS. Another valuable free handbook available from NTIS is *Access EPA* (EPA-220B-92-014), a guide to guides that identifies hotlines, computer bulletin boards, administrative offices, clearing houses, databases, and library-information services within the U.S. EPA.

***Most Commonly Used U.S. EPA References.***

U.S. EPA *AP-42, Compilation of Air Pollution Emission Factors.*[2] Vol. 1 contains emission factors for both stationary point and area sources. Point sources are sources with stacks or vents, while area sources are sources with nonspecific emission points (e.g., regionwide pesticide use). Vol. 2 addresses mobile sources (e.g., autos, trucks). If detailed emission estimates from mobile sources are required, the U.S. EPA mobile-source emissions model, *MOBILE 5.0 User's Guide* should be downloaded from the CHIEF bulletin board. Vol. 2 is designed to document rather than to supplant the MOBILE computer model.

*U.S. EPA Air Emissions Species Manuals.*[12] Species profiles consist of a list of major constituents in an emitted stream (e.g., organic species in flue gas from fuel-oil combustion) broken down by weight percent. This document is bound in at least two volumes and includes species profiles for an array of different industrial processes. Both particulate matter and VOC species profiles are included. Particulate-size-distribution data for several source types also are included. A current version of the available species profiles can be obtained by downloading the database SPECIATE from the CHIEF bulletin board; this database can then be used on a desktop computer.

*Toxic Air Pollutant Emission Factors—A Compilation for Selected Air Toxic Compounds and Sources.* This document (EPA-450/2-88-006a) is a paper-copy compilation of many of the emission factors that are included in the U.S. EPA's emission-factor database program for personal computers, XATEF. (Actual source-test data from the source in question or a similar unit are nearly always preferable to the use of these factors.)

*Estimating Air Emissions From Sources of Polycyclic Organic Matter (POM).* Each document in this series (EPA-450/4-84-007p) focuses on a specific toxic air contaminant and on sources that may emit that contaminant.

U.S. EPA Control Technology Guideline (CTG) Documents [e.g., "Control of Volatile Organic Compound Emissions from Reactor Processes and Distillation Operations Processes in the Synthetic Organic Chemical Manufacturing Industry" (EPA-450/4/91-031)]. CTG documents have been prepared for a variety of industries and contain information on the effectiveness and applicability of various control technologies, as well as emission factors and other information.

*Databases of Emission Factors.* These programs offer quick menu-driven access to collections of published emission factors. The emission factors can be used in hand calculations, spreadsheets, or other software packages to estimate emissions. Several emission-factor databases are available free of charge on the U.S. EPA CHIEF bulletin board. Several software programs that provide quick access to an array of emission factors are available. Computer programs that help estimate emissions and compile inventories are also available.

XATEF. The crosswalk/air toxics emission-factor database contains more than 6,000 emission factors and more than 15,000 pollutant/source associations.

AFSEF. The airs facility subsystem emission-factor database contains criteria-pollutant emission factors for approximately 5,000 industrial processes.

SPECIATE. This database contains approximately 300 VOC and particulate-matter profiles. Appendix T of this document contains printouts of the SPECIATE species profiles relevant to E&P.

***Other Sources.*** Documents are also available from state and local regulatory agencies and industry associations.

A particularly valuable reference for gas-turbine and internal-combustion-engine emissions data is *Compilation of Emissions Data for Stationary Reciprocating Gas Engines and Gas Turbines in Use by the Natural Gas Pipeline Transmission Industry* [AGA Catalog No. L51565, American Gas Assn., Arlington, Virginia (May 1988)]. This document contains an extensive compilation of detailed exhaust ($NO_x$, hydrocarbons, and CO) -emission data for turbines and reciprocating engines used in the gas-pipeline industry. Data are grouped by engine manufacturer and include an array of specific engines and configurations. These data are more applicable to estimating emissions from a specific engine than are the generic emission factors presented in U.S. EPA *AP-42*.

*Proceedings,* 1992 Petroleum Industry Workshop on VOC/HAP Emission Inventory Practices, Hal Taback Co. (March 1992). A compilation of papers presented at a workshop attended by professional environmentalists.

## Nomenclature

$C$ = concentration at the plume centerline, m/L$^3$
$K_e$ = vapor-space-expansion factor, dimensionless
$K_n$ = turnover factor, dimensionless
$K_p$ = working-loss product factor = 0.75 for crude oil and 1.00 for organic liquids
$K_s$ = vented-vapor-saturation factor, dimensionless
$L_f$ = roof-fitting loss, m/t, lbm/yr
$L_r$ = rim-seal losses
$L_s$ = standing or breathing losses, m/t, lbm/yr
$L_t$ = total losses, m/t, lbm/yr
$L_w$ = working losses, m/t, lbm/yr
$L_{wd}$ = withdrawal losses, m/t, lbm/yr
$M_v$ = vapor molecular weight, m, lbm/lbm mol
$n$ = number of turnovers per year, n/t
$p_v$ = storage-temperature true vapor pressure, m/Lt$^2$, psia
$q$ = substance emission rate, m/t, lbm/yr
$Q$ = annual net throughput, L$^3$/t, bbl/yr
$v$ = wind speed, L/t
$V_v$ = vapor-space column
$\rho_v$ = vapor density, m/L$^3$, lbm/ft$^3$
$\sigma_y, \sigma_z$ = dispersion coefficients in the $y$ and $z$ cross-plume directions, respectively

## References


1. *40 Code of Federal Regulations*, Part 50.
2. *AP-42 Compilation of Air Pollution Emission Factors*, U.S. EPA, Washington, DC (January 1995).
3. Bennett, C.O. and Myers, J.E.: *Momentum, Heat, and Mass Transfer*, third edition, McGraw-Hill Inc., New York City (1962) 532–35.
4. "Production Tank Emissions Model" (E&P TANK Vers. 1.0), Publication 4660, API, Washington, DC (October 1997).
5. *Air Pollution Engineering Manual*, A.J. Buonicore and W.T Davis (eds.), Air and Waste Management Assn., Van Nostrand Reinhold, New York City (1992) 341.
6. "TANKS Vers. 3.1," U.S. EPA, Washington, DC (April 1998).
7. "WATER8 Vers. 4," U.S. EPA, Washington, DC (September 1995).
8. "CHEMDAT8 Vers. 1," U.S. EPA, Washington, DC (September 1995).
9. "Fugitive Hydrocarbon Emissions from Petroleum Production Operations," Publication 4322, API, Washington, DC (July 1980).
10. "Reference Method 21, Determination of Volatile Organic Compound Leaks," CFR, Title 40, U.S. Government Printing Office, Washington, DC (22 June 1990) Part 60, Appendix A.

11. "Protocol for Equipment Leak Emission Estimates," U.S. EPA, Research Triangle, North Carolina (1993) 2–7.
12. *Air Emissions Species Manual, Organic Compound Species Profiles*, second edition, U.S. EPA Natl. Technical Information Service, Research Triangle, North Carolina (January 1990). **1.**
13. "Pooled Source Emission Test Report: Oil and Gas Production Combustion Sources, Fresno and Venture Counties, California," Western States Petroleum Assn., Bakersfield, California (December 1990).
14. "SPECIATE Database," CHIEF Bulletin Board, Computer Modem Telephone: 919–541–5742, 1200/2400/9600/bps, U.S. EPA, Washington, DC (1998).
15. Shreve, R.N. and Brink, J.A. Jr.: *Chemical Process Industries*, McGraw-Hill Book Co., New York City (1977) 295.
16. "Sampling and Analysis of Wastes Generated from Natural Gas Industry Operations," topical report, GRI, Chicago (May 1993).
17. Thompson, P.A. *et al.*: *Technical Reference Manual for GRI-GLYCalc: A Program for Estimating Emissions from Glycol Dehydration of Natural Gas, Ver. 3.0*, GRI-96–0091, GRI, Chicago (March 1996).
18. Ferry, K.R. *et al.*: *Technical Reference Manual for GRI-HAPCalc: Software for Estimating Emissions of Hazardous Air Pollutants and Criteria Air Pollutants from Natural Gas Industry Operations*. GRI-96/10346, GRI, Chicago (December 1996).

## General References

*Assessment of Environmental Fate and Effects of Discharges from Offshore Oil and Gas Operations*, EPA-440/4-85/002, U.S. EPA, Washington, DC (August 1985).

Bourgoyne, A.T. Jr. *et al.*: *Applied Drilling Engineering*, Textbook Series, SPE, Richardson, Texas (1986) **2.**

*Bull. E2, Bulletin on Management of Naturally Occurring Radioactive Materials (NORM) in Oil & Gas Production*, first edition, API, Washington, DC (April 1992).

Craft, B.C., Holden, W.R., and Graves, E.D. Jr.: *Well Design: Drilling and Production*, Prentice Hall Inc., Englewood Cliffs, New Jersey (1962).

*Development Document for Effluent Limitations Guidelines and New Source Performance Standards for the Offshore Subcategory of the Oil and Gas Extraction: Point Source Category*, final, EPA-821-R-93-003, U.S. EPA, Washington, DC (November 1991).

Gray, G.R. and Darley, H.C.H: *Composition and Properties of Oil Well Drilling Fluids*, fourth edition, Gulf Publishing Co., Houston (1980).

"Exploration and Production (E&P) Waste Management Guidelines," Report 2.58/196, E&P Forum, London (September 1993).

*Field Handling of Natural Gas*, third edition, Petroleum Extension Service, U. of Texas, Austin, Texas (1972).

*Fundamentals of Petroleum*, third edition, Petroleum Extension Service, U. of Texas, Austin, Texas (1986).

"Keepin' It All Clean in the Oil Patch," training guide, Natl. Environmental Training Assn., Phoenix, Arizona (1995).

"Management of Wastes from the Exploration, Development, and Production of Crude Oil, Natural Gas, and Geothermal Energy," report to Congress, EPA-530-SW-88-003, U.S. EPA,Washington, DC (December 1987).

"Onshore Solid Waste Management in Exploration and Production Operations," environmental guidance document, first edition, API, Washington, DC (January 1989).

"Outer Continental Shelf Environmental Assessment Program," final reports for U.S. Dept. of Commerce, Natl. Oceanic & Atmospheric Admin., Anchorage (December 1984) **27.**

*Plant Processing of Natural Gas*, Petroleum Extension Service, U. of Texas, Austin, Texas (1974).

Stephenson, M.T.: "Components of Produced Water: A Compilation of Industry Studies," *JPT* (May 1992) 548.

*Waste Minimization in the Petroleum Industry: A Compendium of Practices*, Publication 849-00020, API, Washington, DC (November 1991).

# Chapter 7
# Risk Assessment

**John L. Cayais,** Oryx Energy Co.
**Gerald R. Carney,** U.S. Environmental Protection Agency
**June L. Sutherlin,** Louisiana Dept. of Environmental Quality
**Ron Gordon,** Gordon & Assocs.
**Leslie L. Savage,** Railroad Commission of Texas

## 7.1 Introduction

In the oil industry, we are not used to thinking of crude oil or other chemicals used in association with oil and gas E&P activities as potential contaminants. Nevertheless, crude oil is a complex mixture of chemicals, some of which may be hazardous to human health and the environment. Crude oil and industry-related chemicals are not environmental hazards unless they are spilled or otherwise leak into the atmosphere, soil, groundwater, surface water, or ecologically sensitive habitats. There, they may become an environmental hazard that needs to be addressed. The quantitative, scientific method used to determine whether contamination poses a human-health or ecological risk is referred to as risk assessment. Risk assessment also may be used to determine cleanup levels at which contamination is no longer a risk to human health or the ecology.

When faced with a contamination problem, the responsible party has several options. It may clean up to background, clean up to a regulatory limit, or clean up to a level determined through risk assessment. Cleaning up to background may be prohibitively expensive or technically impossible. Cleaning up to a regulatory limit may be expensive and may or may not be protective of human health or the environment. Remedial options evaluated with risk assessment may provide more flexibility when a contaminated site is being cleaned up.

This chapter discusses the following.

1. The risk-assessment process and its importance in determining human-health risks.
2. How risk assessment can be applied to determine cleanup levels for contaminated areas.
3. How risk assessment plays a key role in the overall risk-management process, including environmental planning.

**7.1.1 Overview of Risk-Assessment Process.** People are faced with decisions every day. Some are simple, such as eating or choosing a time to go to bed. Some are arbitrary, such as what food to eat, but carry with the choice a potential danger associated with the type of food chosen. For a person with high blood pressure, consumption of large amounts of sodium may pose a significant health danger. Such situations are risks that an individual should weigh when making decisions.[1] For oil and gas companies addressing environmental issues, a similar evaluation, or risk assessment, can be used to determine the best way to develop, operate, or remediate a location.

***History.*** Methods to evaluate human-health risks from exposure to chemicals in the environment have been studied extensively since the 1960's. The U.S. Environmental Protection Agency (EPA) was formed in 1970. In 1980, the U.S. Congress passed the Comprehensive Environmental Response, Compensation, and Liability Act (CERCLA), or Superfund, to clean up the nation's inventory of hazardous-waste sites. During this period, a vast body of scientific evidence was accumulated that suggested human-health risk from chemicals in the environment could be quantified through an emerging science called risk assessment. In 1983, the Natl. Resource Council (NRC), a group of eminent scientists and engineers, met and formalized the risk-assessment process.[2]

Their report, which became known as the "red book," outlined the principles that formed the scientific basis for risk assessment. In 1985, CERCLA was reauthorized as the Superfund Amendments and Reauthorization Act (SARA) and a decision was made to use risk as a basis for establishing cleanup criteria. Beginning in late 1985, the U.S. EPA published a series of eight risk-assessment guidelines.[3-6] The agency also published the *Superfund Public Health Evaluation Manual*[7] in 1986 and the *Risk Assessment Guidelines for Superfund* (*RAGS*)[8] in 1989. *RAGS* forms the basic guidelines for the risk-assessment process at both U.S. federal and local levels. Nevertheless, *RAGS* is used for all types of risk assessments, not just those performed at Superfund sites.

During this period, various state agencies, including those in California, Washington, Michigan, New Jersey, and Texas, began to develop their own risk-assessment guidelines. In 1995, the American Soc. for Testing and Materials (ASTM) published procedures for performing risk assessment at petroleum- and hydrocarbon-contaminated sites.[9] The ASTM process, although developed primarily for underground storage tanks, is also useful for risk quantification of any hydrocarbon-contaminated site. The general process is referred to as risk-based corrective action.

In the 10 years that followed the NRC report,[2] risk assessment developed vocal proponents and opponents. However, much of the controversy was focused on its policy decisions, not on scientific issues. In 1994, the NRC held a second risk workshop to review the risk-assessment process and concluded that the basic scientific approach to risk assessment was still valid and recommended that no changes be made.[10]

***Scientific Basis of Risk Assessment.*** The risk-assessment process was developed with contributions from many scientific disciplines, including statistics for sampling, data collection, and data evaluation; engineering and geology for fate and transport modeling; chemistry for chemical testing and analysis; and toxicology for developing relationships between chemical dose and human-health response. The process continues to evolve as we become more knowledgeable about the relationship between chemicals, the environment, and human health.

Toxicology is a relatively new science, but its roots are deep in history through the study of poisons.[11] Over time, this study developed into pharmacology and, recently, into the science of toxicology. Simply, toxicology examines the relationship between a chemical dose and a response by the organism. In our everyday life, we are exposed to small doses of chemicals present in our food and in the environment. All chemicals are toxic; the dose determines toxicity. For example, alfatoxin, a toxin produced by a fungus that grows on peanuts, is toxic at the parts-per-billion level, while water is toxic at levels that are physically impossible for the human body to ingest.[12] Toxicology provides an answer to how human-health and ecology receptors are impacted by exposure to small doses of chemicals.

Chemical toxicity is determined by studying chemical exposure in laboratory animals. The animals, commonly rats, are given relatively large doses of chemicals over their lifetimes, and the results are extrapolated to evaluate relatively small dose exposures in human populations. Laboratory test data are used to classify chemicals into two general classes: cancer-causing chemicals (carcinogens) and noncancer-causing chemicals (noncarcinogens). This differentiation is based on the difference in both the mechanism and endpoint of toxicity.

The cancer model is a linear, "nonthreshold" model that assumes that all exposures to a cancer-causing chemical pose a risk and that the risk is cumulative over a lifetime. This model for a carcinogen is referred to as a nonthreshold mechanism, which assumes that no level of exposure to a chemical exists that does not pose a finite probability of generating a carcinogenic response (i.e., no dose is risk free). The model for a noncancer-causing chemical assumes that exposure below a threshold level is not harmful. In the development of acceptable exposure levels for humans, the noncancer model applies uncertainty factors to threshold levels observed in laboratory animals to account for inter- and intraspecies variations.

***Voluntary and Involuntary Risk.*** Risk is defined as the probability of something unwanted happening. Scientists who study risk often divide it into two classes: voluntary and involuntary. Voluntary risks are those where you have choices even if you do not always understand the risk. The risk associated with voluntary activities can be determined from statistical analysis of actuarial data. For example, if you want to travel across country, you can travel by air or automobile. Car travel has an annual risk of death of approximately 1 in 4,000; air travel has a risk of approximately 1 in 20,000. Another voluntary risk is smoking, which has an annual death rate of 1 in 800.

Involuntary risks are harder to define. A relevant example is unwilling exposure to chemicals in the environment. The exposure to ozone and carbon monoxide in urban air pollution is considered an involuntary risk.[13,14] Studies have shown that people are not adept at assessing the level of risk from an activity but that they appear to be willing to accept voluntary risk as high as 1 in 100 to 1 in 1,000 for such activities as mountain climbing, riding a motorcycle, and smoking.[15] Scientists know little about the acceptance of involuntary risk.

***Risk From Exposure to Chemicals in the Environment.*** In 1958, the Delaney Clause of the U.S. Pure Food, Drug, and Cosmetic Act was passed.[16] The clause stated that no residues of cancer-causing chemicals would be allowed in food at any level and is a basis for the so-called zero-tolerance approach to chemicals in the environment. However, advances in technology have led to problems with the Delaney Clause. In 1958, state-of-the-art analytical methods were able to detect chemical contaminants only at the parts-per-thousand level. With advances in instruments and techniques, the level of detection for many chemicals is now on the order of parts per billion and parts per trillion. Future advances are expected to continue to lower detection limits. As a consequence, chemical residues are being found where none were thought to exist. Technology has made environmental zero tolerance a virtually unattainable goal. In response, risk assessment has been developed as a tool that allows for quantification of risk from chemicals to determine "safe" levels of exposure. The next step is to use risk management to define acceptable and unacceptable risk.

***Uncertainty and Risk Assessment.*** There is a high degree of uncertainty in determining risk for something as complex as the interaction between chemicals in the environment and human health. Because of this uncertainty, the risk-assessment process is designed to err on the health-conservative side when site-specific data are not available. Risk assessment frequently focuses on estimating risk associated with a worst-case or "reasonable"-maximum scenario. For example, there is considerable latitude in calculating the amount of drinking water consumed on a daily basis. The default assumption for a person's daily ingestion rate of water is taken as the 95% upper-confidence level of the mean value for water ingestion, or 2.0 L/d, and the mean value is 1.4 L/d. If the risk assessment is complex and has several variables, using default conservative assumptions for each variable may overstate the risk significantly.

An alternative method is to use a Monte Carlo approach to calculate risk and the uncertainty associated with the risk.[17] The Monte Carlo approach takes into account the variability of each input parameter and calculates a risk estimate that is a distribution with confidence levels that define a range of related values rather than a single risk value. This approach is useful for complex problems where the number of risk-related variables is large and where an estimate of uncertainty is required. To use the Monte Carlo approach, the ranges for input data must be available.

***Phased Approach to Risk Assessment.*** An important feature of the risk-assessment process is its flexibility and ability to work on different levels. Risk evaluations can be divided into three general categories or phases relative to available data and their quality and to the level of effort needed to estimate risk. With an understanding of the risk-assessment process and its overall goals, a risk estimate can become the basic tool for environmental planning, determining site-specific remedial goals, and developing general environmental strategy. Risk assessment can be conducted in phases, with each phase increasing in level of complexity. The three general phases of risk assessment are screening risk assessment; intermediate risk assessment; and detailed, site-specific risk assessment. These phases build on each other and provide a long-range view for determining the risk associated with a site. For many sites, a screening risk assessment may be sufficient, while a complete risk assessment may be required for others. Using the risk-assessment process as a planning tool allows for proper collection of data needed at each phase, assists in allocation of resources, and allows for evaluation of the process in light of new information.

*Screening Risk Assessment.* The first phase in risk assessment is the simplest and is referred to as a screening risk assessment. Screening risk assessments often are risk comparisons, engineering evaluations, or modeling or other computer-assisted activities. They frequently are performed with specific portions of complete risk-assessment guidelines or regulatory standards. They usually do not require monitoring studies and can be completed in a matter of days to several weeks. Screening risk assessments may be done with preliminary site-specific data, such as a chemical identification, estimated maximum contaminant levels, and possible exposure pathways. This information can be used to give a worst-case scenario and a qualitative risk estimate that identifies where major uncertainties exist. Understanding the risk process allows for development of data needs and of required environmental sampling goals and objectives. With a screening-risk-assessment approach, the risk estimated is probably a maximum risk that reflects a worst-case scenario. An example of a screening risk assessment is the type of assessment performed in the Resource Conservation and Recovery Act (RCRA) underground-storage-tank program.[18] A similar procedure may be used for spills or leaks of any hydrocarbon as long as chemical-specific data are available.

*Intermediate Risk Assessment.* Intermediate risk assessment builds on preliminary risk assessment and serves as a guide for further data collection and evaluation, with emphasis on developing indirect exposure pathways and locating receptors. In this phase, the broadening knowledge base of the range of chemicals present and their concentrations becomes better defined and leads to refinement of the risk estimates and a better estimate of uncertainty. The risk manager can use the data to establish preliminary cleanup levels, to rank sites by use of risks, and to determine permitting needs under the Natl. Pollutant Discharge Elimination System (NPDES), the Safe Drinking Water Act, or other regulatory programs.

*Site-Specific Risk Assessment.* A site-specific risk assessment is a complete assessment of potential adverse health effects caused by release of contaminants from the site in the absence of actions to control or mitigate the release. The site should be characterized fully, and exposure pathways and potentially affected populations should be understood in detail. Risk managers can use the complete risk assessment to determine whether remedial actions are needed, to set cleanup goals and objectives, and to document the causes and magnitude of risk at the site. A detailed discussion of complete risk-assessment procedures is included later in this chapter.

Detailed risk assessments are usually resource intensive. Their cost may range from tens to hundreds of thousands of dollars, and they may require several months to years of effort and expensive engineering and other scientific expertise. Most risk assessment is done at the screening or intermediate level. Screening assessment is used for leaks, spills, planning, and setting priorities. All risk assessments begin at the screening level and go on to higher levels as the complexity or magnitude of the problem becomes more severe.

*Other Types of Risk Assessment.* Several different categories of risk assessments or analyses can be performed. Examples are human-health, ecological, socioeconomic, and cost/benefit risk assessment. Human-health and ecological risk assessment are discussed in more detail later in this chapter.

Socioeconomic risk analyses are done to protect low-income and minority communities from disproportionate exposure to environmental pollution. The method analyzes household income, minority status, and population data for communities near environmentally contaminated sites.[19]

Cost/benefit analysis is a relatively new type of risk assessment. The basic idea is that the cost of reducing risk needs to be assessed vs. resulting benefits. The U.S. Office of Management and Budget performs cost/benefit risk-assessment analyses for legislation of national significance. These analyses for risk assessments are being proposed at several levels of government. Resources and method questions must be answered before cost/benefit analysis becomes common practice. Limited application of cost/benefit risk analysis is done at Superfund sites for which costs for various remedial actions must be determined before action is taken. Costs and effectiveness are always major considerations when examining remedial options. This usually is considered to be risk management not risk assessment.

**7.1.2 Regulations and Risk Assessment.** Many major environmental regulations are risk based. The U.S. Clean Water, Safe Drinking Water, Natl. Environmental Policy, and Clean Air Acts are based on varying degrees of human-health or environmental risk. Many U.S. state and local regulations are based on risk or on federal statutes and guidelines that are risk based. The U.S. EPA, as well as the states, have used risk as a common basis for understanding the impact that chemical contamination has on human health and the ecology. As previously mentioned, *RAGS* is the basis for risk-assessment guidelines and the regulations that have followed it.[8] The U.S. federal government, in particular, has written a number of risk-based guidelines that examine miscellaneous and esoteric risk areas. An understanding of all the guidelines and documents is not necessary for a basic understanding of risk assessment. However, if the risk assessment is to be used for external purposes, a thorough understanding of all pertaining regulations and guidelines is needed. While the science of risk assessment remains the same, certain issues may be treated differently by various regulatory agencies. Because of its early history, many people in the U.S. tend to equate risk assessments with Superfund projects. While risk-assessment methods are used for Superfund, they may be used to analyze any project where human health or the ecology is adversely affected by emissions, spills, leaks, or accidental releases of contaminants to the environment.

## 7.2 Fundamentals of Human-Health Risk Assessment

This section discusses the fundamentals of a risk assessment. While it focuses on a complete risk assessment, the principles apply to simpler types of risk assessment (e.g., screening and intermediate). A complete health risk assessment is an analysis of potential adverse health effects (current or future) caused by hazardous-substance releases from a site in the absence of any actions to control or mitigate these releases (i.e., under an assumption of no action). Complete risk assessment contributes to site characterization and subsequent development and to evaluation and selection of appropriate response alternatives. Results of a complete risk assessment are used to help determine whether remediation is necessary to protect human health; to modify preliminary remediation goals; to help support the selection of a no-action remedial alternative, where appropriate; and to document the magnitude of risk at a site and the primary cause of that risk. Complete risk assessments are site-specific and therefore may vary in detail and in the extent to which qualitative and quantitative data are used, depending on the complexity and particular circumstances of the site. Health-based risk assessment involves four steps: site characterization, exposure assessment, toxicity assessment, and risk characterization. Ref. 8 discusses each of these steps in detail.

**7.2.1 Site Characterization.** Site characterization involves the gathering and analyzing of site data relevant to the health risk assessment and identifying the contaminants present at the site that are the focus of the risk-assessment process.

***Review of Available Site Data.*** The first step in site characterization is to review available site data. This review serves to determine basic site characteristics; to identify potential exposure pathways and exposure points; and to identify data gaps. Historical or available data that may be relevant to the risk assessment include site maps, aerial photographs, records on past releases or removal actions, information on surface-water-runoff patterns, meteorological data, hydrogeologic data, site-investigation and/or -monitoring data, and information about current and future land use in the vicinity of the site. On the basis of a review of existing data, a conceptual model of the site is formulated that identifies all potential or suspected sources of contamination; types and concentrations of contaminants present at the site; potentially contaminated media; and potential exposure pathways, including receptors. From this model, data needed to assess the health risks associated with contaminants at or migrating from the site can be identified for inclusion in the sampling and analysis work plan. In general, the types of data needed for risk assessment include identities and concentrations of contaminants present; characteristics of primary and secondary contaminant sources, especially information related to release potential; and characteristics of the environmental setting that may affect the fate, transport, and persistence of the contaminants. Other data that may be needed for risk assessment include site-specific parameters for environmental-fate and transport models and naturally occurring background levels of inorganic contaminants.[8]

***Contaminants of Concern for E&P Activities.*** Wastes and contaminated media from E&P activities can contain various organic and inorganic contaminants. In situations where the potential threat to human health and the environment must be determined, knowledge of what contaminants to expect is helpful. Oilfield wastes or contaminants generally include petroleum hydrocarbons, salts, heavy metals, naturally occurring radioactive material (NORM), and various other chemicals. Crude-oil and condensate contamination from spills and leaks at wellheads, gathering lines, gas plants, tank batteries, or transmission lines are other potential oilfield problems. Historical contamination by produced waters from past waste-management practices is a problem in some areas. Also, spills of the various chemicals used in oilfield activities may occur.

*Crude-Oil Chemical Composition.* Crude oil is the source material for all hydrocarbon products. The composition of crude oils is highly variable, depending on their source. This variability is expressed in the different chemical components present and their relative amounts. The principal components of most crude oils are alkanes, aromatics, and asphaltenes. Alkanes are a variety of straight-chain, branched, and cyclic saturated hydrocarbons that contain only carbon and hydrogen. The aromatic (unsaturated cyclic) fraction consists of two general fractions: single-ring compounds, such as benzene, and multiple-ring compounds called polycyclic aromatic hydrocarbons (PAH's). High-molecular-weight materials in crude oil that form a thick, solid or semisolid mass are referred to as asphaltenes. Crude oil may contain heteroatoms (oxygen, sulfur, and nitrogen) and trace metals, such as nickel, arsenic, and vanadium.[20] The most likely toxic components present in crude oils are benzene, toluene, ethyl benzene, xylenes (BTEX) and PAH's.

*Gas-Processing and -Transportation Wastes.* Gas-processing and -transportation wastes can contain constituents from the crude feedstock and other components that are indigenous to the producing formations. The major toxics that may be present are BTEX in condensate, trace metals, and possibly NORM. Arsenic that may come from producing formations is present in some gas-industry wastes. Arsenic and chromium may be present in chemicals used in cooling towers. Mercury contamination also may be a problem because it is indigenous to some gas formations and may concentrate in condensate-tank bottoms or gas-processing wastes. Mercury has also been used widely in gas-flow manometers at wellheads and along pipelines. Mercury spills and leaks may be a soil-contamination problem where mercury gauges are used. Polychlorinated biphenyls (PCB's) may be found on various production and transportation sites either because of leaking transformers or historical contamination from compressor lubricating oils used from the late 1950's through the early 1970's.

*NORM.* Oil and gas wells also may produce water that is contaminated with radium and radon from naturally occurring uranium and thorium radioactive decay. Radium and radon may leave the formation because radium salts are soluble in some forms and radon is a gas. Levels typically found in produced waters generally are not a problem; however, radium is concentrated by precipitation as a sulfate salt and becomes part of scale formation in pipes and other equipment. NORM-scale deposits in separators can range from 1 to 1000 pCi/g. Radium in tank sludges may be as high as 50 pCi/g.[21] Radium 226 has a half-life of 1,600 years, radium 228 has a half-life of 5.75 years, and radon has a relatively short half-life of 3.8 days. Radon gas has a boiling point in the range of $C_2$ to $C_4$ hydrocarbons. Lead 210, a radon decay product, may accumulate as a thin coat in gas-processing equipment.

*Brines.* Produced water, or oilfield brine, is the water produced from the reservoir along with the crude oil or natural gas. Produced water constitutes more than 98% of all oilfield wastes generated in the U.S.[21] While most produced water is reinjected for improved oil recovery, accidental spills and leaks do occur. The chemical composition of produced waters, like crude oil, varies widely and depends on the producing formation; chemical composition of the oil; efficiency of the hydrocarbon-separation process; and chemical additives, such as demulsifiers. Primarily, produced water consists of dissolved solids in concentrations that range from relatively fresh water (1,000 ppm) to heavy brines with compositions several times more saline (100,000 ppm) than common seawater. The dissolved salts in produced water are mainly sodium and chloride but may also include calcium, magnesium, barium, carbonates, bicarbonates, and sulfates. In addition, produced water contains some insoluble hydrocarbons that were not separated and some hydrocarbons that are soluble in water.

*Miscellaneous Chemicals.* A wide variety of other chemicals also are used in the oil and gas industry. These include viscosifiers, corrosion inhibitors, scale inhibitors, biocides, and dispersants and may contain chemicals with known or suspected toxicity.[22] Other chemicals that are used include pH buffers, cement additives, defoamers, deflocculants, explosives, fuels, lost-circulation materials, solvents, surfactants, and weighting agents.[22,23] Additives are often used in low concentrations (up to a few percent by weight of specific fluids) and are diluted even further when they are recovered as part of the produced water.

*Total-Petroleum-Hydrocarbon (TPH) Parameter.* TPH chemical analysis methods provide a measure of the total amount of hydrocarbon present but do not indicate the types of hydrocarbons present or the amount of the individual chemical(s) of concern. TPH data provide useful information regarding the presence and extent of hydrocarbon contamination. However, TPH data cannot be used quantitatively in the risk-assessment process because they do not provide chemical-specific information. Use of risk-based TPH standards implies that the risks from different TPH mixtures are similar. In addition, different analytical methods give varying amounts of TPH in a given sample. TPH methods do not determine the individual chemicals that make up the hydrocarbons in a sample; therefore, they cannot account for changes in chemicals caused by weathering and resulting chemical-specific toxicity. Again, TPH data cannot be used directly in a risk assessment.

***Environmental Fate and Transport for E&P Constituents.*** Generally, water solubility decreases with the complexity and size of the hydrocarbon molecule. $C_6$ through $C_8$ aromatic hydrocarbons are the most-highly-water-soluble constituents of petroleum; they also adsorb poorly to soils. Because of their high mobility, they present the greatest threat to groundwater.

*Benzene.* Benzene in soils is highly volatile and has a high water solubility and relatively low adsorption to soil particles. Thus, benzene is mobile in soils and has the potential to leach into groundwater. Benzene is biodegradable and not expected to bioaccumulate.

*Toluene.* Toluene is highly volatile under ambient aboveground environmental conditions. Its high water solubility and relatively poor adsorption to soils give it the potential to leach into groundwater. Toluene is biodegradable and not expected to bioaccumulate.

*Xylenes.* Xylenes in soils are highly volatile and have moderate water solubility and a moderate capacity to adsorb to soils. Therefore, they tend to be mobile and may leach to groundwater. Xylenes are biodegradable and not expected to bioaccumulate.

*Ethyl Benzene.* Ethyl benzene is volatile in soils and has moderate water solubility and a moderate capacity to adsorb to soils. Therefore, it is moderately mobile in soil and may leach into groundwater. Ethyl benzene is biodegradable and not expected to bioaccumulate.

*PAH's.* PAH's have low volatility and low water solubility and tend to adsorb to soils. Three-ring PAH's, such as anthracene and phenanthrene, have low water solubility (on the order of 1000 $\mu$g/L). The water solubility of this class of compounds decreases with increasing ring number; therefore, PAH's are not highly mobile in the environment. They tend to partition to the soil rather than the groundwater, resulting in minimal movement in the subsurface. PAH's with more than three rings generally have poor biodegradability characteristics and may bioaccumulate in ecological species. Naphthalene and methyl naphthalene are moderately water soluble and moderately adsorbed to soil.[23] Naphthalene is volatile in soil and moderately water soluble and has the potential to migrate to groundwater. Naphthalene is very biodegradable and unlikely to bioaccumulate. Methyl naphthalene in soil may be volatile and exhibit a water solubility similar to that of naphthalene. They are, therefore, slightly immobile in soil. Methyl naphthalenes have the potential to bioaccumulate but do exhibit a moderate degree of biodegradation.

***Data-Collection Considerations.*** Data gathered for site investigations are useful for risk-assessment purposes. Special procedures are necessary to acquire reliable chemical release and exposure data for quantitative human-health risk assessment at hazardous-waste sites. Special considerations for obtaining samples for risk-assessment purposes include sample size, sampling locations, types of samples (grab vs. discrete), temporal factors, and meteorological factors. Other important sampling considerations for risk assessment include definition of data-quality objectives and inclusion of quality-assurance/-control measures to meet those objectives; selection of appropriate analytical methods; determination of the need for special analytical services; and acquisition of sufficient analytical documentation to allow for data evaluation/review. Ref. 8 presents guidance on conducting sampling for risk-assessment purposes, and Ref. 24 presents guidance on data quality and usability for risk assessment. For the purposes of developing a database for risk assessment, the sampling and analysis plan should address the following items.

1. Each medium of concern.
2. Background concentrations.
3. Potential exposure points within each medium.
4. Migration to potential exposure points, including data for environmental-fate and transport models.
5. Potential exposures based on possible future land uses.
6. Sufficient data to satisfy concerns about distributions of sampling data and statistics as well as the number and location of samples.
7. Data-quality objectives.
8. Special analytical requirements.

Crude oils are complex, variable, and subject to weathering and biodegradation. This makes analysis for hydrocarbons in crude-oil-contaminated soils and water a problem. The analytical methods used must be broad in scope but specific enough to capture all compounds involved. Because of the toxicity of some petroleum compounds, a high degree of analytical sensitivity is required. Detection limits in the 1- to 10-$\mu$g/L range per compound may be necessary.[20]

*Soil Sampling.* Soil represents a medium of direct-contact exposure and often is the main source of contaminants released into other media. As such, the number, location, and type of samples collected from soils have a significant effect on the risk assessment. Specific issues related to sampling of soil include the heterogeneous nature of the soils, the designation of hot spots, depth of samples, and the fate and transport properties of the chemicals of concern. The heterogeneous nature of soil makes collecting representative samples difficult. Therefore, a relatively large number of soil samples may be required to obtain sufficient data to estimate exposure concentrations adequately. Hot spots should be characterized because they may have significant impact on direct-contact exposures. Sample depth should be applicable to the exposure pathways and contaminant-transport routes of concern. Samples to assess surface exposures should be collected from the shallowest depth practical. The sampling plan should consider physical and chemical characteristics of soil that are important for evaluating fate and transport.

*Groundwater Sampling.* Groundwater samples should be collected with an approach that adequately defines the contaminant plume with regard to horizontal and vertical extent and with respect to potential exposure points (water supply wells). In addition, the hydrogeologic properties of the impacted aquifers must be characterized to allow for evaluation of contaminant migration and estimation of future exposure concentrations.

*Surface-Water and Sediment Sampling.* Surface-water and sediment samples collected from any nearby surface-water body should have the potential to receive discharge from the site. Samples are needed from a sufficient number of sampling points to characterize exposure pathways and from potential discharge points to the water body to determine whether the site is contributing to surface-water/sediment contamination.

*Air Sampling.* Although air is difficult to sample, sufficient samples should be collected to characterize both temporal and spatial variability. Usually, air-exposure concentrations are estimated by air modeling alone or air modeling in conjunction with monitoring data. The use of air models and monitoring data requires acquisition of site-specific meteorological data. Meteorological data are often available from nearby airports and meteorological stations.

*Biota Sampling.* Biota—commercial and game fish, shellfish, fowl, and terrestrial mammals as well as plants (such as grains, vegetables, and fruit consumed by humans)—may need to be sampled. This is of particular concern with chemicals that are known to bioaccumulate.

***Evaluation of Site-Specific Data.*** The objectives of the data-evaluation step are to identify (1) a set of chemicals that are likely to be site-related, (2) the reported chemical concentrations that are of acceptable quality for use in the quantitative risk assessment, and (3) the chemicals of potential concern that will be carried through the quantitative risk-assessment process. After data have been collected, they should be evaluated carefully to identify reliable, accurate, and verifiable numbers that can be used to quantify risks. Each data set used may be of very different quality. Adequate referencing is of primary importance for any data used in risk assessment. It is imperative to identify sources for all monitoring data, laboratory results, and toxicity data. The risk assessor or site manager is responsible for evaluating data. Data reported from the laboratory should be evaluated to determine its usability for risk assessment. Laboratory data should be evaluated for analytical methods used, sample-quantification limits, qualifiers and codes, blanks, and tentatively identified compounds. Various U.S. EPA documents present guidelines for these evaluations.[8, 24-26]

After data evaluation is complete, the chemicals of potential concern are identified for quantitative risk assessment. These include the following.

1. Chemicals positively detected in at least one sample in a given medium.

2. Chemicals detected at levels significantly above levels of the same chemicals detected in associated blanks.

3. Chemicals detected at levels significantly above naturally occurring levels of the same chemical.

4. Tentatively identified chemicals known to be associated with the site on the basis of historical information or confirmed by special analytical services.

5. Chemicals that are transformation products of chemicals demonstrated to be present.

Use of the criteria presented in Ref. 8, Chap. 5, may reduce further the number of chemicals carried through the risk calculations. These criteria include consideration of historical information; mobility, persistence, and potential for bioaccumulation; special exposure routes; treatability; applicable or relevant and appropriate requirements; frequency of detection; concentrations detected; and toxicity of contaminants present.

**7.2.2 Exposure Assessment.** An exposure assessment is conducted to estimate the magnitude of actual and/or potential human exposures, the frequency and duration of these exposures, and the pathways by which humans potentially are exposed. The exposure assessment addresses the nature and size of the population exposed to a substance, including subpopulations. In the exposure assessment, average and reasonable maximum estimates of exposure are developed for both current and future land-use assumptions. Current exposure estimates are used to determine whether a threat exists on the basis of existing exposure conditions at the site. Future exposure estimates are used to give risk managers an understanding of potential future exposures and threats as well as providing a qualitative estimate of the likelihood of such exposures. Conducting an exposure assessment involves the following.

1. Analysis of contaminant releases.

2. Identification of exposed populations.

3. Identification of all potential pathways of exposure.

4. Estimation of exposure-point concentrations for specific pathways of exposure on the basis of both environmental monitoring data and predictive-chemical-modeling results.

5. Estimation of contaminant intakes for specific pathways.

The results of this assessment are pathway-specific intakes for current and future exposures to individual contaminants. The exposure assessment consists of three steps: characterization of the exposure setting, identification of exposure pathways, and quantification of exposure.

***Characterization of the Exposure Setting.*** The exposure setting should be characterized with respect to the site's general physical characteristics and characteristics of populations on or near the site. Populations are identified and described for those characteristics that influence exposure, such as location relative to the site, activity patterns, and presence of sensitive subpopulations. These data are collected for current and future land uses.

***Identification of Exposure Pathways.*** An exposure pathway describes a unique mechanism by which a receptor population can be exposed to chemical constituents present at or originating from a site. An exposure pathway generally consists of four elements.

1. A source and mechanism of chemical release.

2. A retention or transport medium.

3. A point of potential human contact with the contaminated medium (referred to as the exposure point).

4. An exposure route (i.e., ingestion, inhalation, dermal contact) at the point of exposure.

All four elements must be present for an exposure pathway to be complete and for exposure to occur. Elements of an exposure-pathway analysis include direct contact from an exposure point and indirect exposures from vegetables, milk, meats, and transport of sources (agricultural contaminants); leaching from underground deposits (storage tanks, landfills) or surface stores (lagoons); and volatilization from soil or water. Fates for chemical pollutants include forming complexes with soil particles; bioaccumulation in plant and animal tissue; and chemical degradation in air, water, and soil. Absorption of chemical pollutants may be by oral ingestion of water, food, or soil; by inhalation of contaminated air or fumes; or by dermal absorption from air deposition, soil contact, or water (swimming, bathing). Ex-

| TABLE 7.1—VARIABLES FOR EQ. 7.2 FOR INGESTION OF CHEMICALS IN DRINKING WATER | |
|---|---|
| Variable | How Measured |
| $C_w$ | Site-specific measured or modeled value. |
| $R_i$ | 2 L/d (adult, 90th percentile). |
| | 1.4 L/d (adult, average). |
| | Age-specific values. |
| $f_e$ | Pathway-specific value (for residents, usually daily, 365 d/a). |
| $D_e$ | 70 years (lifetime; by convention). |
| | 30 years [national upper-bound time (90th percentile)] at one residence. |
| | 9 years [national median time (50th percentile)] at one residence. |
| $W$ | 70 kg (adult, average). |
| | Age-specific values. |
| $t_A$ | Pathway-specific period of exposure for noncarcinogenic effects (i.e., $D_e \times 365$ d/a); 70-year lifetime for carcinogenic effects (i.e., $70 \times 365$). |

| TABLE 7.2—VARIABLES FOR EQ. 7.3 FOR INGESTION OF CHEMICALS IN SURFACE WATER WHILE SWIMMING | |
|---|---|
| Variable | How Measured |
| $C_w$ | Site-specific measured or modeled value. |
| $R_c$ | 50 mL/h. |
| $t_e$ | Pathway-specific value in hours per event. |
| $f_e$ | Pathway-specific value (should consider local climatic conditions and age of potentially exposed population). |
| $D_e$ | 70 years (lifetime; by convention). |
| | 30 years [national upper-bound time (90th percentile)] at one residence. |
| | 9 years [national median time (50th percentile)] at one residence. |
| $W$ | 70 kg (adult, average). |
| | Age-specific values. |
| $t_A$ | Pathway-specific period of exposure for noncarcinogenic effects (i.e., $D_e \times 365$ d/a); 70-year lifetime for carcinogenic effects (i.e., $70 \times 365$). |

posure pathway and environmental fate may be addressed together because of their interrelationship. Principal pathways of exposure must be identified. Transport and transformation of the compound should be discussed, including a prediction of potential environmental distribution. With knowledge of the probable exposure pathways, exposed populations can be determined. Key variables are size, age, sex, location, and cultural habits (subsistence fishing, agricultural-pesticide exposure, intense ultraviolet-light exposure).

***Quantification of Exposure.*** Exposure is defined as contact of a human with a chemical agent. The magnitude of exposure is determined by measuring or estimating the amount of an agent available at the exchange boundaries (i.e., lungs, gut, or skin) during a specified time period. Quantification of exposure is conducted in two steps: estimation of exposure concentrations and calculation of chemical intake.

Exposure concentrations are estimated with monitoring data and/or chemical-transport and environmental-fate models. Available data are used to estimate current exposure concentrations. To estimate future exposure concentrations or exposure concentrations a distance from the source, environmental-fate and transport models are commonly used. To ensure that the data are health conservative, the 95% upper-confidence limit of the arithmetic mean is used as the exposure concentration; Ref. 27 provides guidelines for determining this upper-confidence limit. Estimation of chemical intake involves calculation of chemical-specific exposures for each exposure pathway identified. Exposure estimates are expressed in terms of mass of chemical in contact with the body per unit body weight per unit time. These exposure estimates are referred to as intakes and represent normalized exposure rates. Chemical intakes are calculated with equations that include variables for exposure concentration, $C$; contact rate, $R_c$; exposure frequency, $f_e$; exposure duration, $D_e$; body weight, $W$; and exposure averaging time, $t_A$. The values of some of these variables depend on site conditions and the characteristics of the potentially exposed population. Eqs. 7.1 through 7.8 are the standard equations for estimating chemical intake for common exposure pathways.[8]

*General Chemical-Intake Equation.*

$$I = C\left(\frac{R_c f_e D_e}{W}\right)\frac{1}{t_A}, \qquad (7.1)$$

where the chemical-related variable is $C$; the variables that describe the exposed population are $R_c$, $f_e$, $D_e$, and $W$; and the asset-determined variable is $t_A$.

*Ingestion of Chemicals in Drinking Water.*

$$I = \frac{C_w R_i f_e D_e}{W t_A}, \qquad (7.2)$$

where $R_i$ is measured in liters per day. **Table 7.1** gives the variable values for Eq. 7.2.

*Ingestion of Chemicals in Surface Water While Swimming.*

$$I = \frac{C_w R_c t_e f_e D_e}{W t_A}. \qquad (7.3)$$

**Table 7.2** gives the variable values for Eq. 7.3.

*Dermal Contact With Chemicals in Water.*

$$I = \frac{C_w A k_d t_e f_e D_e F_w}{W t_A}. \qquad (7.4)$$

**Table 7.3** gives the variable values for Eq. 7.4.

*Ingestion of Chemicals in Soils.*

$$I = \frac{C_s R_i F \phi_i f_e D_e}{W t_A}. \qquad (7.5)$$

**Table 7.4** gives the variable values for Eq. 7.5.

*Dermal Contact With Chemicals in Soil.*

$$I = \frac{C_s F A F_{ab} F_{d,ab} f_e D_e}{W t_A}, \qquad (7.6)$$

where $f_e$ is measured in events per year. **Table 7.5** gives the variable values for Eq. 7.6.

*Inhalation of Airborne Chemicals.*

$$I = \frac{C_a R_I t_e f_e D_e}{W t_A}, \qquad (7.7)$$

where $R_I$ is measured in cubic meters per hour. **Table 7.6** gives the variable values for Eq. 7.7.

*Food-Pathway Ingestion of Contaminated Fish and Shellfish and Contaminated Fruits and Vegetables.*

$$I = \frac{C_f R_i \phi_i f_e D_e}{W t_A}, \qquad (7.8)$$

where $R_i$ is measured in kilograms per meal and $f_e$ is measured in meals per year. **Table 7.7** gives the variable values for Eq. 7.8 for fish and shellfish and for fruits and vegetables.

Exposure assessments require that many parameters be assumed and default values adopted. Such values include inhalation rate (different for each population analyzed), ingestion rate, exposure duration (hours per day, days per year), and averaging period to be used (lifetime for carcinogens; equal to exposure periods for noncarcinogens). Assessment of all possible exposure scenarios for all potential receptor populations from all known chemicals of concern and their degradation products constitutes an exposure assessment.

**TABLE 7.3—VARIABLES FOR EQ. 7.4 FOR DERMAL CONTACT WITH CHEMICALS IN WATER**

| Variable | How Measured |
|---|---|
| $C_w$ | Site-specific measured or modeled value. |
| $A$ | 1.94 $m^2$ (adult, 50th percentile total body surface area). |
| | 0.23 $m^2$ (adult, 50th percentile surface area—arms). |
| | 0.082 $m^2$ (adult, 50th percentile surface area—hands). |
| | 0.55 $m^2$ (adult, 50th percentile surface area—legs). |
| $k_d$ | Consult Ref. 28 or open literature for values (note that use of $k_d$ values results in an estimate of absorbed dose). |
| $t_e$ | Pathway-specific value (should consider local activity patterns if information is available); U.S. national average for swimming is 2.6 h/d. |
| $f_e$ | Pathway-specific value (should consider local climatic conditions and age of potentially exposed population); U.S. national average for swimming is 7 d/a. |
| $D_e$ | 70 years (lifetime; by convention). |
| | 30 years [national upper-bound time (90th percentile)] at one residence. |
| | 9 years [national median time (50th percentile)] at one residence. |
| $F_w$ | 1 L/1000 cm. |
| $W$ | 70 kg (adult, average). |
| | Age-specific values. |
| $t_A$ | Pathway-specific period of exposure for noncarcinogenic effects (i.e., $D_e \times 365$ d/a); 70-year lifetime for carcinogenic effects (i.e., $70 \times 365$). |

Ref. 8, Chap. 6, and Ref. 29 provide guidance on conducting an exposure assessment. The project summary of Ref. 30 describes principles for exposure assessment. This document provides a summary of available data on various factors used in assessing human exposure, including drinking-water consumption, consumption rates of a variety of foods (i.e., fruits, vegetables, beef, dairy products, fish), soil ingestion, inhalation rate, skin area, lifetime, activity patterns, and body weight. The handbook also recommends standard default values (i.e., daily water intake, soil consumption for adults and children), basic exposure equations, reasonable worst-case scenarios, and procedures for assessing data uncertainties. The outline includes identification of each chemical or mixture, molecular formulas, synonyms, and Chemical Abstract Service numbers. Physical and chemical characteristics of each contaminant should be documented as well as sources of its site-specific uses, its disposal, and summary of its environmental releases.

**TABLE 7.4—VARIABLES FOR EQ. 7.5 FOR INGESTION OF CHEMICALS IN SOILS**

| Variable | How Measured |
|---|---|
| $C_s$ | Site-specific measured value. |
| $R_i$ | 200 mg/d (children, 1 through 6 years old). |
| | 100 mg/d (all age groups greater than 6 years old). |
| $F$ | $10^{-6}$ kg/mg |
| $\phi_i$ | Pathway-specific value (should consider contaminant location and population activity patterns). |
| $f_e$ | 365 d/a. |
| $D_e$ | 70 years (lifetime; by convention). |
| | 30 years [national upper-bound time (90th percentile)] at one residence. |
| | 9 years [national median time (50th percentile)] at one residence. |
| $W$ | 70 kg (adult, average). |
| | Age-specific values. |
| $t_A$ | Pathway-specific period of exposure for noncarcinogenic effects (i.e., $D_e \times 365$ d/a); 70-year lifetime for carcinogenic effects (i.e., $70 \times 365$). |

**7.2.3 Toxicity Assessment.** The toxicity-assessment step of the health risk assessment considers the following points.

1. Types of adverse health effects associated with chemical exposures (hazard identification).
2. Relationship between magnitude of exposure and adverse effects (dose/response evaluation).
3. Related uncertainties, such as weight of evidence for a particular chemical's carcinogenicity in humans.

**TABLE 7.5—VARIABLES FOR EQ. 7.6 FOR DERMAL CONTACT WITH CHEMICALS IN SOIL**

| Variable | How Measured |
|---|---|
| $C_s$ | Site-specific measured value. |
| $F$ | $10^{-6}$ kg/mg |
| $A$ | 1.94 $m^2$ (adult, 50th percentile total body surface area). |
| | 0.23 $m^2$ (adult, 50th percentile surface area—arms). |
| | 0.082 $m^2$ (adult, 50th percentile surface area—hand). |
| | 0.55 $m^2$ (adult, 50th percentile surface area—legs). |
| $F_{ab}$ | 1.45 kg/mg—commercial potting soil (for hands). |
| | 2.77 kg/mg—kaolin clay (for hands). |
| $F_{d,ab}$ | Chemical-specific value (accounts for desorption of chemical from soil matrix and absorption of chemical across skin). |
| $f_e$ | Pathway-specific value (should consider local climatic conditions and age of potentially exposed population). |
| $D_e$ | 70 years (lifetime; by convention). |
| | 30 years [national upper-bound time (90th percentile)] at one residence. |
| | 9 years [national median time (50th percentile)] at one residence. |
| $W$ | 70 kg (adult, average). |
| | Age-specific values. |
| $t_A$ | Pathway-specific period of exposure for noncarcinogenic effects (i.e., $D_e \times 365$ d/a); 70-year lifetime for carcinogenic effects (i.e., $70 \times 365$). |

**TABLE 7.6—VARIABLES FOR EQ. 7.7 FOR INHALATION OF AIRBORNE (VAPOR-PHASE) CHEMICALS**

| Variable | How Measured |
|---|---|
| $C_a$ | Site-specific measured or modeled value. |
| $R_l$ | 20 $m^3$/d (adult, average). |
| | 0.6 $m^3$/h (showering, all age groups). |
| $t_e$ | Pathway-specific value (depends on duration of exposure-related activity). |
| | 12 minutes—showering (90th percentile). |
| | 7 minutes—showering (50th percentile). |
| $f_e$ | Pathway specific value (depends on frequency of showering and other exposure-related activities). |
| $D_e$ | 70 years (lifetime; by convention). |
| | 30 years [national upper-bound time (90th percentile)] at one residence. |
| | 9 years [national median time (50th percentile)] at one residence. |
| $W$ | 70 kg (adult, average). |
| | Age-specific values. |
| $t_A$ | Pathway-specific period of exposure for noncarcinogenic effects (i.e., $D_e \times 365$ d/a); 70-year lifetime for carcinogenic effects (i.e., $70 \times 365$). |

**TABLE 7.7—VARIABLES FOR EQ. 7.8 FOR INGESTION OF CONTAMINATED FISH AND SHELLFISH AND CONTAMINATED FRUITS AND VEGETABLES**

| Variable | How Measured: Fish and Shellfish | How Measured: Fruits and Vegetables |
|---|---|---|
| $C_f$ | Site-specific measured or modeled value. | Site-specific measured value based on soil concentration and plant-to-soil accumulation factor or deposition factors. |
| $R_i$ | 6.5 g/d (daily intake averaged over 1 year). Daily-intake values should be used in conjunction with an exposure rate of 365 d/a. | Specific values for a wide variety of fruits and vegetables are available. |
| | Specific values for age, sex, race, region, and fish species are available. | |
| $\phi_i$ | Pathway-specific value (should consider local usage patterns). | Pathway-specific value (should consider location and size of contaminated area relative to that of residential areas in addition to local usage patterns). |
| $f_e$ | Pathway-specific value (should consider local population patterns if information is available). | Pathway-specific value (should consider anticipated usage patterns). |
| $D_e$ | 70 years (lifetime; by convention). | 70 years (lifetime; by convention). |
| | 30 years [national upper-bound time (90th percentile)] at one residence. | 30 years [national upper-bound time (90th percentile)] at one residence. |
| | 9 years [national median time (50th percentile)] at one residence. | 9 years [national median time (50th percentile)] at one residence. |
| $W$ | 70 kg (adult, average). | 70 kg (adult, average). |
| | Age-specific values. | Age-specific values. |
| $t_A$ | Pathway-specific period of exposure for noncarcinogenic effects (i.e., $D_e \times 365$ d/a); 70-year lifetime for carcinogenic effects (i.e., $70 \times 365$). | Pathway-specific period of exposure for noncarcinogenic effects (i.e., $D_e \times 365$ d/a); 70-year lifetime for carcinogenic effects (i.e., $70 \times 365$). |

***Hazard Identification.*** Hazard identification is the process of determining whether exposure to a chemical agent can cause an increase in the incidence of an adverse health effect. Hazard identification also involves characterizing the nature and strength of the evidence of causation. Hazard identification involves data on the types of health injury and disease that may be produced by a chemical and the exposure conditions that may produce injury or disease. The question often asked is whether it is scientifically correct to infer that toxic effects observed in one setting will occur in other settings (i.e., laboratory animals and industry workers). Most chemical-toxicity data used in risk assessments are either animal or epidemiological (human) data. Typically, the toxicity assessment of a health risk assessment is based on existing toxicity information developed by the U.S. EPA about specific chemicals.

***Dose/Response Evaluation.*** Dose/response evaluation is the process of quantitatively evaluating the relationship between the dose of the contaminant received and the incidence of adverse health effects in the exposed population. From this quantitative dose/response relationship, toxicity values are derived that can be used to estimate the incidence of adverse effects occurring in humans at different exposure levels. Evaluating exposure data becomes critical because many dose/response relationships may exist for a substance if it produces different toxic effects under different exposure conditions. Dose/response data must be quantitative if any degree of confidence is to be assigned to the risk evaluation. When developing toxicity values, adverse health effects are distinguished as noncarcinogenic or carcinogenic because of assumed differences in the mechanisms of action of noncarcinogens and carcinogens.

***Noncarcinogens.*** Toxicity values used to assess noncarcinogens are referred to as reference doses. A reference dose, $d_r$, is defined as an estimate of a daily exposure level for the human population, including sensitive subpopulations, that is likely to be without risk of adverse health effects. $d_r$ is derived by dividing the no-observed-adverse-effect level identified for a critical toxic effect by uncertainty factors. These uncertainty factors are used to adjust for interspecies and intrahuman variations and differences in exposure duration or routes of exposure. $d_r$ is expressed in units of milligrams of chemical per kilogram of body weight per day.

***Carcinogens.*** To assess potential carcinogenic effects, a two-part evaluation is used: (1) a weight-of-evidence classification (**Table 7.8**), which characterizes the likelihood that a chemical is a human carcinogen, and (2) a cancer slope factor, which is a plausible upper-bound estimate of the probability of an individual developing cancer as a result of exposure to a unit intake of the substance over a lifetime. The cancer slope factor converts estimated daily chemical intakes averaged over a lifetime of exposure directly into incremental risk of an individual developing cancer. Cancer slope factors are expressed in terms of risk per milligram of chemical per kilogram of body weight per day. The slope factor is often a 95th percentile upper-confidence limit of the probability of response based on experimental animal data and is considered an upper-bound estimate of carcinogenic risk.

***Toxicity Assessment of Chemicals Associated With E&P Activities.*** An indicator/surrogate approach generally is used to assess chemical components of crude oil, natural gas, condensate, and other oily waste associated with E&P activities. The indicator/surrogate method uses toxicity data from individual constituents and designated hydrocarbon fractions of a crude-oil mixture to determine soil-cleanup levels. Indicator compounds generally are selected after consideration of toxicity, pervasiveness, and mobility.[8] The most commonly used indicators for petroleum hydrocarbons include BTEX, PAH's, and, if applicable, metals. The remainder of the hydrocarbon mass is evaluated with a surrogate method. Specific hydrocarbon fractions, based on chemical structure and carbon number, are evaluated with a surrogate toxicity value identified on the basis of the toxicity of the hydrocarbon constituents making up the fraction. In general, the most conservative toxic component of a fraction is chosen as the surrogate compound. The indicator compounds, BTEX and

**TABLE 7.8—U.S. EPA WEIGHT-OF-EVIDENCE CLASSIFICATION SYSTEM FOR CARCINOGENICITY[8]**

| Group | Description |
|---|---|
| A | Human carcinogen. |
| B1 or B2 | Probable human carcinogen. |
| | B1 indicates that limited human data are available. |
| | B2 indicates sufficient evidence in animals and inadequate evidence in humans. |
| C | Possible human carcinogens. |
| D | Not classifiable as to human carcinogenicity. |
| E | Evidence of noncarcinogenicity for humans. |

PAH's, are used to evaluate carcinogenic toxicity, and the surrogates are used to evaluate noncarcinogenic toxicity of the remaining mass.

For very recent releases, use of a whole-mixture method that uses toxicity data from the whole petroleum product to determine clean-up levels may be feasible. Toxicity data for gasoline, diesel, mineral oil, and jet fuels have been determined. However, toxicity data for crude oil as a whole mixture have not been developed because of its extreme variability. Whole-mixture methods are appropriate only for "fresh" releases because they do not account for degradation or weathering. Toxicity values derived for fresh crude may not be appropriate for a weathered crude because the toxicity of the mixture changes as the relative percent weight of distillate fractions change with weathering. As expected, fractions containing lighter hydrocarbons are lost over time.[31]

Ref. 8, Chap. 7, presents guidelines on conducting a toxicity assessment. After extensive peer review, a U.S. EPA on-line database, Integrated Risk Information System (IRIS),[32] makes available toxicity values and the information used in their development. This database is the preferred source of toxicity information and supersedes all other sources. However, if information is not available in IRIS, other sources of toxicological data (e.g., Ref. 33) may be consulted. Ref. 34 provides guidance on addressing the TPH parameter.

**7.2.4 Risk Characterization.** The risk-characterization step summarizes and integrates the outputs of the exposure and toxicity assessments to characterize complete risk, both in quantitative expressions and qualitative statements.

***Noncarcinogenic Effects.*** To characterize potential noncarcinogenic effects, comparisons are made between an exposure level over a specified time period with a chemical-specific, route-specific reference dose derived for a similar exposure period. This ratio of exposure to toxicity is called a hazard quotient.

$$Q_{nc} = I/d_r. \quad (7.9)$$

The nonhazard quotient assumes that a level of exposure, $d_r$, exists below which it is unlikely for even sensitive populations to experience adverse health effects. If the exposure level exceeds this threshold (i.e., $I/d_r$ exceeds unity), there may be concern for potential noncancer effects. As a rule, the more the $I/d_r$ value exceeds unity, the greater the level of concern. It is important to note that the ratios of $I/d_r$ are not statistical probabilities.

A hazard-index approach is used to assess overall potential for noncarcinogenic effects posed by more than one chemical. This approach assumes that simultaneous subthreshold exposures to several chemicals could result in cumulative adverse health effects. The approach assumes that the magnitude of adverse effects is proportional to the sum of the ratios of the subthreshold exposures to acceptable exposures and that no synergism occurs. The hazard index is equal to the sum of the hazard quotients.

$$i_{nc} = I_1/d_{r1} + I_2/d_{r2} + ... + I_i/d_{ri}, \quad (7.10)$$

where Subscript $i$ = the $i$th toxicant.

When the hazard index exceeds unity, there may be concern for potential health effects. This approach assumes that doses are additive. This assumption is applied most properly to chemicals that induce the same effect by the same mechanism of action. Therefore, if the hazard index is greater than unity as a consequence of summing several hazard quotients, it may be appropriate to segregate the chemicals of concern by health effect and by mechanism of action to derive a hazard index for each group.

At some sites, exposure to a chemical or combination of chemicals may be through several pathways. Therefore, to assess the overall potential for noncarcinogenic effects posed by several exposure pathways, the total hazard index, $i_{nc,t}$, is calculated by summing the hazard indices of all exposure pathways, $P$.

$$i_{nc,t} = P_1 + P_2 + ... + I_i. \quad (7.11)$$

To assess total cancer risks, $i_{c,t}$, the cancer risks for each exposure pathway contributing to exposure of the same individual are

$$i_{c,t} = P_{R1} + P_{R2} + ... + P_{Ri}. \quad (7.12)$$

Cancer risks from various exposure pathways are assumed to be additive as long as the risks are for the same individual and the same time period. Ref. 8, Chap. 8, gives guidelines for assessing potential cancer risks.

**7.2.5 Uncertainties in the Risk-Assessment Process.** Although U.S. EPA risk-assessment guidelines provide a systematic approach for evaluating health risks, uncertainties and limitations are inherent in the risk-assessment process. Uncertainties are associated with site-characterization, exposure assessment, toxicological assessment, and risk characterization. In all risk assessments, uncertainties need to be addressed quantitatively and qualitatively. Each individual data evaluation affects the overall quality of the risk assessment because of the multiplicity process used.

***Uncertainty in Site Characterization.*** Uncertainty associated with site-characterization data is dependent on the quantity and quality of the historical site data and any new data generated during sampling and analysis programs for the risk assessment. Techniques to reduce uncertainty in data, such as taking a sufficient number of samples, may increase the cost of the risk assessment. In general, the more uncertain the data, the more conservative the assumptions have to be to ensure that risk calculations are health conservative. Determining acceptable levels of uncertainty is an important part of the risk-management process.

***Uncertainty in Exposure Assessment.*** Potential sources of uncertainty identified for the exposure-assessment step include (1) use of monitoring data that may not be representative of actual site conditions, (2) use of exposure models with assumptions and input variables used to estimate exposure concentrations that may not be applicable to the site, and (3) values for intake variables used to calculate chemical intakes.

***Uncertainty in Toxicological Assessment.*** Human toxicity associated with a chemical is inferred from data derived in experiments on animals. However, substantial interspecies variation in terms of susceptibility to the toxic effects of some chemicals has been documented. In addition, considerable variation occurs within a species. These inter- and intraspecies variations introduce a major source of uncertainty into the risk-assessment process. Other sources of variation in the toxicity assessment include extrapolations between routes of exposure (e.g., from oral to dermal route) and exposure conditions (e.g., from acute to chronic exposure duration). Toxicological databases are often deficient; consequently, toxicity values derived from these sources have inherent uncertainty. Also, using uncertainty factors to account for potential inter- and intraspecies variations, differences in exposure conditions, and other deficiencies in the databases introduces another source of uncertainty. Cancer slope factors represent the 95% upper-confidence limit of the probability of a response per unit intake of a chemical over a lifetime, thereby representing another source of uncertainty. Ironically, the default assumptions used to ensure that the risk calculations are health conservative are a major source of uncertainty in risk assessment.

***Uncertainty in Risk Characterization.*** Risk measures used in standard risk-assessment methods are not really probabilistic estimates of risk, but rather conditional estimates driven by the conservative assumptions made about exposure and toxicity. To place risk estimates in a proper perspective, the assumptions and uncertainties inherent in the risk-assessment process must be specified. An important potential source of uncertainty in assessment of health risks involves evaluation of simultaneous exposure to multiple constituents. Chemicals may interact to produce additive, synergistic, or antagonistic toxic responses. These interactions are not well-understood for most chemicals. For risk-assessment purposes, dose additivity is assumed; however, this may not be applicable when the specific mechanisms of toxicity and the target organs differ between chemicals. Therefore, this assumption may under- or overestimate potential health risks, depending on the chemicals involved.

### 7.3 Ecological Risk Assessment

Ecological risk assessment is defined as the process of assigning probabilities to the adverse effects that human activities have on the environment and their impact on the ecology.

**7.3.1 Background.** Risk-based regulations, guidelines, and reports have focused primarily on human-health concerns. Nevertheless, since 1987, the U.S. EPA has worked to develop similar guidelines for ecological risk assessment. In fact, Superfund legislation was the basis for both human-health- and ecological-risk-assessment guidelines. In 1990, the U.S. EPA recommended that ecological risk assessment be put on a level equal to human-health-based risk assessment.[35] To support establishment of guidelines, the U.S. EPA developed a number of ecological-risk-assessment case studies[36-39] and guideline documents.[40-42]

**7.3.2 Methods.** In the general approach, ecological risk assessments are similar to human-health risk evaluations. Ecological risk assessments may be quantitative, qualitative, or a combination of both. Ecological risk depends on a stressor (contaminant) contacting a biological component for sufficient time and dosage to cause an adverse effect (acute or chronic). Ecosystems may be compared with components of the human body. For example, rivers and streams, like the vascular system, transport nutrients and contaminants. Groundwater is similar to interstitial fluids (water between cells). Lake and river sediments and organic materials in soil are similar to fatty tissue in their ability to retain and concentrate fat-soluble contaminants, such as dichlorodiphenyltrichloroethane (DDT) and PCB. Both human and ecological wildlife toxicological data are derived from laboratory-animal studies.

Ref. 40 defines the process for ecological risk assessment. The method evaluates the likelihood that adverse ecological effects are occurring or may occur as a result of exposure to one or more stressors. A stressor is defined as a physical, chemical, or biological entity that can induce an adverse ecological response. Adverse responses may range from sublethal chronic effects in an individual organism to loss of an entire ecosystem function. Evaluation procedures exist for plants and animals that exclude domesticated species.

As with human-health risk assessments, several levels and categories of ecological assessment can exist. Guidelines[43] under the Natl. Environmental Policy Act (NEPA) may require submission of an environmental impact statement (EIS) as a condition to receive an NPDES permit to discharge wastewater. The EIS must address several site-specific issues as part of the environmental assessment. Some environmental-assessment issues include describing project processes and the affected environment and predicting environmental impacts. The EIS may include surface-water resources; groundwater resources; air quality and odors; noise; public health; land-use changes; socioeconomics; cultural resources; flood-plain/wetland resources; endangered species; impacts to soil, vegetation, and landscape; and cumulative effects.

An ecological assessment may have a format similar to that of a complete risk assessment. After the problem is identified, an ecological assessment has two phases: analysis and risk characterization.[40]

***Analysis.*** The analysis phase combines characterization of exposure and ecological effects. This phase combines the hazard-identification, dose/response, and exposure-assessment components of human-health risk assessment except that dose response is replaced by stressor response. This change emphasizes the physical changes that may stress ecosystems as severely as chemical stressors. Physical stressors usually are not measured in terms of dose.

***Risk Characterization.*** Risk characterization, as with human-health risk assessment, is the final product of the ecological assessment. This phase integrates the results of the exposure and ecological-effects analyses to evaluate the likelihood of adverse ecological effects. The risk-characterization phase considers types and magnitudes of effects, their spatial and temporal patterns, and the likelihood of ecological recovery. Information in this phase is presented to risk managers along with assumptions, limitations of the analytical process, and estimates of uncertainty.

While a complete discussion of an ecological risk assessment is beyond the scope of this chapter, the following general questions should be answered when collecting data and planning to do an ecological risk assessment.

1. What is the environmental setting in terms of upland forest, onsite stream, wildlife, manmade areas, potential contamination, offsite impacts from pollutant migration, habitat used by endangered species, and sensitive habitats?
2. What may be the contaminant's fate and transport (e.g., windblown dust, surface-water runoff, erosion, drainage ditches, deposition in sediment downstream, and photochemical-oxidation and hydrolysis potential)?
3. What are the ecotoxicities and potential receptors in terms of toxic mechanisms of the pollutants, target species, species organ systems affected, reproductive effects, and food-chain concerns?
4. What are the potential exposure routes of animal inhalation, ingestion, or dermal exposure (e.g., plant-root absorption, leaf deposition, direct contact to fish gills, and direct contact with aquatic plants) and chemical form of the contaminant (acid, base, salt, organic, metal, and combinations)?
5. What are the lowest exposure concentrations shown to produce adverse effects (e.g., benchmark doses for species other than the species in question)?
6. What are the appropriate conservative exposure assumptions (i.e., use the following: 100% for area-factor home range of species of concern, 100% for bioavailability of contaminant, most sensitive life stage, and minimum body weight to maximum ingestion rate)?

## 7.4 Risk Management

Risk management, as the words suggest, is management of risks that have been identified through the risk-assessment process. Risk management is applied to the overall management of environmental risk. Many people confuse risk management with risk assessment; however, a distinct difference exists between managing risks and determining them. The reader should now be sufficiently familiar with risk assessment to understand the scientific methods used to identify and quantify risks. Risk-assessment information can be used to make better-informed decisions and to set priorities to obtain the most effective use of available resources in a process known as risk management. Risk assessment is an important part of risk management, but not the only part. Risk management considers project economics, technical feasibility, remedial options, environmental planning, risk communications, and regulatory compliance when making decisions.

Risk communications are an important part of today's information age, where the slightest comments can be transmitted around the world in following or documenting an event. Specific attention must be given to the communication phase of risk management to have a successful and effective program.

**7.4.1 Remedial Options.** When a remediation project is considered, risk assessment may be used to provide an estimate of current risk and post-cleanup risk. Risk assessment allows for a discussion of the advantages and disadvantages of each remedial approach in terms of technology, economic feasibility, and risk reduction. Risk and risk-related parameters, such as contaminant levels and potential contaminant mobility, are evaluated for each remedial alternative. Before a risk assessment is done, preliminary remediation goals are established on the basis of chemical-specific information, such as maximum contaminant levels for groundwater. The preliminary risk assessment redefines risk in chemical- and site-specific terms. Final remediation goals may be redefined or adjusted after results from more detailed risk assessment become available. Final remediation options and target cleanup levels usually are part of a complete risk assessment.

The general approach is to develop remedial alternatives, determine the state-of-the-art technology, look at short- and long-term risk reduction, and develop preliminary cost projections for each alternative. Next, the alternatives are screened to determine which ones meet preselected evaluation criteria. Several things should be considered in developing evaluation criteria, including the following.

1. Overall protection of human health and the ecology.
2. Application of general regulatory cleanup guidelines, such as maximum contaminant level for groundwater.
3. Long- and short-term effectiveness of remedial approach.
4. Reduction of various health-related parameters, such as toxicity, mobility, and volume of contaminant.
5. Availability of state-of-the-art remediation technology.

6. Applicable permits required and regulation involved.
7. Technical and economic feasibility.

**7.4.2 Risk Planning.** Risk assessments, especially at the screening level, are often used in environmental planning. This allows risk to be used as a means to assign environmental priorities to areas where contamination exists or may exist in the future. Using risk to rank environmental problems allows for allocation of time and effort where it will have the greatest impact on risk reductions. Ref. 44, which ranks environmental issues by their human-health and ecological risk, shows (from an overall point of view) that the most important issues for human-health risks are pesticides, air pollution, and contaminated drinking-water supplies. The most important ecological risks are water pollution and physical degradation of terrestrial and wetland ecosystems.

A primary purpose of comparative-risk studies is to allocate resources. The concept is to focus resources on the problems that pose the greatest risk. Risk can be used to set priorities for cleaning up sites, to initiate preventive action at sites that pose significant potential risk, and to rank cleanup options to assist in cost-efficiency. Risk also can be used to compare pollution-prevention technology options and to evaluate chemical processes and treatment options.

**7.4.3 Risk Communication.** We live in an information age; consequently, risk communication is a very important aspect of managing and operating a company in today's environment. Most companies have a public-relations group or office to handle their communications with the public and the news media. Unfortunately, environmental issues need to be handled with knowledge and accuracy. Accurate risk information and good intentions sometimes are perceived as foolish because of the manner in which they are presented. Social scientists, media reporters, government officials, environmental advocacy groups, and industry public-relations personnel all participate in the art of risk communication. For the past decade, much has been published by different groups attempting to relate success and failure, rules and cautions. In 1988, the U.S. EPA published cardinal rules for risk communication.[45] The following are summaries of these rules.

1. Accept and involve the public as a legitimate partner. In a democracy, people and communities have a right to participate in decisions that affect their lives, their property, and the things they value. Demonstrate your respect for the public and underscore the sincerity of your effort by involving the community early, before important decisions are made.

2. Plan carefully and evaluate your efforts. Begin with clear, explicit risk-communication objectives. Evaluate the information and know its strengths and weaknesses. Realize that there is no one "public," but many individuals with many different interests.

3. Listen to the public's specific concerns. If you do not listen to people, you cannot expect them to listen to you. Let them know you are listening. Do not make assumptions about what people know, think, or want done about risks. Recognize emotions, hidden agendas, and broader economic or political concerns.

4. Be honest, frank, and open. Trust and credibility are your most precious assets. They are difficult to obtain. Once lost, they are almost impossible to regain completely. If you do not know an answer or are uncertain, say so. Admit mistakes. Disclose risk information as soon as possible. Discuss data strengths and weaknesses.

5. Coordinate and collaborate with other credible sources. Consult with others to determine whether you or others are best able to answer questions about risk. Try to issue communications jointly with academic contacts, physicians, or local officials. Conflicts or public disagreements with other credible sources make risk communication very difficult.

6. Meet the needs of the media. The media play a critical role in setting agendas and determining outcomes. Be open with and accessible to reporters. Do not hesitate to follow up on stories with praise or criticism, as warranted. Try to establish long-term relationships with specific editors and reporters.

**TABLE 7.9—WORD PAIRS THAT INFLUENCE PUBLIC PERCEPTION OF RISK**

| Less Risky | More Risky |
|---|---|
| Voluntary activities | Involuntary activities |
| Familiar | Unfamiliar |
| Controllable | Uncontrollable |
| Controlled by self | Controlled by someone else |
| Fair | Unfair |
| Not memorable | Memorable |
| Not dreaded | Dreaded |
| Chronic | Acute |
| Diffuse in time and space | Focused in time and space |
| Not fatal | Fatal |
| Immediate | Delayed |
| Natural | Artificial |
| Individual mitigation possible | Individual mitigation impossible |
| Detectable | Undetectable |

7. Speak clearly and with compassion. Use simple, nontechnical language. Acknowledge and respond to emotions that people express. Respond to distinctions that the public views as important (e.g., involuntary risk situations, benefits, fairness, familiarity with the stressor). Promise only what you can do, and be sure to do what you promise.

Refs. 46 through 48 discuss the public perception of risk. If risk communicators understand not just the technical point of view but also the public's, better communications result. These references identified the word pairs in **Table 7.9** as important factors in public perception of risk.

While explaining risk to both the media and the public is a matter best left to experts, risk-communication training is becoming more common for technical and management personnel. Often, personnel properly trained for communication are not available to handle all risk-communication duties. Understanding risk-communication concepts may be useful for briefing nonrisk personnel on risk concepts.

## 7.5 Summary and Conclusions

Crude oil is a complex mixture of chemicals. Some constituents, such as benzene and PAH's, are hazardous to human health and the environment. In normal operations, these chemicals are not released to the environment, so no human-health or ecological risk occurs. However, when accidental spills or other releases to the environment do occur, risk assessment may be used to provide a quantitative determination of the resulting risk. Knowing the magnitude of the risk allows for establishment of cleanup levels based on human-health and environmental concerns as well as on regulatory guidelines.

The risk-assessment process is science based, flexible, and accepted by regulatory agencies. Risk assessment may be used in environmental planning, especially for establishment of priorities and allocation of funds based on risk. The risk-assessment process has four principal steps: hazard identification, toxicity assessment, exposure assessment, and risk characterization. The process may be used at various levels from screening, to intermediate, to a complete risk assessment. The same basic process may be used to establish ecological, socioeconomic, and cost/benefit risks. Information from risk assessment is an important input to the overall process of risk management.

While a large body of scientific information exists about the risk-assessment process, the process is dynamic and continues to change as more is understood about the complex interactions between chemicals, the environment, and human health. Research continues on the role that chemicals play in cancer formation and development. Ecological risk assessment is not as advanced as human-health risk assessment, but, over time, as more information is developed, it may become as important. Risk communication is an important part of risk assessment and risk management.

**7.5.1 Need for Expertise—Consultants and Regulatory Agencies.** A fundamental understanding of the risk-assessment process allows an engineer/manager to perform preliminary risk calculations and to assist in planning data-gathering strategies. Risk assessment for internal purposes could require the services of a consultant, depending on the purpose of the assessment and the experience of the person or team performing the risk assessment. For external risk assessments, the state or federal environmental authority with jurisdiction should be contacted to determine which guidelines or special conditions might apply. For detailed, site-specific risk assessments, especially when liability concerns exist, an outside consultant may be needed.

With risk assessment, as with many scientific methods, there is a need to know how and when to seek expert help. The information in this chapter is an introduction to risk assessment. As such, it is a first step toward providing an overview of the process and a familiarity with and definition of the terms particular to the process. Requesting a detailed, site-specific risk assessment when a screening risk assessment is sufficient is a costly and time-consuming error.

The U.S. EPA and many U.S. state regulators have considerable experience in risk assessment and are accustomed to thinking of environmental regulations in terms of risk. Understanding the basic concepts of risk assessments allows for a common basis for communication and negotiation. Consultants usually should be hired for a detailed, site-specific risk assessment because few organizations have the in-house expertise needed; however, screening risk assessments often are the basis for preliminary discussion and project reviews with regulators.

**7.5.2 Risk-Based Software.** ***American Petroleum Inst. (API).*** API has written software for the petroleum industry called API-DSS for Decision Support System. API-DSS is written in KnowledgePro under the Microsoft Windows environment. The interface uses FORTRAN code for computation and Microsoft's Excel spreadsheet for graphics and data presentation. The software has a limited database that is directed toward hydrocarbons and metals, common chemicals of concern in the petroleum industry. The database consists of two parts: toxicity data and chemical data. The database can be modified and new chemical data can be added with an editor. Version 1.0 has four modules. The first module allows the user to develop a conceptual model that is used to estimate risk. The second module is the fate and transport module that allows the user to estimate receptor-point concentrations using site-specific data that is used by hydrogeological and meteorological models. The third module is a chemical-intake and risk-calculation module. The last module for risk presentation takes advantage of graphic and data-handling qualities of the Excel spread sheet. The software also can be used to estimate uncertainty by use of Monte Carlo simulation, albeit with a limited number of iterations.

The API software is relatively easy to use for people familiar with the risk-assessment process. Other software packages are available and undoubtedly more will be developed over time as risk assessment enjoys wider usage. A word of caution about risk software. Commercially available packages are deceptively easy to use and may tempt inexperienced users to begin to develop risk assessments with little real knowledge of the process. The API-DSS software package has an excellent manual that explains the fate and transport models in detail and is a good overview of risk assessment from a petroleum-industry point of view.

***Other Sources.*** CalTox is a multimedia, total-exposure model for hazardous-waste sites developed by the State of California's EPA Dept. of Toxic Substances Control. Risk Assistant is a microcomputer-based software system developed by the U.S. EPA with Hampshire Research Inst. for risk assessment. RiskQ is from Lawrence Livermore Laboratory and the Regents of the U. of California. Crystal Ball is from Decisioneering in Denver, Colorado, and is used for Monte Carlo simulations.

### 7.6 Definitions

**Chemicals of Potential Concern.** Chemicals that are potentially site-related for which data are of sufficient quality to use in quantitative risk assessment.

**Contact Rate.** Amount of medium (e.g., groundwater, soil) contacted per unit time or event (e.g., liters of water ingested per day).

**Chronic Reference Dose.** An estimate (with uncertainty spanning perhaps an order of magnitude or greater) of a daily exposure level for the human population, including sensitive subpopulations, that is likely to be without an appreciable risk of deleterious effects during a lifetime. Chronic reference doses are specifically developed to be protective for long-term exposure to a compound (7 years to lifetime).

**Data-Quality Objectives.** Qualitative and quantitative statements that ensure that data of known and documented quality are obtained during site-investigation activities to support decisions concerning protection of human health.

**Dose/Response Evaluation.** The process of quantitatively evaluating toxicity information and characterizing the relationship between dose of a contaminant administered or received and incidence of adverse health effects in the exposed population. From the quantitative dose/response relationship, toxicity values are derived that are used in the risk-characterization step to estimate the likelihood of adverse effects occurring in humans at different exposure levels.

**Exposure.** Contact of an organism with a chemical or physical agent. Exposure is quantified as the amount of the agent available at the exchange boundaries of the organism (e.g., skin, lungs, gut) and available for absorption.

**Exposure Assessment.** Determination or estimation (qualitative or quantitative) of the magnitude, frequency, duration, and route of exposure.

**Exposure Event.** Incident of contact with a chemical or physical agent. An exposure event can be defined by time (e.g., days, hours) or by incident (e.g., eating a single meal of contaminated fish).

**Exposure Pathway.** Course that a chemical or physical agent takes from a source to an exposed organism. An exposure pathway describes a unique mechanism by which an individual or population is exposed to chemicals or physical agents at or originating from a site. Each exposure pathway includes a source or release from a source, an exposure point, and an exposure route. If the exposure point differs from the source, a transport/exposure medium (e.g., air)—or media in cases of intermediate transfer—also is included.

**Exposure Point.** Location of potential contact between an organism and a chemical or physical agent.

**Exposure Route.** Way a chemical or physical agent comes in contact with an organism (e.g., ingestion, inhalation, dermal contact).

**Hazard Identification.** Process of determining whether exposure to an agent can cause an increase in the incidence of a particular adverse health effect (e.g., cancer, birth defect) and whether the adverse health effect is likely to occur in humans.

**Hazard Index.** Sum of more than one hazard quotient for multiple substances and/or multiple exposure pathways. The hazard index is calculated separately for chronic, subchronic, and shorter-duration exposures.

**Hazard Quotient.** Ratio of a single-substance exposure level over a specified time period (e.g., subchronic) to a reference dose for that substance derived from a similar exposure period.

**Intake.** Measure of exposure expressed as the mass of a substance in contact with the exchange boundary per unit body weight per unit time (e.g., milligram chemical per kilogram body weight per day). Also called normalized exposure rate; equivalent to administered dose.

**IRIS.**[32] U.S. EPA database containing verified reference doses and slope factors and up-to-date health-risk and U.S. EPA regulatory information for numerous chemicals. IRIS is the U.S. EPA's preferred source for toxicity information for risk assessment.

**Lifetime Average Daily Intake.** Exposure expressed as mass of a substance contacted per unit body weight per unit time, averaged over a lifetime.

**Lowest-Observed-Adverse-Effect Level.** In dose/response experiments, the lowest exposure level at which there are statistically or biologically significant increases in frequency or severity of adverse effects between the exposed population and its appropriate control group.

**NORM.** Ambient concentrations of radioactive chemicals that are present in the environment naturally and have not been influenced by humans (e.g., aluminum, manganese).

**No-Observed-Adverse-Effect Level.** In dose/response experiments, exposure level at which there are no statistically or biologically significant increases in frequency or severity of adverse effects between the exposed population and its appropriate control. Some effects may be produced at this level, but they are not considered to be adverse or precursors to specific adverse effects. In an experiment with more than one no-observed-adverse-effect level, the regulatory focus is primarily on the highest one, leading to common usage of the term no-observed-adverse-effect level to mean the highest exposure level without adverse effect.

**Reference Dose.** Preferred toxicity value for evaluating noncarcinogenic effects resulting from exposures at hazardous-waste sites. Reference dose, when used without other modifiers refers either generically to all types of reference doses or specifically to chronic reference doses; it never refers specifically to subchronic or developmental reference doses.

**Slope Factor.** Plausible upper-bound estimate of the probability of a response per unit intake of a chemical over a lifetime. Slope factor is used to estimate an upper-bound probability of an individual developing cancer as a result of a lifetime of exposure to a particular level of a potential carcinogen.

**Special Analytical Services.** Nonstandardized analyses conducted to meet user requirements that cannot be met with routine analytical services, such as shorter analytical turnaround time, lower detection limits, and analysis of nonstandard matrices or compounds.

**Toxicity Value.** Numerical expression of a substance's dose/response relationship used in risk assessments. The toxicity values most commonly used in risk assessments are reference doses (for noncarcinogenic effects) and slope factors (for carcinogenic effects).

**Weight-of-Evidence Classification.** U.S. EPA classification system for characterizing the extent to which available data indicate that an agent is a human carcinogen. Recently, the agency has developed weight-of-evidence classification systems for some other kinds of toxic effects, such as developmental effects.

## Nomenclature

$A$ = skin surface area available for contact per event, $L^2$, $m^2$
$C$ = chemical concentration, m/m, mg/kg
$C_a$ = contaminant concentration in air, m/m, mg/kg
$C_f$ = contaminant concentration in food, m/m, mg/kg
$C_s$ = chemical concentration in soil, $m/L^3$, mg/L water
$C_w$ = chemical concentration in water, $m/L^3$, mg/L water
$d_r$ = reference dose, m/m-t, mg/kg-d
$D_e$ = exposure duration, t, years
$f_e$ = exposure frequency, t/t, d/a
$F$ = conversion factor, m/m, mg/kg
$F_{ab}$ = absorption factor, m/m, mg/kg
$F_{d,ab}$ = desorption/absorption factor
$F_w$ = volumetric conversion factor for water, $L^3$/L, L/cm
$i_{nc}$ = noncancer-hazard index
$i_c$ = cancer-hazard index
$I$ = intake or exposure level, m/m-t, mg/kg-d
$k_d$ = chemical-specific dermal permeability constant, L/t, cm/h
$P$ = exposure pathway
$P_R$ = risk-exposure pathway
$Q_{nc}$ = noncancer-hazard quotient
$R_c$ = contact rate, $L^3$/t, L/h
$R_i$ = ingestion rate, m/t, mg/d for soil, kg/meal for food
$R_I$ = inhalation rate, $L^3$/t, $m^3$/h
$t_A$ = averaging time per event = period over which exposure is averaged = $D_e \times 365$ d/a, t, days
$t_e$ = exposure time per event, t, hours
$W$ = body weight, m, kg
$\phi_i$ = fraction ingested from contaminated source, dimensionless

## References


1. Carson R., *Silent Spring*, Houghton Mifflin Co., Boston, Massachusetts (1962).
2. *Risk Assessment in the Federal Government, Managing the Process*, Natl. Research Council, Natl. Academy Press, Washington, DC (1983).
3. "Guidelines for Mutagenicity Risk Assessment," U.S. EPA, *Federal Register* (24 September 1986) **51,** 34006.
4. "Guidelines for the Health Risk Assessment of Chemical Mixtures," U.S. EPA, *Federal Register* (24 September 1986) **51,** 34014.
5. "Guidelines for the Health Risk Assessment of Suspect Developmental Toxicants," U.S. EPA, *Federal Register* (24 September 1986) **51,** 34028.
6. "Guidelines for Exposure Assessment," U.S. EPA, *Federal Register* (24 September 1986) **51,** 34042.
7. *Superfund Public Health Evaluation Manual*, EPA 540/1-86/060, U.S. EPA, Washington, DC (October 1986).
8. *Risk Assessment Guidelines for Superfund, Human Health Evaluation Manual,* interim final, EPA 540/1-89/002, U.S. EPA, Washington, DC (December 1989) **1,** Part A.
9. *ES-1739-95, Standard Guide for Risk Based Corrective Action Applied at Petroleum Release Sites*, ASTM, West Conshohocken, Pennsylvania (November 1995).
10. *Science and Judgment in Risk Assessment*, Natl. Research Council, Natl. Academy Press, Washington, DC (1994).
11. Casarett, L.J. and Doull, J.: *Casarett and Doull's Toxicology: The Basic Science of Poisons*, third edition, C.D. Klassen, M.O. Amdor, and J. Doull (eds.), McMillan Publishing Co., New York City (1986).
12. Rodericks J.V.: *Calculated Risks*, Cambridge U. Press, New York City (1992).
13. Lewis, H.W.: *Technological Risk*, W.W. Norton & Co., New York City (1990).
14. Lowrance, W.: *Of Acceptable Risk: Science and the Determination of Safety,* William Kaufman, Los Altos, California (1979).
15. Fischhoff, B. *et. al.*: *Acceptable Risk*, Cambridge U. Press, New York City (1989).
16. The 1938 Federal Food, Drug, and Cosmetic Act, Public Law No. 85-929, 72 Statute 1784, 1785, Codified at 21 USC 348 (1958).
17. McGrath, E.J. and Irving, D.C.: "Techniques for Efficient Monte Carlo Simulation," *Variance Reduction*, Office of Naval Research, Arlington, Virginia (1973) **3,** Chap. 1.
18. The Resource Conservation and Recovery Act, 40 CFR (1976) Part 280, Subtitle I.
19. Carney, G.R. and Morrison, C.A: *Computer Assisted Environmental Justice Index Methodology,* Office of Planning and Analysis, Region 6 U.S. EPA, Dallas (July 1994).
20. Spreight, J.C.: *The Chemistry and Technology of Petroleum,* Marcel Decker, New York City (1980).
21. *Risk Assessment Guidance For Superfund, Human Health Evaluation Manual, Development of Risk-Based Preliminary Remediation Goals,* EPA/540/R-92/003, Office of R&D, U.S. EPA, Washington, DC (1991) **1,** Part B.
22. Shepard, M.C. *et al.*: "Characterization of Produced Waters from Natural Gas Production and Storage Operations: Regulatory Analysis of a Complex Matrix," *Proc.*, API Produced Water Symposium, San Diego, California (1992) 120.
23. Brendehaug, J. *et al.*: "Toxicity and Chemical Characterization of Produced Water—A Preliminary Study," *Proc.*, API Produced Water Symposium, San Diego, California (1992) 96.
24. "Guidance for Data Usability in Risk Assessment (Superfund)," Publication 9285.7-09A, PB92-963356, Hazardous Site Evaluation Div., Office of Emergency and Remedial Response, U.S. EPA, Washington, DC (1992) Part A.
25. "Laboratory Data Validation Functional Guidelines for Evaluating Organics Analysis," EPA/540/R-94/082, U.S. EPA, Washington, DC (1988).
26. "National Functional Guidelines for Organic Data Review," EPA/540/R-94/013, U.S. EPA, Washington, DC (1991).

27. "Supplemental Guidance to RAGS: Calculating the Concentration Term," OSWER Directive 9285.7-081, Office of Health and Environmental Assessment, U.S. EPA, Washington, DC (1992).
28. "Dermal Exposure Assessment: Principles and Applications," EPA/600/8-91/011-B, Exposure Assessment Group, Office of Health and Environmental Assessment, U.S. EPA, Washington, DC (January 1992).
29. "Guidelines for Exposure Assessment," EPA/600/Z8-92/001, U.S. EPA, Washington, DC (1992).
30. Schaum, J.: *Exposure Factors Handbook, Project Summary*, EPA/600/S8-89/043, Office of Health and Environmental Assessment, U.S. EPA, Washington, DC (1989).
31. MacKay, D.: "The Chemistry and Modeling of Soils Contaminated with Petroleum," *Soils Contaminated with Petroleum: Environmental and Public Health Defects*, E.J. Calbrese and P.T. Kostecki (eds.), John Wiley & Sons, New York City (1988).
32. "Integrated Risk Information System (IRIS)," database, Office of R&D, Natl. Center for Environmental Assessment, U.S. EPA, Washington, DC (October 1993).
33. "Health Effects Summary Tables Annual Update, FY 1995," annual report, EPA 540/R-95-036, PB95-921199, Office of Emergency & Remedial Response, Office of R&D, U.S. EPA, Washington, DC (May 1995).
34. "Development of Health-Based Alternative to the Petroleum Hydrocarbon (TPH) Parameter," interim final petroleum report, Massachusetts Dept. of Environmental Protection, Boston, Massachusetts (August 1994).
35. "Reducing Risk: Setting Priorities and Strategies for Protection," SAB-EC-021, Science Advisory Board A-101, U.S. EPA, Washington, DC (September 1990).
36. "A Review of Ecological Assessment Case Studies from a Risk Assessment Perspective," EPA 630/R-92/005, U.S. EPA, Washington, DC (1993).
37. "Ecological Assessment of Hazardous Waste Sites: A Field and Laboratory Reference," EPA 600/3-89/013, U.S. EPA, Corvallis, Oregon (1989).
38. "Summary of Ecological Risks, Assessment Methods, and Risk Management Decisions in Superfund and RCRA," EPA-230-03-89-046, Office of Policy Analysis, U.S. EPA, Washington, DC (1989).
39. "Review of Ecological Risk Assessment Methods," EPA/230/10-88/041, U.S. EPA, Washington, DC (1988).
40. "Framework for Ecological Risk Assessment," EPA/630/R-92/001, U.S. EPA, Washington, DC (1992).
41. "Ecological Risk: A Primer for Risk Managers," EPA/734/R-95-001, Office of Pesticides & Toxic Substances, U.S. EPA, Washington, DC (1995).
42. "Managing Ecological Risk at EPA: Issues and Recommendations for Progress," EPA/600/R-94/184/3, Office of R&D, U.S. EPA, Cincinnati, Ohio, and Office of Policy, Planning and Evaluation, U.S. EPA, Washington, DC (1994).
43. *Environmental Information Document (EID) Handbook: Guidance to Applicants for New Source NPDES Permits* Office of Planning and Coordination, U.S. EPA Region 6, Dallas (October 1995).
44. *Unfinished Business: A Comparative Assessment of Environmental Problems*, NTIS Stock No. PB88-127030, Office of Policy, Planning, and Evaluation, U.S. EPA, Washington, DC (1987) 1.
45. Corvello, V.T. and Allen, F.W.: "Seven Cardinal Rules of Risk Communication," EPA 230/09-88/001, U.S. EPA, Washington, DC (April 1988).
46. Sandman, P.M.: "Explaining Environmental Risk," No. 15561443, TSCA Assistance Office, Office of Toxic Substances, U.S. EPA, Washington, DC (November 1986).
47. Slovic, P.: "Perception of Risk," *Science* (February 1987) **236,** 280.
48. Covello, V.T., Von Winterfeldt, D., and Slovic, P.: "Communicating Scientific Information about Health and Environmental Risks: Problems and Opportunities from a Social and Behavioral Perspective," *Uncertainties in Risk Assessment and Risk Management,* John Wiley & Sons, New York City (1987) 180–90.

## General References

*Access EPA*, Information Access Branch, Office of Information Resources Management, U.S. EPA, Washington, DC (1993).

Adamus, C.L.: "Environmental Risk Assessment: Dealing with Uncertainty in Environmental Impact Assessment," environment paper 7, Office of the Environment, Asian Development Bank, Manila, Philippines (1990).

Adamus, C.L.: *Guidelines for the Health Impact Assessment of Development Projects*, Office of the Environment, Asian Development Bank, Manila, Philippines (1992).

Adamus, C.L. and Bergman, M.J.: "Estimating Non-Point Source Pollution Loads with a GIS Screening Model," *Water Resources Bull.* (1995) **31,** No. 4, 6.

Allen, B. *et al.*: "Investigation of Cancer Risk Assessment Methods," project summary, EPA/600/S6-87/007, Office of Health and Environmental Assessment, U.S. EPA, Washington, DC (1988).

Allen, F.W.: "What the Public Believes; How the Experts See It," *EPA J.* (1987) **13,** No. 9, 9.

Alm, A.: "Why We Didn't Use Risk Before," *EPA J.* (1991) **17,** No. 2, 12.

Antenucci, J.C.: *Geographic Information Systems: A Guide To The Technology*, Van Nostrand Reinhold, New York City (1991).

"Approaches to Acceptable Risk," *Proc.*, American Soc. of Civil Engineers Engineering Foundation Conference on Risk-Based Decision Making in Water Resources, New York City (1985) 188.

Armente, P.M.: *Contingency Planning For Industrial Emergencies*, Van Nostrand Reinhold, New York City (1991).

Arquiaga, M.C., Canter, L.W., and Nelson, D.I.: "Integration of Health Impact Considerations in Environmental Impact Studies," *Impact Assessment* (1994) **12,** No. 2, 175.

Barnthouse, L.W. *et al.*: *User's Manual for Ecological Risk Assessment*, Publication No. 2679, Environmental Sciences Div., Oak Ridge Natl. Laboratory, Oak Ridge, Tennessee (1986).

Batt, S.C. and Peterson, P.J.: "Risk Assessment Techniques for Carcinogenic Chemicals," *Risk Assessment of Chemicals in the Environment*, Royal Soc. of Chemistry, London (1988).

Bierman, V.J. *et al.*: "Research Strategy for Ocean Disposal: Conceptual Framework and Case Study," *Environmental Hazard Assessment Of Effluents*, Pergamon Press, Elmsford, New York (1986).

Bull, R.J.: "Cancer Risk Assessment: Importance of Identifying Mechanisms of Action," *J. Amer. Water Works Assn.* (1990) **82,** No. 10, 31.

Burke, B.: "Some ABCs in Addressing Risk," *EPA J.* (1984) **10,** No. 10, 10.

Byrne, C.D.: "Selection of Substances Requiring Priority Action," *Risk Assessment of Chemicals in the Environment*, M.L. Richardson (ed.), Royal Soc. of Chemistry, London (1988) Chap. 23.

Calabrese, E.J. and Baldwin, L.A.: *Performing Ecological Risk Assessments*, Lewis Publishers, Boca Raton, Florida (1993).

Canter, L.W.: *Health Risk Impacts in Environmental Impact Statements*, Environmental and Groundwater Inst., U. of Oklahoma, Norman, Oklahoma (1990).

Canter, L.W.: "Pragmatic Suggestions for Incorporating Risk Assessment Principles in EIA Studies," *Environmental Professional* (1993) **15,** No. 1, 125.

Canter, L.W. and Canty, G.A.: "Impact Significance Determination—Basic Considerations and a Sequenced Approach," *Environmental Impact Assessment Rev.* (1993) **13,** No. 5, 54.

Canter, L.W., Nelson, D.I., and Everett, J.W.: "Public Perception of Water Quality Risks—Influencing Factors and Enhancement Opportunities," *J. Environmental Systems* (1993) **22,** No. 2, 163.

Carney, G.R.: "Computer Assisted Environmental Justice Index Methodology," *Environmental Justice, Human Health Risk Index (HRI),* Region 6 Comparative Risk Project, Office of Planning and Analysis, U.S. EPA Region 6, Dallas (1993) Chap. 4.

Carney, G.R.: "Overview Report: Region 6 Comparative Risk Project," Office of Planning and Analysis, U.S. EPA Region 6, Dallas (1990).

Carpenter, R.A. and Maragos, J.E.: "How to Assess Environmental Impacts on Tropical Islands and Coastal Areas," *Training Manual for South Pacific Regional Environmental Programme (SPREP)*, Asian Development Bank program, Environmental and Policy Inst., East–West Center, Honolulu, Hawaii (1989).

Carpenter, R.A.: "Communicating Environmental Science Uncertainties," *Environmental Professional* (1995) **17,** No. 2.

*Catalog of Digital Geographic Data, New Mexico Program*, Resource Geographic Information System, U. of New Mexico, Albuquerque, New Mexico (1995).

*Catalog of Publications, Office of Science and Technology*, EPA-820-B-94-002, U.S. EPA, Washington, DC (1994).

"Chemical Hazard Evaluation for Management Strategies: A Method for Ranking and Scoring Chemicals by Potential Human Health and Environmental Impacts," EPA/600/R-94/177, Risk Reduction Engineering Laboratory, Office of R&D, U.S. EPA, Cincinnati, Ohio; U. of Tennessee, Center for Clean Products and Clean Technologies, Knoxville, Tennessee (1994).

Chowdhury, A.K.M. and Canter, L.W.: *Expert Systems Related to Groundwater Protection and Management*, Environmental and Groundwater Inst., U. of Oklahoma, Norman, Oklahoma (1993).

Cirone, P. and Coco, M.: "The Lessons of Commencement Bay," *EPA J.* (1993) **19,** No. 1, 33.

Clay, D.R.: "In Toxics Control," *EPA J.* (1985) **11,** No. 5, 3.

Cleland-Hamnett, W.: "The Role of Comparative Risk Analysis," *EPA J.* (1993) **19,** No. 1, 18.

Cohrssen, J.J. and Covello, V.T.: "Risk Analysis: A Guide to Principles and Methods for Analyzing Health and Environmental Risks," NTIS PB 89-137772, U.S. Council on Environmental Quality, Washington, DC (1989).
"Comparative Risk Assessment: A Tool for Remedial Action Planning," *Risk/Decision Analysis*, Environmental Intl. Corp., Washington, DC (1983).
Conway, R.A.: "Introduction to Environmental Risk Analysis," *Environmental Risk Analysis for Chemicals*, Van Nostrand Reinhold Co., New York City (1982) Chap. 1.
Cothern, C.R. *et al.*: "Estimating Risk to Human Health," *Environmental Sci. & Tech.* (1986) **20,** No. 2, 111.
Cox, L.A.: "Risk, Uncertainty, and Causation: Quantifying Human Health Risks," *The Risk Assessment of Environmental and Human Health Hazards: A Textbook of Case Studies*, D.J. Paustenbach (ed.), John Wiley & Sons, New York City (1989).
Cox, L.A. and Ricci, P.F.: "Legal and Philosophical Aspects of Risk Analysis," *The Risk Assessment of Environmental and Human Health Hazards: A Textbook of Case Studies*, D.J. Paustenbach (ed.), John Wiley & Sons, New York City (1989).
Cutter, S.L.: *Living with Risk: The Geography of Technological Hazards*, Edward Arnold Publishers, London (1993).
Davies, J.C., Gusman, S., and Irwin, F.: "Determining Unreasonable Risk Under the Toxic Substances Control Act," issue report, Conservation Foundation, Washington, DC (1979).
Dellarco, V.L. and Kimmel, C.A.: "Update on Noncancer Assessments," *EPA J.* (1993) **19,** No. 1, 30.
"Development of Risk Assessment Methodology for Municipal Sludge Landfilling," EPA/600/6-90/008, U.S. EPA, Cincinnati, Ohio (1989).
*Digital Cartography/GIS Data Catalog*, Texas Natural Resources Information System, Austin, Texas (1995).
"Drinking Water, Public Health, and Wastewater Recycling," report, City of El Paso Public Service Board and the U.S. EPA Region 6, Lee Wilson & Assocs., Santa Fe, New Mexico (1980).
Durenberger, D.: "A Dissenting Voice," *EPA J.* (1991) **17,** No. 2, 49.
Durenburger, D. *et al.*: "Debating EPA's New Chemical Program: A Forum," *EPA J.* (1985) **11,** No. 5, 12.
Dwyer, J.P.: "Limits of Environmental Risk Assessment," *J. Energy Engineering* (1990) **116,** No. 3, 231.
"Ecological Assessment of Superfund Sites: An Overview, ECO Update," intermittent bulletin, PB92-963335, Office of Emergency and Remedial Response, U.S. EPA, Washington, DC (1991) **1,** No. 2.
Elkins, C.L.: "Right-To-Know [SARA Title III]: What It Means for EPA," *EPA J.* (1989) **15,** No. 3, 17.
Ellis, D.: *Environments at Risk: Case Histories of Impact Assessment*, Springer-Verlag, Berlin, Germany (1989).
"EMAP Project Descriptions," EPA/620/R-93/009, Environmental Monitoring and Assessment Program, U.S. EPA, American Management Systems, Arlington, Virginia (1993).
"Environmental Assessment, New Source NPDES General Permit, Pig Improvement Co. Inc., CAFO Operations, Kingfisher, Major and Woodward Counties, Oklahoma," review draft, U.S. EPA Region 6, Lee Wilson & Assocs. Inc., Santa Fe, New Mexico (1995).
"Environmental Equity, Reducing Risk for all Communities: Workgroup Report to Administrator," EPA 230-R-92-008, Environmental Equity Work Group, Office of Policy, Planning and Evaluation, U.S. EPA, Washington, DC (1992) **1.**
"Environmental Equity, Reducing Risk for all Communities: Supporting Document," EPA 230-R-92-008A, Environmental Equity Work Group, Office of Policy, Planning, and Evaluation, U.S. EPA, Washington, DC (1992) **2.**
*Environmental Impact Assessment*, Monitoring and Assessment Research Center, King's College, London (1987).
Erickson, P.A.: *Environmental Impact Assessment: Principles and Applications*, Academic Press, New York City (1979).
"Final Environmental Impact Statement, Cummins Creek Project, Fayette County, Texas," EPA 906/9-85-002, U.S. EPA Region 6, Dallas (1985).
Finkel, A.M. and Golding, D.: "Alternative Paradigms," *EPA J.* (1993) **19,** No. 1, 50.
Fischhoff, B., Solvicand, P., and Lichtenstein, S.: "Which Risks are Acceptable?," *Environment* (1979) **21,** No. 4, 17.
Fitchen, J.M. *et al.*: "Risk Perception in Community Context: A Case Study," *Social and Cultural Construction of Risk*, D. Reidel Publishing Co., Dordrecht, The Netherlands (1987).
Flagstad, K.: "Risk Management: FIFRA and the Dinoseb Case," *EPA J.* (1987) **13,** No. 9, 16.
Gilad, A.: "The Health Component of the Environmental Health Impact Assessment Process," *Perspectives on Environmental Impact Assessment*, D. Reidel Publishing Co., Dordrecht, The Netherlands (1984) 130.
Go, F.C.: "Environmental Impact Assessment, Monitoring and Assessment" Report 41, Research Centre, U. of London, London (1990).
Goldstein, B.D.: "Strengthening the Assessment of Risk," *EPA J.* (1984) **10,** No. 10, 5.
Gori, G.B.: "The Regulation of Carcinogenic Hazards," *Science* (1980) **208,** 256.
"Guidance Manual for Assessing Human Health Risks from Chemically Contaminated Fish and Shellfish," Draft Report C737-01, U.S. EPA, PTI Environmental Services Inc., Bellevue, Washington (1987).
"Guidance Manual for Risk Assessment, Contaminant Fate, and Transport Modeling," Report RG-91, Texas Natural Resource Conservation Commission, Woodward-Clyde Inc., Houston (1994).
"Guide for Risk Based Corrective Action Applied at Petroleum Release Sites—New Standard," draft document, Committee on Environmental Assessment, ASTM, West Conshohocken, Pennsylvania (19 September 1994).
Habicht, F.H.: "The Road to Innovation," *EPA J.* (1991) **17,** No. 2, 44.
Hallenbeck, W.H. and Cunningham, K.M.: *Quantitative Risk Assessment for Environmental and Occupational Health*, Lewis Publishers, Boca Raton, Florida (1986).
Hance, B.J. *et al.*: *Industry Risk Communication Manual: Improving Dialogue with Communities*, Lewis Publishers, Boca Raton, Florida (1990).
Harvey, P.D.: "Educated Guesses: Health Risk Assessment in Environmental Impact Statements," *Amer. J. Law & Medicine* (1990) **16,** No. 3, 399.
Hassan, H.M.: "Implementing Geographic Information Systems in Environmental Assessment, Update No. 9," *Environmental Assessment Sourcebook*, Environment Dept., World Bank, Washington, DC (1995).
"Health Guidelines for the Use of Wastewater in Agriculture and Aquaculture," technical report series, No. 778, World Health Organization (WHO), Geneva, Switzerland (1989).
*Improving Risk Communication*, Natl. Academy Press, Washington, DC (1989).
Inhaber, H.: *Energy Risk Assessment*, Gordon & Breach Science Publishers, New York City (1982).
"Joint WHO/FAO/UNEP Panel of Experts on Environmental Management for Vector Control," report of third meeting, WHO, Geneva, Switzerland (1983).
Jowa, L.: "The Role of Structure Activity Relationships in Risk Assessment," *The Analysis, Communication and Perception of Risk*, Plenum Press, New York City (1991).
Kates, R.W.: *Risk Assessment of Environmental Hazard: Scope 8*, John Wiley & Sons, New York City (1978).
Keoleian, G.A. *et al.*: "Product Life-Cycle Assessment to Reduce Health Risks and Environment Impacts," Noyes Data Corp., Park Ridge, New Jersey (1994).
Khoury, G.: "Risk-Based Concentrations Associated with 1E-06 Cancer Risk and HI = 1 for Noncancer Risk," draft report, Superfund Sec., U.S. EPA Region 6, Dallas (1992).
Kim, N.K. and Stone, D.W.: "Organic Chemicals and Drinking Water," New York State Dept. of Health, Albany, New York (undated).
Klapp, M.: "Risk Assessment, Risk Management and Informed Bargaining," *Environmental Impact Assessment Rev.* (1987) **7,** 23.
Kramer, K.A.: *Summary of Comparative Risk Projects, By State, City and Tribe*, Western Center for Comparative Risk, Boulder, Colorado (1995).
Lave, L.B.: "Health and Safety Risk Analyses: Information for Better Decisions," *Science* (April 1987) **236,** 291.
"LEAP to 2000," project report, Louisiana Dept. of Environmental Quality, Baton Rouge, Louisiana (1991).
Lewis, J.: "Reviewing New Chemicals," *EPA J.* (1985) **11,** No. 5, 6.
Lister, M.: "EPA Region 6 GIS User's Guide, Version 5.0," U.S. EPA Region 6, Dallas, Computer Data Systems Inc. (1994).
Loehr, R.: "What Raised the Issue [Of Risk]?," *EPA J.* (1991) **17,** No. 2, 6.
Macdonnell, M.M. *et al.*: "Strategy for Integrated CERCLA/NEPA Risk Assessments," Publication DE92-003379, Argonne Natl. Laboratory, Argonne, Illinois (1991).
"Making Decisions About Risk," *EPA J.* (1984) **10,** No. 10, 8.
Mccallum, D.B. and Covello, V.T.: "What the Public Thinks About Environmental Data," *EPA J.* (1989) **15,** No. 3, 22.
"Measuring for Environmental Results," *EPA J.* (1989) **15,** No. 3, 2.
Messer, J.J.: "Keeping a Closer Watch on Ecological Risks," *EPA J.* (1989) **15,** No. 3, 34.
Mhaisalkar, V.A.: *Environmental Risk Assessment and Management with Specific Reference to Industrial Hazards—A Review*, Environmental and Groundwater Inst., U. of Oklahoma, Norman, Oklahoma (1990).
Miller, I.C.: "Health Effects Assessment Summary Tables (HEAST), Annual Update, FY-1994," EPA 540-R-94-020, U.S. EPA Environmental Criteria and Assessment Office, Cincinnati, Ohio (1994).
Miller, T.A.W. and Keller, E.B.: "What the Public Thinks [About Risk]," *EPA J.* (1991) **17,** No. 2, 40.
Mittelman, A. *et al.*: "Methodology for Assessing Environmental Releases of and Exposure to Municipal Solid Waste Combustor Residuals," EPA/600/8-91/031, U.S. EPA, Washington, DC (1991).
Moore, J.A.: "Speaking of Data: The Alar Controversy," *EPA J.* (1989) **15,** No. 3, 5.

Morrison, A.: "Risk Analysis: Is EPA Changing the Rules?," *Civil Engineering* (October 1982) 60.

Moynihan, D.P. *et al.*: "The Debate: Two Forums—The Policy and the Science, Forum One: Should We Set Priorities Based on Risk Analysis?," *EPA J.* (1991) **17,** No. 2, 17.

Munns, W.R., Walker, H.A., and Paul, J.F.: "An Ecological Risk Assessment Framework for Examining the Effects of Oceanic Disposal," U.S. EPA/NTIS No. PB91-137000, SAIC Corp., Narragansett, Rhode Island (1990).

Murthy, K.S.: *National Environmental Policy Act (NEPA) Process*, CRC Press, Boca Raton, Florida (1988).

Nesnow, S.: "Breakthroughs in Cancer Risk Assessment," *EPA J.* (1993) **19,** No. 1, 27.

North, D.W. *et al.*: "The Debate: Two Forums—The Policy and the Science, Forum Two: Do We Know Enough to Take a Risk-Based Approach?," *EPA J.* (1991) **17,** No. 2, 31.

North, W. and Yosie, T.F.: "Risk Assessment: What It Is; How It Works," *EPA J.* (1987) **13,** No. 9, 13.

Okrent, D.: "Comment on Societal Risk," *Science* (April 1980) **208,** 372.

Olivieri, A.W.: "Risk-Benefit Analysis: On-Site Waste Treatment and Disposal Options," PhD dissertation, U. of California, Berkeley, California (1982).

O'Neill, R.V. *et al.*: "Ecosystem Risk Analysis," CONF-8109103-1, Oak Ridge Natl. Laboratory, Oak Ridge, Tennessee (1981).

Pandian, M.D.: "Evaluation of Existing Total Human Exposure Models," project summary, EPA/600/S4-87/044, Environmental Monitoring Systems Laboratory, U.S. EPA, Las Vegas, Nevada (1988).

*Pathogen Risk Assessment for Land Application of Municipal Sludge, Methodology and Computer Model*, EPA/600/6-90/002A, U.S. EPA, Cincinnati, Ohio (1989) **1.**

*Pathogen Risk Assessment for Land Application of Municipal Sludge, User's Manual*, EPA/600/6-90/002B, U.S. EPA, Cincinnati, Ohio (1989) **2.**

Patton, D.E.: "The ABC's of Risk Assessment," *EPA J.* (1993) **19,** No. 1, 10.

Petts, J. and Eduljee, G.: *Environmental Impact Assessment for Waste Treatment and Disposal Facilities*, John Wiley & Sons, Chichester, England (1994).

"Pollution Prevention/Environmental Impact Reduction Checklists for NEPA/309 Reviewers," report, Office of Federal Activities, U.S. EPA, Science Applications Intl. Corp., Falls Church, Virginia (1995).

Popkin, R.: "Keeping Tabs on Air Quality," *EPA J.* (1989) **15,** No. 3, 24.

"Prediction In EIA," report, The Netherlands Ministries of Public Housing and of Agriculture, Environmental Resources Ltd., London (1984).

"Preliminary Risk Assessment for Bacteria in Municipal Sewage Sludge Applied to Land," EPA/600/6-91/006, U.S. EPA, Cincinnati, Ohio (1991).

*Public Health Assessment Guidance Manual*, Agency for Toxic Substances and Disease Registry, U.S. Dept. of Health and Human Services, Atlanta, Georgia (1992).

Raider, R.: "A Solid Approach to Risk Assessment," *Water Environment & Technology* (October 1994) 40.

Reilly, W.K.: "Why I Propose a National Debate on Risk," *EPA J.* (1991) **17,** No. 2, 2.

"Report on the Workshop on Cancer Risk Assessment Guidelines Issues," EPA/630/R-94/005a, Risk Assessment Forum, U.S. EPA, Washington, DC (1994).

"A Review of Ecological Assessment Guidelines from a Risk Assessment Perspective," EPA/630/R-94/003, U.S. EPA, Washington, DC (1994) **2.**

*Risk Assessment Guidance for Superfund, Human Health Evaluation Manual—Risk Evaluation of Remedial Alternatives*, EPA/540/R-92/004, Office of R&D, U.S. EPA, Washington, DC (1991) **1,** Part C.

*Risk Assessment Guidance for Superfund, Human Health Evaluation Manual—Supplemental Guidance, Standard Default Exposure Factors*, Toxics Integration Branch, Office of Emergency and Remedial Response, U.S. EPA, Washington, DC (1991) **1.**

*Risk Assessment Guidelines and Information Directory*, Government Insts. Inc., Rockville, Maryland (1988).

*Risk Assessment in the Federal Government: Managing the Process*, Natl. Academy of Sciences Press, Washington, DC (1983).

"Risk Assessment, Management and Communication of Drinking Water Contamination," EPA/625/4-89/024, Office of Drinking Water, U.S. EPA, Washington, DC (1990).

"Risk Reduction Rules: Hazardous Waste Seminar for TSD Operators," Texas Natural Resource Conservation Commission, Houston, Texas (1994).

Roberts, L.: "Risk Assessors Taken to Task," *Science* (1990) **249,** 1173.

Rowe, M.D.: "National Energy Strategy: Recent Studies Comparing the Health Impacts of Energy Technologies," Brookhaven Natl. Laboratory, Upton, New York (1990).

Ruckelshaus, W.D.: "EPA's Strategy to Reduce Risks," *EPA J.* (1984) **10,** No. 10, 2.

Sample, B.E. and Suter, G.W.: "Estimating Exposure to Terrestrial Wildlife to Contaminants," ES/ER/TM-125, Office of Environmental Restoration and Waste Management, Environmental Sciences Div., U.S. Dept. of Energy, Oak Ridge Natl. Laboratory, Oak Ridge, Tennessee (1994).

Sandman, P.M.: "Risk Communication: Facing Public Outrage," *EPA J.* (1987) **13,** No. 9, 21.

Saunders, J.: "Region 6 Comparative Risk Project," ecological report, Office of Planning and Analysis, U.S. EPA Region 6, Dallas (1990) Appendix A.

Schaum, J.: *Exposure Factors Handbook*, project summary, EPA/600/S8-89/043, Office of Health and Environmental Assessment, U.S. EPA, Washington, DC (1990).

Scheuplein, R.J.: "Uncertainty and the 'Flavors' of Risk," *EPA J.* (1993) **19,** No. 1, 16.

*Science and Judgment in Risk Assessment*, Natl. Academy of Sciences Press, Washington, DC (1994).

"Setting Priorities, Getting Results: A New Direction for the EPA," report to U.S. Congress, Natl. Academy of Public Administration, Washington, DC (1995).

Shovlin, M.G. and Tanaka, S.S.: "Risk Communication in Los Angeles: A Case Study, Los Angeles Department of Water and Power," *J. American Water Works Assn.* (November 1990) 40.

Slovic, P., Fischhoff, B., and Lichtenstein, S.: "Rating the Risks," *Environment* (1979) **21,** No. 3, 14.

Smith, K.: *Environmental Hazards: Assessing Risk and Reducing Disaster*, Routledge, Chapman, & Hall, London (1992).

Smith, R.L.: "Risk-Based Concentration Table, January–June 1995," Technical Support Sec. (3HW13), U.S. EPA Region 3, Philadelphia, Pennsylvania (1995).

Snider, J. and Jover, T.: "On the Scene with CAMEO," *EPA J.* (1989) **15,** No. 3, 20.

Stone, R.: "California Report Sets Standard for Comparing Risks," *Science* (October 1994) **266,** 214.

"Summary Report on Issues in Ecological Risk Assessment," EPA/625/3-91/018, Risk Assessment Forum, U.S. EPA, Washington, DC (1991).

*Superfund Exposure Assessment Manual*, EPA/540/1-88/001, U.S. EPA, Washington, DC (1988).

Thomas, L.M.: "Environmental Decision-Making Today," *EPA J.* (1987) **13,** No. 9, 2.

Tolba, M.K. and El-Kholy, O.A.: *The World Environment 1972–1992*, Chapman & Hall, London (1992).

*Toward the 21st Century: Planning for the Protection of California's Environment*, California Comparative Risk Project, California Office of Environmental Health Hazard Assessment, Sacramento, California (1994).

"The Use of the Benchmark Dose Approach in Health Risk Assessment," EPA/630/R-94/007, Office of Risk Assessment Forum, U.S. EPA, Washington, DC (1995).

Varshney, L.: "Risk Assessment in Remedial Action Decision-Making for Hazardous Waste Sites," special report, Dept. of Civil Engineering, U. of Oklahoma, Norman, Oklahoma (1990).

Whelan, E.M., Roe, D., and Black, R.F.: "What Kind of Data Does the Public Need?," *EPA J.* (1989) **15,** No. 3, 10.

Wicke, H.D.: "Risk Assessment for Water Quality Management," PhD dissertation, U. of Michigan, Ann Arbor, Michigan (1983).

Williams, M.: "Regulating Toxic Substances: An Overview," *EPA J.* (1985) **11,** No. 5, 2.

Wilson, A.R.: *Environmental Risk: Identification and Management*, Lewis Publishers, Chelsea, Michigan (1991).

Wilson, R. and Crouch, E.A.C.: "Risk Assessment and Comparisons: An Introduction," *Science* (April 1987) **236,** 267.

Zamuda, C.: "The Nature and Extent of Ecological Risks at Superfund Sites and RCRA Facilities," EPA 230-03-89-043, U.S. EPA Office of Policy, Planning and Evaluation, Washington, DC (1989).

Zaporozec, A.: "Groundwater Protection Principles and Alternatives for Rock County, Wisconsin," special report 8, Wisconsin Geological & Natural History Survey, Madison, Wisconsin (1985).

Zeckhauser, R.J. and Viscusi, W.K.: "Risk within Reason," *Science* (May 1990) 559.

## SI Metric Conversion Factors

| | | |
|---|---|---|
| curie | × 3.7* | E + 10 = Bq |
| ft | × 3.048* | E − 01 = m |
| $ft^2$ | × 9.290 304* | E − 02 = $m^2$ |
| $ft^3$ | × 2.831 685 | E − 02 = $m^3$ |
| in. | × 2.54* | E + 00 = cm |
| $in.^2$ | × 6.451 6* | E + 00 = $cm^2$ |
| $in.^3$ | × 1.638 706 | E + 01 = $cm^3$ |
| lbm | × 4.535 924 | E − 01 = kg |

*Conversion factor is exact.

# Chapter 8
# Oilfield-Pit Management, Closure, and Remediation

**Andrew Wojtanowicz,** Louisiana State U.

## 8.1 Introduction

The purpose of this chapter is to provide the reader with an understanding of the depth of knowledge required for remediation, focusing on oilfield pits. The literature should be consulted for specific types of remediation efforts. Waste material that has been discharged from drilling or production operations typically is stored on site in earthen pits. Oilfield pits usually are excavated directly adjacent to the site of operation so that they can be used for final disposal of all or part of wastes or for temporary storage before off-site disposal. This chapter should be read in conjunction with Chap. 2.

## 8.2 Types of Oilfield Pits

On-site pits include reserve pits for drilling operations, workover pits, and production pits. Production pits are classified according to the type of waste (such as produced-water and basic-sediment pits) or function (such as blowdown/emergency, skimming, percolation, and evaporation pits).

Reserve pits are an essential design component in the great majority of well drilling operations and are used to accumulate, store, and, to a large extent, dispose of spent drilling fluids, cuttings, and associated drillsite wastes generated during drilling and completion operations. In the past, reserve pits were used both to remove and dispose of drilled solids and cuttings and to store the active mud system for later recycling in the well being drilled. As more advanced solids-control and drilling-fluid technology has become available, mud tanks have replaced reserve pits as the storage and processing area for the active mud system and the reserve pit is used to dispose of waste mud and cuttings. Reserve pits, however, continue to be the principal method for storage and management of waste drilling fluids.

Workover pits used to contain workover fluids typically are open only for the duration of workover operations. Siting of these pits is based on whether usable groundwater and surface waters are present. If site-specific considerations indicate that these waters could be endangered, workover pits are lined or replaced with tanks. Typically, the toxicity of workover drilling fluids is higher than that for drilling fluids. Workover fluids may contain total dissolved solids (TDS) in excess of 3,000 ppm (approximately 4-m℧/cm conductivity) in addition to hydrocarbons or potentially toxic additives or compounds. [The mho (℧), the unit of measurement for conductivity, is the inverse of the ohm, the unit commonly used to express resistance. Values of specific conductance for groundwater are expressed in millionths of mhos.]

Produced-water (collecting) pits are used to store produced water before sea disposal at a coastal (tidal) disposal facility or to store produced water or other oil and gas (O&G) wastes before disposal at a fluid-injection well. In some cases, the same pit is both a collecting and a skimming pit. Produced-water pits also have been used in lieu of tankage. Produced-water pits should be lined and operated only as a substitute for water-treatment process vessels before the surface-water discharge or injection-well-regulated disposal of the produced water. These pits should not be used for permanent water disposal.

Basic sediment pits, also called burn pits, are used in conjunction with a tank battery to store basic sediment removed from a production vessel or from the bottom of an oil-storage tank. Unlined burn pits were used in the past for temporary storage of oily wastes, such as paraffin and vessel bottoms. Currently, basic sediment pits are either lined or replaced with tankage. Their contents are periodically cleaned out or burned.

Blowdown/emergency pits are used to store produced water for limited periods of time. Use of these pits is necessitated by a temporary shutdown of a disposal or fluid-injection well and/or associated equipment; by temporary overflow of produced-water storage tanks on a producing lease; or by the loading of a producing well with formation fluids, which possibly leads to the death of the well. Currently, blowdown/emergency pits (flare, pressure-vessel-relief, and fluid-overflow pits) are not used for storage or disposal. Fluids diverted to emergency pits are removed as quickly as is practical and in accordance with local regulations.

Skimming pits are used to skim oil off produced water before disposal of the water at a tidal disposal facility, disposal well, or fluid-injection well. Also, unlined skimming/settling pits are used in association with discharges to provide additional retention time, which allows more complete residual oil separation. Unlined skimming/settling pits are used only with permitted discharges.

Percolation pits allow liquid contents to drain or seep through the bottom and sides of the pit into surrounding soils. Percolation pits are used for disposal of produced waters only where permitted by regulatory agencies and only where underground freshwater aquifers are not present or endangered. Percolation pits are unlined.

Evaporation pits, defined as surface impoundments that are lined with clay or synthetics, are used in areas where small volumes of waste waters are generated. Disposal of waste water by evaporation concentrates salts and residual hydrocarbons.

TABLE 8.1—pH VALUES OF OILFIELD WASTE STORED ON SITE[2]

| Waste | Sample Size | Average Value* | Minimum Value | Maximum Value | Standard Deviation |
|---|---|---|---|---|---|
| Produced brine (pit or tank) | 17 | 6.8 to 6.8 | 3.0 to 3.4 | 8.1 to 8.7 | 1.1 to 1.2 |
| Pit liquids | 15 | 8.2 to 8.4 | 6.4 to 6.5 | 12.7 to 13.0 | 1.8 to 1.9 |
| Pit solids | 19 | 9.5 to 10.2 | 6.9 to 7.9 | 12.3 to 12.8 | 1.2 to 1.7 |

*Two values of each parameter show discrepancies between the independent assessments by the U.S. Environmental Protection Agency (EPA) and American Petroleum Inst. (API).[2]

TABLE 8.2—O&G DISTRIBUTION IN OILFIELD PITS[4]

| O&G Concentration Range (wt%) | Frequency Distribution | Percent of Total | Cumulative Percent |
|---|---|---|---|
| <1.0 | 75 | 92.6 | 92.6 |
| 1.0 to 2.9 | 4 | 4.9 | 97.5 |
| 3.0 to 5.9 | 2 | 2.5 | 100.0 |
| Total | 81 | 100 | — |

### 8.3 Pit Pollutants and Environmental Risk

Common types of pollutants found in oilfield pits include high-/low-pH contaminants, salts, organics (O&G), and dissolved heavy metals. High-pH contamination is a result of the standard practice of adding caustic (NaOH) to drilling fluids and generating excessive volumes of cement slurry from well completions. Sources of low-pH contamination are spent acidic fluids in workover pits. Sources of soluble-salt contamination include formation waters, downhole salt layers, and chemical additives. Sources of organic contamination include lubricating oil from equipment leaks; well-pressure-testing fluids; oil-based spotting fluids; and, in some cases, oil-based mud cuttings. Sources of potential heavy-metal contamination include drilling-fluid additives, drilled solids, weighting materials, pipe dope, and spilled chemicals.

Low-pH contamination (acidity) elevates the activity of metals in the soil solution by both increasing solubility and decreasing surface adsorption. The most important aspect of metals activity is reduction of soil fertility and disturbances in plant metabolism. Surface reactions involve changes in soil clay-mineral compositions, which lower the cation-exchange capacity (CEC). High-pH contamination, on the other hand, alters soil alkalinity, reducing micronutrients and causing loss of nitrogen. The suggested pH range (loading factor) is 6 to 9 for wetlands and 6 to 8 for uplands.[1]

**Table 8.1** shows the results of a 1987 survey of oilfield pits in the U.S. that revealed that the average pH for brines collected from produced-water pits was nearly neutral (pH of 6.8). This value resulted primarily from high concentrations of sodium chloride (NaCl) in the brines. Some brines were slightly acidic (pH of 3.0 to 3.4) as a result of oxidation of reduced metallic species during brine storage in pits. Slurries from other oilfield pits surveyed had pH values ranging from neutral to extremely alkaline (12.8 and 13.0 for solids and liquids, respectively). Because these pits initially were constructed as drilling-mud reserve pits and converted later to production pits, most of their contents comprised waste drilling fluids, which are basic in nature.

The presence of organics in soil, measured as O&G, may severely limit revegetation efforts after oilfield-pit closure. Usually, revegetation should be accomplished in one season. It has been established that, for most soils, an O&G concentration of 1% is an acceptable maximum.[3] **Table 8.2** shows statistical data on O&G concentration in oilfield pits closed in Louisiana and Mississippi during 1984–85 by Shell Western E&P Inc. The data show that 92.6% of the pits had organic concentrations below the soil-loading level. The remaining 7.4% pits required some dilution mixing of the waste with soil to reduce the O&G concentration to 1 wt%.

The salinity hazard of oilfield-pit waste has been assessed on the basis of several parameters of the solid and liquid phases of soils. These parameters include electrical conductivity (EC), sodium adsorption ratio (SAR), CEC, and exchangeable sodium percentage (ESP). Typically, an EC value greater than 4 m℧/cm, which corresponds to a TDS content of 3,000 ppm, is considered a maximum loading factor for soils in the uplands and freshwater wetlands.[5] For brackish wetlands, no salinity hazard exists because the wetland system has a natural influx of salt caused by tidal activity. SAR can be related empirically to poor physical soil characteristics and plant toxicities and is defined as the ratio of soluble sodium to calcium and magnesium in the soil water. The remaining two parameters, CEC and ESP, characterize cation distribution in the solid phase of soils. CEC is the sum total of exchangeable cations that a soil can absorb; ESP is the ratio of exchangeable sodium to CEC. When the ESP is greater than 15, the soil structure generally lacks stability and water infiltration is impeded. The result is slow percolation of the water necessary to move excess salts out of the plant-root zone.

Unlike pH, organics, and salts, the environmental risk of metals contained in oilfield pits has been evaluated taking human-health criteria into consideration. This health-based approach and method were developed initially for municipal sewage sludges[6] and applied by analogy to oilfield pits.[7] The presence of metals in soils after oilfield-pit closure constitutes a potential health hazard to higher-risk human populations, such as highly exposed or most exposed (the worst-case scenario) individuals. Several possible pathways of exposure (from soil to an individual) are considered for each metal. Then, a maximum soil concentration (soil-loading factor) is calculated by use of a so-called soil-ingestion rate (i.e., estimated amount of soil ingested by an individual per day). Of 14 possible exposure pathways for sewage sludge, four have been identified as most likely to apply to oilfield pits (**Table 8.3**).

**Table 8.4** lists maximum loading factors for 12 metals of concern in soils associated with oilfield pits. The table also shows the most likely exposure pathway for each metal and its maximum concentration detected in oilfield waste. The values of maximum soil loading for Exposure Pathways 2 through 4 have been calculated assuming a loaded depth of topsoil of 6 in. from the surface, which is an average for disking. For Exposure Pathway 1, the risk is associated with the ingestion of whole, undiluted waste in the oilfield pit.

Table 8.4 also compares soil-loading factors recommended by the API guidelines[5] with those from Louisiana Statewide Order 29-B[3] and Canadian Interim Soil Remediation Criteria.[8] The Louisiana 29-B criteria were developed primarily from early works on metals

TABLE 8.3—EXPOSURE PATHWAYS FOR OILFIELD PITS IDENTIFIED BY API[1]

| Pathway | Description of Oilfield-Pit Sludge | Description of Highly Exposed Individual |
|---|---|---|
| 1 | Human | Children ingesting sewage sludge |
| 2 | Soil/plant/animal | Livestock ingesting crops grown on sewage-sludge-amended soil |
| 3 | Soil/plant | Plants grown in sewage-sludge-amended soil |
| 4 | Soil/soil organism/soil-organism predator | Animals eating soil organisms living in sewage-sludge-amended soil. |

**TABLE 8.4—MAXIMUM SOIL LOADING FOR OILFIELD-PIT METALS[1]**

| | Exposure Pathway | API Guidance[5] | Louisiana 29-B* | Canadian Agriculture[8] | Maximum Concentrations Detected** |
|---|---|---|---|---|---|
| Arsenic | 1 | 41 | 10 | 20 | 29/27.9/140 |
| Barium* | 1 | 180,000 | 20,000<br>40,000<br>100,000 | 750 | 56,200<br>24,500<br>10,700 |
| Boron | 3 | 2 mg/L | — | 2 mg/L | 290/73.6 |
| Cadmium | 4 | 26 | 10 | 3 | 14/1.5/3 |
| Chromium | 3 | 1,500 | 500 | 750 | 368/145/54 |
| Copper | 3 | 750 | — | 150 | 82/124/210 |
| Lead | 1 | 300 | 500 | 375 | 446/302/970 |
| Mercury | 1 | 17 | 10 | 0.8 | 2.1/1.1/1.4 |
| Molybdenum | 2 | — | — | 5 | 16/9 |
| Nickel | 3 | 210 | — | 150 | 61/40.6/100 |
| Selenium | 1 | — | 10 | 2 | 3/0.6/1.4 |
| Zinc | 3 | 1,400 | 500 | 600 | 823/413/400 |

All concentrations are in mg/kg unless specified.
*Louisiana 29-B barium values for wetlands, uplands, and commercial land farming facilities, respectively.[3]
**Independent evaluations by API in 1987 and 1995.[2,7]

in sewage sludge (before 1980) and were confirmed to be protective in field studies. These early works were later superseded by the research supporting the API guidelines. For example, the Louisiana 29-B criteria for barium were based on studies that used barium analysis inconsistent with that used for the API guidelines. The Canadian Agriculture values for maximum loading have been adopted by the Canadian Council of Ministers for the Environment (CCME) from values that were currently in use in various jurisdictions across Canada in response to its urgent need to begin remediation of high-priority "orphan" contaminated sites. Many of these criteria do not have complete scientific rationale and, therefore, are considered to be only interim. On the basis of experience and professional judgment, the CCME considered the criteria to be generally protective of human and environmental health for agricultural uses of soil. On the other hand, the API guidance criteria, which provide less conservative guidelines than those proposed by CCME, resulted from a quantitative risk assessment in combination with the best available data.

### 8.4 Pit Construction

Major environmentally related changes in oilfield-pit design and construction involve reduction of pit area; increase of pit depth; and replacement of a single, multipurpose pit with an array of small, specialized pits. Also, the cleared area for oilfield locations has become smaller and more compact. **Fig. 8.1,** which shows a sized floor plan of a modern drilling location, depicts these changes. Note that the drillsite is almost a square (400 × 430 ft) and has an area of 3.95 acres. The drillsite is hydraulically isolated from surrounding land by a levee and drainage ditch. The site is not leveled but has a two-sided slope of 1 to 2° so that the runoff waste water can be collected quickly in the ditch. In addition, the drillsite is divided by a levee and ditch into management/storage, drilling-pad, and earthen-pit sections. Four specialized pits surrounded by pit berms are situated in a row along one side of the drilling location. Also, adjacent to the "cuttings pit" is a 100 × 100-ft cuttings-processing area designated for land treatment of the drill cuttings.

Oilfield locations may be located in a variety of environments and land configurations, so their layout must be site-specific. The following are the general rules for oilfield-pit construction.[3]

1. Pits should be protected from surface waters by levees or walls and by drainage ditches (where needed), and no siphons or openings should be placed in or over levees or walls that would permit the contents from escaping and causing pollution or contamination.

2. Pits should be sized to ensure adequate storage during oilfield operations, with historical rainfall patterns taken into account. Liquid levels should not be permitted to rise to within 2 ft of the top of the pit levees or walls. Pit levees or walls should be maintained at all times to prevent deterioration, subsequent overfill, and leakage of waste to the environment.

3. Pits should be constructed so that the pit bottom does not penetrate usable groundwater or, if in wetland areas, pit contents do not impact surface waters significantly.

4. Pits should be lined if the drilling fluids are saltwater or oil-based muds. Liners should be made of natural clay or synthetic material that will withstand normal operating conditions. Liners should protect the sides and bottom of the pit and provide a sufficient barrier to flow for the period during which the pit is in use. For complete isolation, liners should exhibit a hydraulic conductivity of less than $10^{-7}$ cm/sec. (In permafrost areas, lining of only dike walls is required because permafrost provides an effective permanent barrier to downward movement of fluid.)

5. Completion fluids containing TDS concentrations greater than 3,000 ppm (approximately 4-m℧/cm conductivity), produced water, or liquid hydrocarbons should not be stored in unlined pits where groundwater requires protection.

6. Reserve pits should be strategically sited on drilling locations to collect rig-washdown water; spills and leaks from drilling equipment; and, where practical, storm-water runoff from the drilling pad. The pad should be constructed with runon/-off controls to minimize entry of nondrilling-pad storm water into the reserve pit.

7. Separate pits should be designated to store different waste materials, particularly those with drastically different toxicity levels. For example, drill cuttings from oil-based muds should not be stored in a base-mud reserve pit.

In the U.S., where some oilfield waste is exempted from hazardous-waste provisions, Principle 7 is based on a so-called "mixing rule." This rule states that, when small amounts of hazardous waste get mixed with large volumes of exempted oilfield waste, the whole mixture becomes a hazardous waste and, hence, becomes extremely costly to dispose of.

### 8.5 Pit Assessment Before Closure

The most undesirable and expensive method for oilfield-pit closure is off-site transportation and disposal of waste from the pits. For pit liquids, the alternatives to off-site disposal are on-site pit-closure methods, such as land spreading, discharging to surface waters under a permit, evaporation, annular injection, or injection into an injection well. For pit solids, the alternatives are land treatment (land spreading), burial, and encapsulation and burial.

The decision regarding which method of pit closure should be used is based on sampling and testing of the pit contents and the

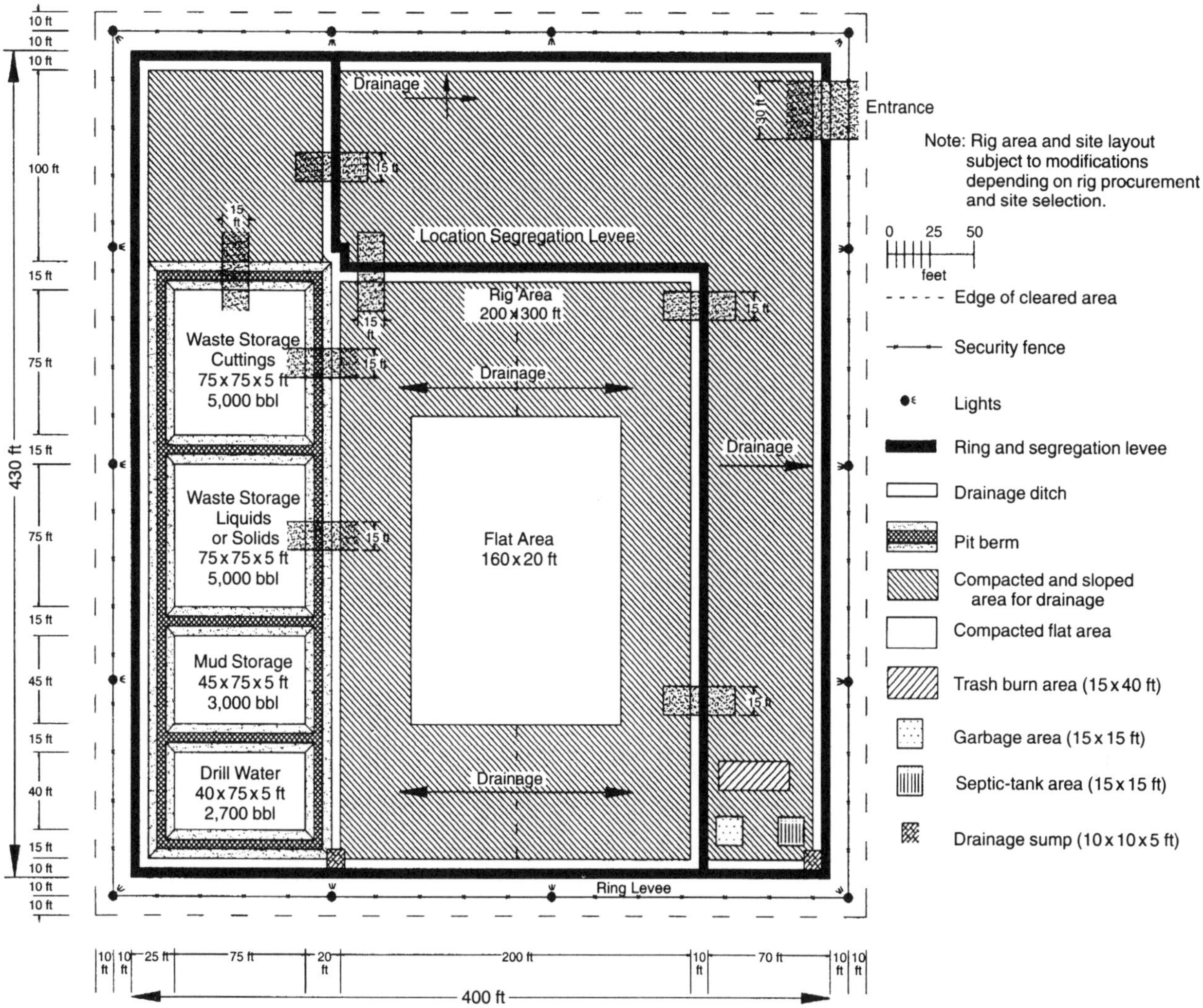


**Fig. 8.1—Layout of environmentally controlled drillsite.**

background soils. Samples of background soils should be collected from the area designated for potential land treatment. The area should be well drained and out of flood plains and wetlands. Additional criteria for land-treatment sites may be enforced by regulations. Background-soil samples should be collected from the "A" soil horizon, or upper 1 ft, and should be composited from a number of nearby locations. Refs. 5 and 9 provide details for designing and executing a soil-sampling plan.

Oilfield-pit samples must represent fully the concentration of pollutants in the pit waste material. Ref. 10 provides methods for representative sampling that use grid networks and composite samples. For example, sampling can be performed at the 50 × 50-ft-grid level with subsamples collected over 2-ft intervals and the lowermost sample taken below the waste bottom. Then, at each sampling point (not necessarily a gridpoint), the subsamples are combined into a single composite for that point. Detailed testing procedures have been developed for environmental analysis of oilfield waste.[3] Particularly important in these procedures are measurement of true total barium[11] and hot-water-soluble boron.[12]

Optimization of the sampling plan is an important issue because, theoretically, the cost of taking and analyzing samples at each gridpoint multiplied by the number of gridpoints is prohibitive. Usually, the number of sampling points can be much smaller than the number of gridpoints. However, sufficient samples should be collected for the analysis to describe pit content statistically with a high degree of confidence. The optimal number of samples depends on both pit size and the variability of pollutants. As an example, **Table 8.5** shows the variability of the concentration of four metals in oilfield reserve pits. The data indicate that metals are not uniformly mixed into the pit slurry; the coefficient of variation ranges from 11 to 54%.

An analytical method for determining the minimum required number of pit samples was developed by use of the variability of metals from the four oilfield reserve pits in Table 8.5. The method provides procedures for two types of sampling problems.

1. Pit assessment. Determine the minimum number of samples necessary to make sure that land treatment is necessary (i.e., at least one pollutant in the pit significantly exceeds the maximum permitted concentration).

2. Land-treatment design. Specify sampling requirements to ensure that pollutant concentrations in the waste-amended soil zone will not exceed loading factors.

**TABLE 8.5—VARIATION COEFFICIENT* FOR METALS IN OILFIELD RESERVE PITS[13]**

| | Pit | | | |
|---|---|---|---|---|
| | 1 | 2 | 3 | 4 |
| Barium | 0.23 | 0.26 | 0.25 | 0.47 |
| Chromium | 0.41 | 0.14 | 1.32 | 0.30 |
| Lead | 0.38 | 0.46 | 0.54 | 0.11 |
| Zinc | 0.16 | 0.31 | 0.47 | 0.16 |

*Variation coefficient is a ratio of standard deviation to the mean value.

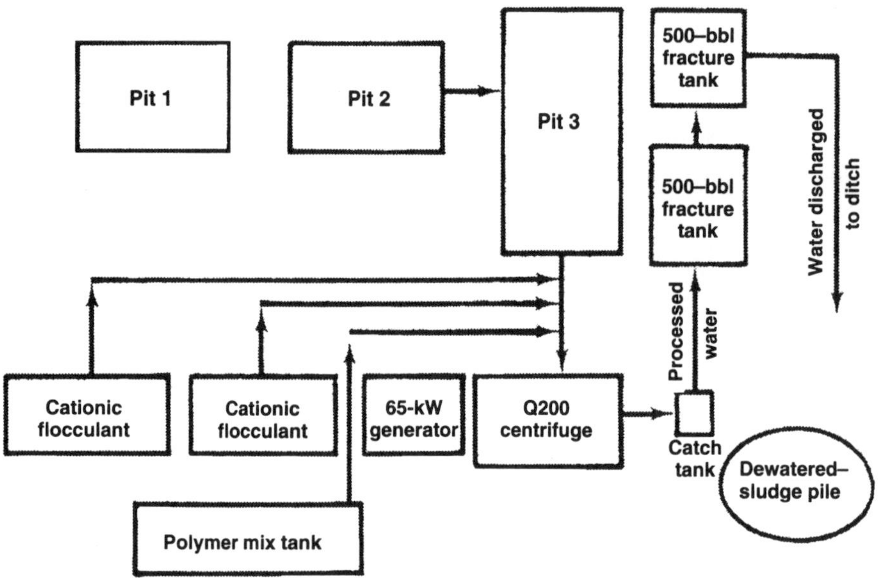


**Fig. 8.2—Layout of portable process for oilfield-pit dewatering.**

## 8.6 Pit Closure

**8.6.1 Liquid Phase.** Oilfield pits are closed by segregating the liquid phase from the solid phase and disposing of each phase separately. The liquid phase can be broadly defined as an aqueous layer that usually contains some suspended solids and is situated above settled solids. The solid phase comprises the settled solids and significant amounts of liquids remaining in the pit after pumping the liquid phase out. Usually, pumping continues until the remaining mixture becomes nonpumpable.

Three options for on-site disposal of the liquid phase are disposal to surface waters, land spreading, or subsurface injection (annular injection or injection well). Disposal to surface waters requires dewatering the oilfield pit. The dewatering process can be accomplished in situ by either evaporation or chemical flocculation followed by gravity settling inside the pit or on site by use of a portable process of chemically enhanced mechanical expression. Evaporation methods are not discussed in detail here because they do not truly constitute treatment and are, in most cases, of limited acceptability. In many areas, high precipitation and low evaporation rates preclude this option. Another problem is a regulatory time limit that dictates how long a pit may be left untreated, which could result in migration of contaminants. Methods have been applied to improve evaporation rates, such as spray-nozzle atomizers.

One method of in-situ pit dewatering with chemical treatment is use of alum (aluminum sulfate) and polymers to flocculate and remove suspended solids. In this method, liquid waste is pumped from the pit through a hopper, where the flocculating agents are added. The slurry is then either discharged to a second pit or returned to the same pit for gravity settling. The separated water can be discharged to the environment in some instances or recycled for rig washing and drilling-mud mixing. The reported cost of this method is U.S. $0.30/bbl, with 60,000 to 80,000 bbl treated within a few days.[14]

The portable process for pit dewatering includes three stages: chemical treatment, mechanical expression, and water treatment. **Fig. 8.2** shows an example layout of the portable dewatering process.[15] First, the pit slurry is homogenized and pumped to the chemical conditioning unit for neutralization, coagulation, and flocculation. Neutralization is necessary to remove organics, heavy metals, and other contaminants. The pH is reduced to precipitate organics, while reactive solids are neutralized electrochemically and begin to coagulate. Further addition of flocculant converts solids into larger particles (flocs) that settle quickly.

Flocculated slurry is then processed by a solid/liquid separator that expresses the free water from the spongy mass of the flocculated solids. Typically, a decanting centrifuge is used for expression. Other separators, such as recess plate filter presses, static clarifiers, screw presses, and belt presses, also have been used for expression.[16, 17]

The objective of the third stage, water treatment, is to reduce the concentration of the remaining contaminants below the permitted levels for surface discharge. **Table 8.6** shows an example requirement for disposal of oilfield-pit liquids to surface waters. In a simple case, this stage (water polishing) may involve increasing pH to the neutral range to precipitate metals followed by addition of a low-molecular-weight polymer to remove residual suspended solids. (In Fig. 8.2, the polishing is performed in the first fracture tank.) In a complex case, polishing the centrifuge centrate may include adjustment of pH followed by coagulation and electroflotation.[18]

If the liquid phase cannot meet the requirements for surface-water disposal, the only two options for disposal are subsurface injection or land spreading. Most typically, a decision in this case is based solely on the EC of pit liquids.[19] For an EC greater than 4 m℧/cm, liquids should be injected underground. Note that this EC threshold value applies mostly to agricultural soil. This criterion may be relaxed if the native soils or freshwater wetlands are of poorer quality than agricultural land. Also, this threshold value does not apply to brackish wetlands, where there is no salinity hazard. **Table 8.7** summarizes disposal criteria for the pit liquid phase.

The design of land spreading of pit liquids requires calculation of the minimum land area needed for liquid application to reduce values of constituents to less than those in Table 8.7. One criterion is water-infiltration rate. For native soils, water-infiltration rates decrease with increasing water saturation in the soil. Water-infiltration rates are used to determine the minimum required land-spreading area that would not cause liquid-phase runoff during the land-spreading operation. Another design criterion relates the minimum land area to the required ESP value of 12% after the pit liquid phase infiltrates the soil. The usual assumption here is that the depth of infiltration is 6 in. Calculations of the minimum area of land for land spreading of pit liquids require knowledge of the sodium content in the pit liquid phase together with measurements of bulk density, moisture content, CEC, ESP, and liquid-runoff limit of the background soil.

**8.6.2 Solid Phase.** Land-treatment technology provides consistent and proven methods for rendering waste materials harmless by soil incorporation. The technology uses dilution, chemical-alteration, and biodegradation mechanisms to reduce concentrations of pollutants to acceptable levels consistent with intended land use. The technology has been used for many years and has proved to be a viable waste-disposal technique for oilfield-pit solids. Land treatment of pit solids can be performed with land spreading, burial (trenching or landfill), or solidification and burial.

**TABLE 8.6—EFFLUENT LIMITATIONS (MAXIMUM ALLOWABLE CONCENTRATION) FOR RESERVE-PIT-WATER DISCHARGE FOR U.S. GULF COAST STATES**

| Analysis | Texas | Louisiana | Mississippi |
|---|---|---|---|
| pH | 6 to 9 | 6 to 9 | 6 to 9 |
| O&G, mg/L | 15.0 | 15.0 | — |
| Chloride, mg/L | | | |
| Inland | 500 | 500 | 500 |
| Coast | 1,000 | | |
| EC, m℧/cm | | | 1,000 |
| Total solids, mg/L | — | — | — |
| Total suspended solids, mg/L | 50.0 | 50.0 | 100 |
| TDS, mg/L | 3,000 | — | — |
| Chemical oxygen demand, mg/L | 200.0 | 125.0 | 250 |
| Total organic carbon, mg/L | — | — | — |
| Metals, mg/L | | | |
| Arsenic | 0.1 | — | — |
| Barium | 1.0 | — | — |
| Cadmium | 0.05 | — | — |
| Chromium | 0.5 | 0.5 | 0.5 |
| Copper | 0.5 | — | — |
| Iron | — | — | — |
| Lead | 0.5 | — | — |
| Mercury | 0.005 | — | — |
| Nickel | 1.0 | — | — |
| Selenium | 0.05 | — | — |
| Zinc | 1.0 | 5.0 | 5.0 |
| Phenol, ppm | | | 0.1 |

**TABLE 8.7—ON-SITE DISPOSAL OF OILFIELD LIQUID PHASE[19]**

| | Criteria | | | |
|---|---|---|---|---|
| Disposal Option | EC (m℧/cm) | SAR (ratio) | ESP (ratio) | O&G (wt%) |
| Subsurface injection | ≧4 | — | — | — |
| Land spreading | <4 | ≦12 | ≦15 | ≦1 |

SAR, ESP, and O&G are for liquid phase only

Solids contaminated with petroleum hydrocarbons are best managed on site by the land-spreading technique. In this technique, pit-waste solids are added to the receiving soil, then disked to an appropriate depth so that the final waste/soil mixture meets the constituent threshold criteria.

The most limiting constituent for land spreading is salt (NaCl). Organics degrade, but salts require leaching by rainfall to move them out of the intended root zone. The sodicity property, measured as SAR for pore liquids and ESP for solids, is a major concern but can be managed easily by calcium treatment (i.e., additions of gypsum) if the total salt is kept in control. The calcium-amendment treatment requires significant soil moisture for cation exchanges to occur and displace desorbed sodium ions with calcium ions. Therefore, land spreading is best suited for more humid, warmer climates (precipitation >20 in./yr).

In submerged wetlands, the pit area will be inundated with water after the pit is closed. This type of closure design is more difficult to complete because the dilution of pit wastes with surface soils is limited by the fact that the only surface soil available for mixing is the pit-levee material. Consequently, reduction of pollutant concentrations to the maximum loading values through surface-soil dilution is limited.

In elevated wetlands, the pit area will be inundated after closure but some adjacent soil is available for mixing. Reduction of pollutant concentrations can be achieved in most cases through surface-soil dilution. The waste fluids can be mixed thoroughly with pit levees and adjacent soil at the well location. However, when an insufficient area of surface soil is available for dilution, concentrations of pollutants can exceed maximum loading values after closure. In such a case, some of the pit solids must be transported off site for disposal.

In upland areas, constituents can be reduced easily with surface-soil mixing. However, land spreading of pit solids may be difficult to do in upland areas because landowners are reluctant to give permission to spread wastes on adjacent property. Burial of pit solids in trenches may be the only on-site disposal option available.

In the burial technique, a layer of pit solids mixed with background soil is placed either at the pit bottom (landfill) or inside a trench. Then, the solid layer is covered with at least 5 ft of soil to protect the mixture from rapid leaching by rainfall waters percolating through the ground. A further requirement is that the bottom of the buried mixture be at least 5 ft above the average groundwater table in the area. Burial or landfill techniques are best suited for semiarid (rainfall of <20 in./yr) climates with no potential for leaching to the subsurface. The recommended criteria could be relaxed in semiarid regions after evaluation of the site for any potential environmental impact.

The technique of solidification and burial involves mixing solidifying agents, such as commercial cement, flash, and lime kiln dust with pit sediments to produce a relatively insoluble concrete matrix. Then, the solidified concrete is buried in the pit by use of the levee material or in trenches by use of a protective liner. Solidification can be accomplished in two ways: (1) by injecting a solidifying agent into the reserve pit, followed by burying the pit contents with the levee material, or (2) by mixing the pit sediment with a solidifying agent in a mixer and pumping the slurry into trenches (topsoil removed). Then, the slurry is cured for several weeks and the topsoil is restored.

The oldest and cheapest technique for pit closure is backfilling. This technique involves removing pit liquids by pumping or breaking the pit berm and spreading the fluids over adjacent land. Then, the pit berms are thoroughly mixed with pit solids inside the pit. In elevated wetland or upland areas, the pit mixture is treated with 50

**TABLE 8.8—LIMITS FOR OILFIELD-PIT CLOSURE AND ON-SITE DISPOSAL**

| | Land Treatment for Waste/Soil Mixtures | | | |
|---|---|---|---|---|
| | Uplands | | Freshwater Wetlands | |
| Waste-Material Parameters | Limit | Test Procedure | Limit | Test Procedure |
| pH | 6 to 9 | | 6 to 9 | |
| EC, m℧/cm solid phase | <8 | 1 | <4 | 1 |
| SAR, ratio in solution phase | <14 | 1 | <12 | 1 |
| ESP, % solid phase | <25 | 1 | <15 | |
| CEC, meq/100 g soil | | 5 | | 5 |
| O&G, % dry weight | <1 | 1 | <1 | 1 |
| Metals, ppm | | | | |
| Arsenic | <10 | 1 | <10 | 1 |
| Barium | | | | |
| Elevated wetlands | | | <20,000 | 2 |
| Uplands | <40,000 | 2 | | |
| Cadmium | | 1 | | 1 |
| Chromium | <10 | 1 | <10 | |
| Lead | <500 | 1 | <500 | 1 |
| Mercury | | 1 | | 1 |
| Selenium | <500 | 1 | <500 | |
| Silver | <10 | 1 | <10 | 1 |
| Zinc | <10 | 1 | <10 | |
| | <200 | | <200 | |
| | <500 | | <500 | |
| Radioisotopes in coastal areas after 20 October 1990 | | 5 | | 5 |

Testing procedure
1. Standard soil testing.
2. True total testing.
3. Leachate testing.
4. Testing must be done according to American Soc. for Testing and Materials (ASTM).
5. Mentioned as parameters to analyze for but no limitations are given.

lbm/acre of balanced fertilizer (10-10-10 or 12-12-12) to complete closure. Finally, the backfilled area is graded to natural elevation.

The backfilling technique can be used safely when pit solids and liquids meet land-spreading requirements. If surface waters (streams, creeks, pools, lakes, rivers) are present near the pit location, a risk exists of runoff of the pit liquids after they are spread in the ground. In such a case, the liquids must be treated to comply with surface-discharge limitations. Suspended solids, metals, and pH

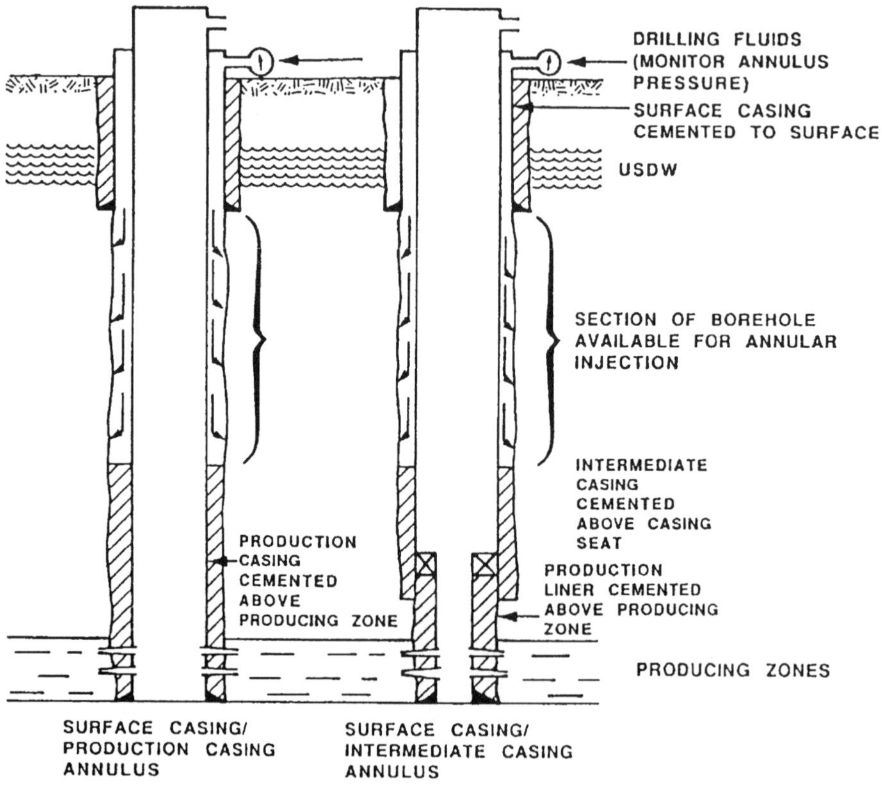


**Fig. 8.3—Well configurations for annular injection; USDW = underground source of drinking water.**

| TABLE 8.8—LIMITS FOR OILFIELD-PIT CLOSURE AND ON-SITE DISPOSAL (continued) | | | | |
|---|---|---|---|---|
| | Burial or Trenching of Waste/Soil Mixtures | | Solidification and Burial of Solidified Materials | |
| Waste Material Parameters | Limit | Test Procedure | Limit | Test Procedure |
| pH | 6 to 9 | | 6 to 12 | |
| EC, m℧/cm solid phase | <12 | 1 | | |
| SAR, ratio—solution phase | | | | |
| ESP, % solid phase | | | | |
| CEC, meq/100 g soil | | 5 | | 5 |
| O&G, % dry weight | <3 | 1 | | |
| Metals, ppm or mg/L | | | | |
| Arsenic | <10 ppm | 1 | <10 mg/L | 3 |
| Barium | | | | |
| Elevated wetlands | <20,000 ppm | 2 | <0.5 mg/L | 3 |
| Uplands | <40,000 ppm | 2 | <10 mg/L | 3 |
| Cadmium | <10 ppm | 1 | <0.1 mg/L | 3 |
| Chromium | <500 ppm | 1 | <0.5 mg/L | 3 |
| Lead | <500 ppm | 1 | <0.5 mg/L | 3 |
| Mercury | <10 ppm | 1 | <0.02 mg/L | 3 |
| Selenium | <10 ppm | 1 | <0.1 mg/L | 3 |
| Silver | <200 ppm | 1 | <0.5 mg/L | 3 |
| Zinc | <500 ppm | 1 | <5.0 mg/L | 3 |
| Soluble anions | | | | |
| Chlorides | | | <500 mg/L | 1 |
| Radioisotopes in coastal areas after 20 October 1990 | | 5 | | 5 |
| Other requirements | | | | |
| Moisture content, wt% | <50 | | | |
| Top of buried mixture, ft | >5 ft below ground level with 5 ft of soil on top | | >5 ft below ground level with 5 ft of soil on top | |
| Bottom of burial cell, ft | >5 ft above high-water table | | >5 ft above high-water table | |
| Unconfined compressive strength, psi | | | >20 | 4 |
| Permeability, cm/sec | | | $<1\times10^{-6}$ | 4 |
| Wet/dry durability, cycles to failure | | | >10 | 4 |

Testing procedure
1. Standard soil testing.
2. True total testing.
3. Leachate testing.
4. Testing must be done according to ASTM.
5. Mentioned as parameters to analyze for but no limitations are given.

usually can be altered to meet the limitations by use of lime treatments. Lime treatments are used to flocculate suspended solids and to adjust pH to desired levels. Furthermore, pH adjustment also removes metals from liquids through precipitation.

**Table 8.8** lists the limiting requirements for oilfield-pit closures related to on-site disposal options in Louisiana.[3] Note that O&G limitations are more relaxed than those recommended by API, while limitations for metals are more conservative.[20]

## 8.7 Annular Injection of Pit Solids and Liquids

Annular disposal of slurrified solids and liquids from drilling-mud reserve pits has been practiced for years.[21] As **Fig. 8.3** shows, annular injection is the injection of fluids into the annulus created by the space between the surface and intermediate casings or between the surface and production casings. The surface casing is cemented all the way to the surface to protect fresh waters, and its setting depth may range from approximately 300 to 2,000 ft. The intermediate casing is cemented below the depth at which the surface casing is set so there is an openhole annulus below the surface casing shoe. The annular space that has an openhole exposure allows fluids to go down between the surface casing and the intermediate casing and out into the permeable formation. In wells with no intermediate casing strings, the fluid will go down below the surface casing and above the top of the cement on the production casing and out into the zones of least resistance which usually are low-pressure nonproductive sands.

In the past, application of annular injection typically involved the following procedure. The pit-fluid-injection contractor connected the injection-pump-discharge line to the proper valve at the wellhead leading to the annulus. Then, waste drilling mud from the pit was pumped into the annulus to fill it up. (Settling of the mud sometimes caused some void space in the annulus.) Next, pumping pressure was increased to "break the formation down." This breakdown pressure usually exceeded the average pumping pressure by 200 to 500 psi. The process of formation breakdown is believed to have been, in fact, a fracturing treatment because gelled and thick mud was pushed out of the annulus and into the permeable rock.

After pumping for a few minutes, the pumping pressures were returned to normal. In most cases, pumping was begun with water, then gradually changed from water to pit slurry, often with a corresponding increase in pressure. Most contractors injected the entire contents of the pit; therefore, at the end of injection, the pit usually was almost empty. Crowding (pushing) the pit levee with bulldozers ensured that most of the slurry was removed from the pit. By the time the pumping was finished, the bulldozers would have covered and closed the pit and graded the surface back to its original elevation. During the reserve-pit injection, the wellhead pressure typically ranged from 500 to 1,000 psi in most areas. For shallow wells, the average injection pressure ranged from 500 to 700 psi. For deep wells (Anadarko basin, for example), the injection pressure was from 1,000 to 5,000 psi. The waste volume injected from a well depended on well depth and pit volume and ranged from 15,000 to 60,000 bbl. The rates of injection varied from 2 to 10 bbl/min, depending on the contractor's equipment. The equipment used in this technology was a type of centrifugal pump, known as a "trash" pump, that homogenized the contents of the pit by circulating and stirring the pit and mixing the mud, cuttings, and water together.

Since the late 1980's, the technology of annular injection of pit slurry has improved greatly and is now applied to disposal of drill cuttings from oil-based muds on- and offshore in the U.S. gulf coast area.[22,23] With this technology, hydraulic fracturing of the disposal zones is not a concern. The disposal zones in this area are shallow (3,600 to 4,600 ft), unconsolidated-sand strata with extremely high permeabilities because of the presence of shell deposits. The high permeability of these formations allows successful disposal of such materials as slurrified, drilled-out cement; shredded paper waste (mud sacks and cardboard boxes); shredded visqueen plastic; and ground wood with plastics (shredded wooden pallets and crates).[23]

## 8.8 Remediation After Pit Closure

Site remediation is a follow-up treatment after pit closure. Treatment depends on the type of contamination that controlled the pit-closure design [i.e., whether it was hydrocarbon, salt, or pH (alkalinity/acidity) contamination].

For hydrocarbon-controlled pit closure (organic management), bioremediation usually is accomplished with naturally occurring microflora in native soil that biodegrade hydrocarbons to the background level in one season. When natural-species diversity is poor and its population limited, the time required for biodegradation sometimes (but not frequently) is shortened by inoculation with commercial microorganisms. However, the performance of these microbes is difficult to measure when compared with regulatory standards, such as O&G or total petroleum hydrocarbons.[24] In any case, controlled bioremediation programs require treatment with fertilizers, aeration, and moisture. Treatments with fertilizer provide nitrogen, phosphorus, and potassium. Disking distributes the application and aerates the system. Refs. 4 and 24 give example calculations of nitrogen requirements.

For salinity-controlled pit closures (salt management), freshwater dilution resulting from rainfall and infiltration in upland areas or from inundation in wetland areas is the only mechanism of remediation. Therefore, good structural conditions and hydraulic conductivity of soil must be maintained through disking and treatment with calcium. Native soils mixed with oilfield wastes contaminated mostly with NaCl become saturated with sodium, or sodic. Gypsum or sulfur usually are added to sodic soils to improve their poor physical characteristics (i.e., lack of soil structure and impermeability to water) resulting from ESP values greater than 15%. Each milliequivalent of exchangeable sodium in excess of 15% requires approximately 1.7 tons of gypsum or 0.3 tons sulfur/acre-ft soil, allowing the total treatment to be designed easily.[4,24]

When oilfield-pit-waste acidity (or alkalinity) dominates the pit-closure design (metal management), lime is used to neutralize acidity and control metals in the waste/soil mixture. The excessive alkalinity usually associated with waste drilling fluids can be controlled with aluminum sulfate. The amount of lime or sulfur required to raise or lower pH is determined empirically and is described in literature.[24]

## References


1. *Criteria for pH in Onshore Solid Waste Management in Exploration and Production Operations*, Publication 4595, API, Washington, DC (March 1995).
2. *Oil and Gas Industry Exploration and Production Wastes*, Document 471-01-09, API, Washington, DC (1987).
3. "Amendment to Statewide Order No. 29-B," Louisiana Dept. of Natural Resources, Office of Conservation, Baton Rouge, Louisiana (1990).
4. Freeman, B.D. and Deuel, L.E.: "Guidelines for Closing Drilling Waste Fluid Pits in Wetland and Upland Areas," *Proc.*, Petroleum Div. of the American Soc. of Mechanical Engineers (ASME) Seventh Annual Energy Source Technology Conference and Exhibition, New Orleans (1984) 190.
5. *Environmental Guidance Document: Onshore Solid Waste Management in Exploration and Production Operations*, API, Washington, DC (15 January 1989).
6. *Standards for the Use or Disposal of Sewage Sludge*, 40 CFR, U.S. EPA, Washington, DC (1993) Parts 257, 403, and 503.
7. *Metals Criteria for Land Management of Exploration and Production Wastes*, technical support document, Publication 4600, API, Washington, DC (January 1995).
8. "Review and Recommendations for Canadian Interim Environmental Quality Criteria for Contaminated Sites," Canada IWD, CAI EP 75691597, Scientific Series 197, Toronto, Canada (1991) 10–15.
9. Peterson, R.G. and Calvin, L.D.: "Sampling," *Methods of Soil Analysis, Part 1: Physical and Mineralogical Methods*, A. Klute *et al.* (eds.), monograph, American Soc. of Agronomy, Madison, Wisconsin (1986) **9,** No. 1, 33–50.
10. Crepin, J. and Johnson, R.L.: "Soil Sampling for Environmental Assessment," *Soil Sampling and Methods for Analysis*, M.R. Carter (ed.), Canadian Soc. of Soil Science, Lewis Publishers, Ann Arbor, Michigan (1993) 18–36.
11. Deuel, L.E. Jr. and Freeman, B.D.: "Amendment to Louisiana Statewide Order 29-B Suggested Modifications for Barium Criteria," paper SPE 18673 presented at the 1989 SPE/IADC Drilling Conference, New Orleans, February 28–3 March.
12. Carter, M.R.: *Soil Sampling and Methods of Analysis*, Lewis Publishers, Boca Raton, Florida (1993) 91–93.
13. Wojtanowicz, A.K. *et al.*: "Statistical Assessment and Sampling of Drilling-Fluid Reserve Pits," *SPEDE* (June 1989) 162.
14. Dover, V.L.: "Land Restoration Following Oil Well Drilling and Products Useful Therefore," U.S. Patent No. 4,353,803 (12 October 1982).
15. "Conditioning Centrifuge, a Key to Cost-Effective Pit Water Cleanup," *Oil & Gas J.* (9 February 1987) 36.
16. Wojtanowicz, A.K. *et al.*: "Comparison Study of Solid/Liquid Separation Techniques for Oilfield Pit Closures," *JPT* (July 1987) 845.
17. Osterman, M.C., Wojtanowicz, A.K., and Field, S.D.: "The Development Study of a Centrifuging Technique for Oilfield Production Pit Closure," *Trans.*, ASME Energy Resources Technology (December 1987) 51.
18. Wentz, C.A. and Shiver, C.M.: "Cleanup of Oilfield Wastes in an Environmentally Acceptable Manner," paper SPE 11402 presented at the 1983 SPE/IADC Drilling Conference, 20–23 February.
19. *Evaluation of Limiting Constituents Suggested for the Land Disposal of Exploration and Production Wastes*, Publication 4527, API, Washington, DC (August 1993).
20. *Laboratory Procedures for Analysis of Oilfield Waste*, Louisiana Dept. of Natural Resources, Office of Conservation, Baton Rouge, Louisiana (January 1989).
21. McCaskill, C.: "Well Annulus Disposal of Drilling Waste," *Proc.*, Natl. Conference on Disposal of Drilling Waste, U. of Oklahoma, Environmental and Groundwater Inst., Norman, Oklahoma (May 1985) 248.
22. Louviere, R.J. and Reddoch, J.A.: "On-Site Disposal of Rig-Generated Waste Via Slurrification and Annular Injection," paper SPE 25755 presented at the 1993 SPE/IADC Drilling Conference, Amsterdam, The Netherlands, 23–25 February.
23. Anderson, E.E., Louviere, R.J., and Witt, D.E.: "Guidelines for Designing Safe, Environmentally Acceptable Downhole Injection Operations," paper SPE 25964 presented at the 1993 SPE/EPA Exploration and Production Environmental Conference, San Antonio, Texas, 7–10 March.
24. Deuel, L.E. Jr. and Holliday G.E.: *Soil Remediation for the Petroleum Extraction Industry*, Pennwell, Tulsa, Oklahoma (1994).


## SI Metric Conversion Factors

| | | | |
|---|---|---|---|
| acre | × 4.046 873 | E − 01 | = ha |
| acre-ft | × 1.233 489 | E − 03 | = ha |
| bbl | × 1.589 873 | E − 01 | = $m^3$ |
| ft | × 3.048* | E − 01 | = m |
| in. | × 2.54* | E + 00 | = cm |
| lbm | × 4.535 924 | E − 01 | = kg |
| ℧ | × 1.0* | E + 00 | = S |
| psi | × 6.894 757 | E + 00 | = kPa |

*Conversion factor is exact.

# Chapter 9
# Legal and Regulatory Structure for Environmental Protection

**Kit E. Armstrong,** Chevron Overseas Petroleum Inc.
**Michael L. Godec,** ICF Resources Inc.
**Ronald Morris,** Texaco Inc.
**William C. Hutton,** Wm. C. Hutton Consultants

## 9.1 Introduction

The objective of this chapter is to provide basic guidance to petroleum professionals about international environmental regulatory structures that affect oil and gas E&P activities. The material in the chapter provides an overview of the typical basic elements of environmental regulatory structures; discusses alternative approaches to pollution control that can be incorporated into regulatory requirements; summarizes steps company personnel should take to identify environmental requirements and standards; describes internal and external drivers that can influence development of standards and guidelines, either by national governments in regulations or by companies for specific projects; and discusses some of the responsibilities that petroleum professionals may have in helping governments establish and implement regulatory programs that function rationally and predictably. Finally, case studies are presented that illustrate examples of some of the principles, guidelines, and recommendations discussed elsewhere in the chapter.

## 9.2 Major Drivers Influencing International Environmental Regulatory Structures

Public awareness and concern about the environment has been growing worldwide, causing governments to develop or revise their laws and regulations to ensure environmentally sound industry operations. Policies, laws, regulations, guidelines, and performance standards that require the oil and gas E&P industry to operate with proper regard for the environment are increasing.

At the same time, governments in many parts of the world are transforming their petroleum sectors. The change is from predominantly nationalized and monopolistic to more decentralized, private, and competitive systems. Governments are having to use more varied types of regulatory mechanisms to ensure that oil and gas companies satisfy government expectations for environmental protection while accomplishing their economic objectives.

The concept of "sustainable development" is being endorsed by many governmental and nongovernmental organizations (NGO's) around the world as a framework for promoting economic development that is environmentally sustainable and socially equitable.[1-3] National governments are developing policies and implementing regulations consistent with this concept. It is also being incorporated into policies and programs of major development aid organizations, such as the World Bank, which now requires compliance with specified environmental standards and guidelines as a condition of its project loans and grants for development projects.[2,4]

An expanding body of international environmental technical and management standards and guidelines is also being developed by organizations like the Intl. Standards Organization (ISO) and by industry organizations. Most multinational petroleum companies are establishing environmental performance standards within their companies to ensure a consistent, predictable approach to environmental protection for their activities worldwide.

These trends all have considerable significance for petroleum engineers and other petroleum professionals. Oil and gas companies must ensure compliance of their operations with environmental laws and regulations while maintaining productivity. They must guard against both catastrophic and chronic risks that their activities pose to the environment, to the health and safety of their fellow employees, and to the public, whether those risks are real or perceived. They also must ensure that proper consideration of environmental issues is incorporated throughout the entire project-planning and decision-making process. This starts with formulation of the bid on a particular prospect and continues through project costing and scheduling, construction, the life of project operations, to final project closure.[5]

As part of their day-to-day activities, petroleum engineers play an integral role in helping companies achieve these objectives. As such, they must perform their jobs with a heightened concern for the environment and work to make sure that companies conduct their operations in a manner that ensures proper environmental stewardship.

## 9.3 Basic Elements of a National Regulatory Structure

**9.3.1 Typical Components.** Government has the primary responsibility to set policy and provide a framework for appropriate industrial development while ensuring that the environment is protected adequately. Petroleum professionals must work to make sure that the operations for which they are responsible are conducted in a sound environmental manner that complies with the laws and regulations of the country. To do this well, they should understand the basic environmental regulatory structure of the country and how it may evolve over the lifetime of a project.

National laws for the protection of the environment are varied and have many purposes. National laws provide a framework for managing industrial and other activities within a country to provide an appropriate level of environmental protection. They may also address international treaty obligations and protection of the environment of other countries from transboundary pollution or protection of the global environment.[6]

Conventional regulatory structures that affect oil and gas E&P operations typically have common elements. They include provisions on environmental impact assessments (EIA's), pollution control, and prevention; spill and other emergency prevention and control; operational discharges and waste management; decommissioning and abandonment; and violations and fines. More developed regulatory programs may also address chemical use and control, ambient environmental monitoring, and construction certification in addition to a host of other requirements.

Responsibility for developing and implementing environmental legal requirements can rest with one or more government agencies at the national level and also at the regional/state and/or local level. At the national level, responsible agencies might be a ministry-level environmental department and/or a ministry-level petroleum, energy, interior, or natural resources department. Other government agencies, industry, environmental and other nongovernmental organizations, academics, and consultants may formally or informally advise or consult with relevant ministries.

**9.3.2 Evolution of Regulatory Systems.** A nation's regulatory cycle consists of four basic and interdependent elements.[7]

1. Policy setting. This involves establishing the basic principles and approaches for environmental protection, generally codified in laws and regulations.

2. Implementation. This involves establishing and implementing a permitting or licensing system that defines who is authorized to pursue specific activities and the terms under which those authorized activities can be undertaken.

3. Verification and monitoring. This involves establishing and implementing a system to ensure compliance with required environmental practices and standards as specified by laws and regulations and by permits and licenses.

4. Enforcement. This involves establishing and implementing procedures and penalties to punish those who do not adhere to the established environmental requirements.

The scope and detail of requirements in each component and how they are implemented typically evolve over time within a country. In some countries, governments have recognized the need for environmental legislation only recently and have just begun to put a basic framework in place. However, others have attempted to put highly detailed regulatory programs in place very quickly, sometimes exceeding the abilities of both government and regulated entities to implement them effectively.

Development and implementation of a comprehensive, effective, rational, and fair national environmental regulatory system is a gradual process. A number of countries have started with a general framework to which they have added over time on the basis of need, experience, and the stage of their national economic and social development. This incremental approach provides the opportunity to develop experience and expertise within government and the regulated community. In addition, it allows for essential evaluation and for feedback of "lessons learned" into decisions so that modifications can be made. The process can move slowly or be greatly expedited by politically influential events, such as a major oil spill or other catastrophic incident.

Understanding the stage of regulatory development that a particular country has reached, the drivers that have influenced the process, and the steps that have occurred can be very valuable to petroleum professionals working for companies in that country. It can help them anticipate and define appropriate environmental protection requirements for operations at a particular location and can also help predict likely permitting problems that will affect project siting, scheduling, planning, and economics. Equally important, it can give insights into where regulatory requirements may be heading. These can then be incorporated into project definition, evaluation, and implementation, reducing the potential for future exposure and liability.

## 9.4 Alternative Regulatory Approaches

**9.4.1 Types of Approaches.** Traditionally, approaches for environmental regulation of industrial activities, including the oil and gas sector, have focused on "end-of-the-pipe" or "environmental quality" standards applied to discharges into water, air, and soil. These types of requirements specify either the composition of waste streams that are allowed to be discharged to the environment or the composition of the receiving medium (e.g., water, soil, air) into which the waste stream is discharged.

Technology-based standards are one type of end-of-the-pipe approach. These specify the processes and technologies that must be used to achieve an established level of environmental performance. They often are expressed in terms of "best available technology" that must be used and sometimes are used to encourage the development of more protective technologies.

Prescriptive command-and-control approaches to environmental protection have the advantage of providing specific, consistent, and predictable requirements. This can facilitate monitoring and enforcement by government agencies, provide a "level playing field" for the regulated community, and be relatively easy for the public to understand how they are expected to be applied. However, numerous problems exist with the command-and-control approach. While generally effective in achieving a predictable level of environmental performance, lack of flexibility often results in unnecessary expenditures by operators to achieve the desired level and can also be very expensive for governments to administer and enforce. In some cases, technology-based standards have had little apparent relationship to environmental risks. Prescriptive standards can also create an adversarial relationship between regulatory agencies and the regulated industry and discourage innovations to reduce emissions and discharges below mandated levels.

Some of these concerns are causing regulatory agencies in the U.S., such as the U.S. Environmental Protection Agency (U.S. EPA), and other countries to consider alternatives to prescriptive approaches to environmental regulation. A number of potential alternatives can be implemented to overcome some of the problems. One example is goal-setting approaches, which, when appropriately implemented, can ensure that the responsibility for environmental protection lies with those who create, best understand, and work most closely with the environmental risks associated with specific activities. These approaches generally provide operators with more opportunity to develop their own solutions to address relevant risks most cost-effectively.

At the same time, goal-setting regulatory structures must be developed in a manner that achieves both the flexibility desired by industry and a specified level of environmental protection.[8] Goal-setting approaches can require significant technical expertise, judgment, and cooperation on the part of both regulators and regulated entities. They often are difficult to monitor and evaluate and therefore may not be trusted or accepted by NGO's and the general public.

Application of economic mechanisms to encourage a desired objective is another approach to achieving a predetermined level of environmental performance. For example, pollution taxes may require facilities emitting pollutants above a defined level to pay a tax for emissions above that level. These taxes are a form of economic disincentive that can be used to achieve pollution-control objectives. Financial incentives typically consist of trading systems for credits for emissions reductions. These systems allow companies that achieve levels of pollution reduction below an established level to trade or sell the resulting "pollution-reduction credits" to companies that are having greater difficulty achieving prescribed reductions. Such a trading system was established in the U.S. by the 1990 Amendments to the Clean Air Act.[9]

Other approaches for achieving environmental objectives include the use of negotiated agreements between regulators and the regulated community (such as those envisioned by the U.S. EPA under its Common Sense Initiative[10]). Voluntary approaches to compliance (e.g., the Gas Star Program[11] initiated by the U.S. EPA to reduce emissions of methane in the natural gas industry) and industry-initiated self-enforcing approaches, are also possibilities.

**9.4.2 Factors Affecting Choice of Approaches.** The stage of evolution of a country's regulatory structure is important in understanding the types of regulatory approaches used in that country. Regulatory structures and systems typically change over time. The type of approach government or industry prefers or considers in a given situation can be a function of approaches previously attempted, what worked and what did not, and what other countries in a similar situation may be considering or are implementing. Equally important is that the emphasis on or attitude toward a regulatory system is partially a function of a country's culture and legal system. For example, in countries where negotiation and consensus are the primary means to resolve conflicts, stringent command-and-control approaches that use highly legalistic implementation and enforcement mechanisms may be unappealing or difficult to carry out effectively.

Occurrence of a catastrophic environmental event often can influence the course of development of environmental regulations in a country. Political pressures force lawmakers and regulators to act in response to environmental crises, even when such action is unnecessary or inappropriate. In actuality, the responses to these events in many cases do little to address the problems that exist and to reduce the risks that such events will occur again. Nonetheless, the potential impacts of these types of events on regulatory development cannot be ignored.

## 9.5 Steps To Identify Environmental Requirements and Standards

Company personnel need to identify and be prepared to work within the legal structure and culture of the country where the company is undertaking activities. A number of basic steps are involved in this process.

**9.5.1 Identify Applicable Laws, Regulations, and Other Legal Requirements.** The domestic law of the host country where a company conducts activities is the primary source of environmental laws that apply to those activities. Environmental requirements may be found at the international, national, regional/state, and/or local levels.

Treaties, conventions, protocols, and other international governmental legal agreements may establish mandates for national governments to regulate industrial activities. The national government must first sign the agreement and take other steps required by national law to commit the country to adhering to the agreement. The government may also adopt specific legislative requirements at the national level to implement the agreement. The following are some international agreements of particular potential interest to oil and gas E&P operations.

1. Convention on the Prevention of Marine Pollution by Dumping of Wastes and Other Matter (London Convention).[12] Regulates marine disposal of a wide range of materials.
2. Regional seas conventions, such as the Convention for the Protection of the Marine Environment of the North-East Atlantic.[13]
3. United Nations (U.N.) Framework Convention on Climate Change.[14] Provides for national commitments to reduce emissions of so-called "greenhouse gases" (e.g., $CO_2$ and methane) that result, in part, from burning of fossil fuels, such as petroleum.
4. Basel Convention on the Control of Transboundary Movements of Hazardous Wastes and Their Disposal.[15] Strictly regulates movements of hazardous wastes among countries for disposal and recycling.
5. U.N. Framework Convention on Biological Diversity.[16] Mandates countries to establish national plans to conserve biodiversity, including, where necessary, implementation of new laws and regulations and establishment of protected areas.

National environmental legal requirements come in many forms and can be found in many places in a country's laws. Sometimes, they take the form of general environmental protection laws, laws specific to environmental media (e.g., air, water, waste), or laws regarding specific activities (e.g., EIA's). Environmental requirements of potential application to oil and gas activities also can be contained in laws relating to energy or hydrocarbon development, forests, wildlife, protection of indigenous people, or other subjects. Important provisions may be in civil, administrative, or criminal codes. Such requirements may not be found in laws or regulations at all. Host governments may impose environmental performance conditions through provisions of the basic license, production-sharing, and other required agreements.

**9.5.2 Identify Responsible Government Agencies.** Legal requirements may be established by the executive authority, legislative body, or specific government agencies at the national, regional/state, or local level. Judges and the courts may have either a strong or very weak role in establishing or interpreting environmental legal requirements. Numerous and diverse governmental bodies may have direct or indirect authority for or involvement in some aspect of environmental legal requirements relevant to a company's activities.

**9.5.3 Obtain Copies of Laws and Regulations.** Obtaining copies of applicable environmental laws and regulations often can be a challenge, especially in countries that do not routinely publish or otherwise make such documents available to the public. Getting high-quality translations of such materials can be especially difficult, so a company should be prepared to have such translations made. Company personnel may need to work hard and be creative in this effort. Potential sources of copies of laws and regulations are government agencies themselves, local or international consultants or law firms, other companies doing business in the country, and local or regional industry associations.

**9.5.4 Identify Local Cultural Customs and Practices.** Cultural and other nonlegal norms may dictate what is considered acceptable practice, particularly where indigenous people occupy the area of the oil and gas activity. These customs can be unknown to the national government. Considerable advance planning and contacts in the local area may be required to determine accepted practice.

**9.5.5 Assess Need for Supplemental Environmental Standards.** A company should take the position that compliance with requirements of applicable environmental laws and regulations in a country constitutes the minimum acceptable level of company performance. A responsible analysis of such requirements may lead to the conclusion that the baseline level of legal compliance is not sufficient to constitute responsible environmental performance by the company, be acceptable to other stakeholders, or protect the company's interests. Therefore, companies should be committed to do the following.

1. Carry out responsible environmental stewardship within the framework of total resource management.
2. Be proactive in anticipating and addressing environmental concerns, and avoid creating concerns and risks wherever feasible.
3. Consider both short- and long-term environmental implications of its activities.
4. Follow good (and preferably best) international practices.
5. Be a responsible corporate citizen to maintain society's "permission to operate."
6. Conduct audits (annually or more frequently) to document performance for the operator and the country.

Consistent with these principles, the company may need to identify specific internal and external nonregulatory environmental policies, standards, or guidelines to apply to its activities to supplement applicable laws and regulations. Such policies and guidelines may be provided by industry and other organizations or developed within the company. In most cases, they also are likely to be influenced by concerns of local communities and other stakeholders.

## 9.6 External Sources of Standards and Practices

The domestic law of the host country is the primary source of environmental legal requirements that apply to E&P operations in that country. However, despite diligent efforts, a company may be unable to identify a clear set of regulatory standards that sets forth the operational requirements for oil and gas E&P operations in a given country or local site. At the same time, various external pressures and drivers are likely to influence or even dictate what is considered to be environmentally responsible operations, even when specific regulatory standards exist.

**9.6.1 Public Concerns.** A major driver is any general public concern associated with E&P operations. For example, public and media attention to past E&P activities can influence a local community's expectations for E&P operations in its area. The needs, concerns, and opinions of other commercial interests in the area (such as fishing, agriculture, forestry, and tourism) and of other local stakeholders influence what is environmentally acceptable E&P operations at the location, regardless of governmental requirements.

As mentioned earlier, many public institutions and nongovernmental organizations around the world are promoting the concept of "sustainable development." This concept encourages approaches for environmental protection that do not compromise the economic viability of E&P operations unnecessarily and *vice versa*. Two major texts were issued as part of the 1992 U.N.'s global Conference on Environment and Development, The Rio Declaration on Environment and Development and Agenda 21—A Call for Action.[17] These texts focus on the interdependence between development and the environment and emphasize the need for partnership between different nations and sectors of society, including industry, to tackle the complex issues involved. As a result, environmental aspects of E&P operations may be influenced and measured by stakeholders vs. the sustainable development concept.

**9.6.2 International Governmental Organizations.** Other important external drivers can be standards and guidelines adopted by major intergovernmental development aid organizations, such as the World Bank and the Intl. Finance Corp.; regional development banks, such as the Asian Development Bank; and private lending institutions. (Ref. 2 provides a description of the environmental concerns and corresponding activities of the World Bank.) Projects seeking financial support from these organizations must meet the specified environmental standards to qualify.

For example, some of the World Bank environmental requirements are contained in the bank's *Industrial Pollution Prevention and Abatement Handbook*.[18] It contains standards for specific industrial sectors and operations, and provides guidance to bank staff and policy makers about what constitutes acceptable performance with respect to the activities and operations in each sector. The bank has adopted standards for emissions and discharges associated with on- and offshore oil and gas activities. Other World Bank guidelines contain detailed requirements for preparing environmental impact statements for projects funded by the bank.

Environmental performance standards and guidelines are also being defined by other international organizations. For example, the U.N. Environmental Program (UNEP) has joined with the Oil Industry Intl. E&P Forum to publish environmental management standards for oil and gas E&P activities.[19] UNEP has also sponsored the Regional Seas Program,[20] under which a number of conventions have been developed to protect various marine areas of the world from pollution, some of which include provisions relating to control of oil and gas E&P activities. The Intl. Maritime Organization (IMO) also has been actively developing standards for oil-spill contingency planning and response by industry and national governments.[21]

**9.6.3 International Standards Organizations.** An expanding body of environmental technical and management standards is being developed by such organizations as the ISO and regional and national standards organizations. These standards are intended to promote a common level of environmental performance for industrial operations worldwide. ISO's Technical Committee 207 has developed what is likely to be the most far-reaching of these standards, the ISO 14000 Series. This series establishes international standards for environmental management systems, environmental auditing, environmental performance monitoring, environmental labeling, and life-cycle assessments. (Although the ISO standards are voluntary, they are likely to find their way into regulatory requirements in some countries.)

**9.6.4 Industry Organizations.** General industry and E&P-industry-specific guidelines and statements of "best industry practices" are proliferating in efforts to encourage voluntary efforts toward improving environmental performance. For example, the Intl. Chamber of Commerce has adopted a set of environmental management principles.[22] Guidelines most relevant to the E&P industry are those developed by the E&P Forum, the Intl. Assn. of Geophysical Contractors (IAGC), and the Intl. Petroleum Industry Environmental Conservation Assn. (IPIECA).

**9.6.5 NGO's.** In addition to the predominantly international drivers just described, NGO's in many countries exercise considerable influence on prospective oil and gas E&P activities. They include both international and regional environmental organizations in addition to local community groups concerned about how a company's project may affect their local environment, culture, traditions, and economic livelihoods. NGO's may also be organized by other economic interests in the region.

Interaction with NGO's and the communities they represent often focuses on preparation of EIA's and social-impact assessments associated with a project. The primary challenge for an oil and gas company attempting to initiate a project often involves efforts to get diverse, and sometimes competing, stakeholders to understand and accept the project. This requires open and effective dissemination of information about the project and its anticipated effects. Usually, it also involves the further step of active consultation and dialogue with local communities and NGO's in which the company listens to their needs and concerns and incorporates their input into relevant project decisions. Increasingly, a key factor in successful projects is the need to ensure that the project includes real and meaningful benefits identified by and delivered to the local community.

**9.6.6 Laws of Other Countries.** Policies, laws, and regulations of a neighboring or like-minded country may also influence operational standards to be applied to a project in a country without a well-established regulatory regime. Laws of a country with a well-established tradition of environmental regulation, such as the U.S. (or even key U.S. states, like California or Texas), may also be used as standards. In some cases, an operator is expected to conduct its activities in a manner consistent with other operations in the region or with "best practice in the international petroleum industry."

**9.6.7 Historical Experience.** In some cases, past practices of oil and gas companies in a country or region or the historical occurrence of a catastrophic environmental event may dictate expectations for environmental performance in a particular country. Where past environmental performance resulted in damages and discontent, new operators in the area may not only have to prove themselves environmentally responsible for their own operations but also may have to mitigate damages associated with operations that preceded them.

**9.6.8 Partner Companies.** Oil and gas producer companies increasingly are pursuing project opportunities in partnership with other producer companies and, in some instances, with service companies. The policies, principles, and practices of partner companies may be a very significant driver of how environmental considerations are addressed in a particular project. With increasing frequency, environmental considerations are included as a specific component of an alliance charter or agreement.[23]

## 9.7 Internal Sources of Environmental Standards and Practices

As discussed previously, company personnel should be prepared to identify and use a variety of company and industry sources to form a company's environmental protection programs and its approaches in a country. Some should be considered essential components of the company's business program, regardless of how extensive external regulatory requirements may be. For example, a company should have a written policy that expresses the basic philosophy and key principles by which it conducts its activities with respect to the environment. Such a policy is a primary element of a company's broader integrated health, safety, and environmental (HSE) management system. The HSE management system is a structured, systematic process that a company uses to maintain a specified level of environmental performance. ISO 14001 and E&P Forum guidelines provide two similar models for an HSE management system.

A company's HSE management system should identify appropriate objectives to be achieved by the system. To be complete, it also should incorporate specific environmental operating guidelines and standards that the company applies in its activities when it concludes that external regulatory standards are not sufficient. A company may develop such standards and guidelines with its internal resources. It may also use guidelines developed by relevant organizations, such as the E&P Forum, IPIECA, IAGC, and national petroleum associations. A company may find such guidelines very useful in identifying what can be considered good or best international industry practice. Guidance for identifying potential environmental standards and approaches may also be found in industry literature, previous company projects, and advice of internal company technical experts and outside consultants.

## 9.8 Establishment of Project-Specific Environmental Standards and Practices

Determination of specific environmental standards and practices to be applied to a project generally involves certain fundamental activities. As early as possible in the project, a company needs to collect and analyze all external and internal regulatory standards or other environmental norms and guidelines and determine their application to the project or activity. The objective should be to identify all environmental standards, criteria, and practices that may be needed or are appropriate to ensure that the project complies with applicable legal requirements and is the most environmentally responsible as appropriate for local conditions. Assessment of all environmental issues and consequences of the project or activity should be part of this process. This may be an internally driven, an externally required EIA, or some other such process.

Whatever its form, the process should be designed and implemented to produce all information necessary to prepare specific plans and programs to monitor and manage all potentially significant environmental aspects of the project effectively. Plans should cover all project phases, from final design through construction, operation, decommissioning, and rehabilitation, with provision for revision over time based on new needs, information, and experience.

Remembering that a project may create or be a catalyst for social and cultural changes in local communities is particularly important. These can come from direct environmental or health effects (e.g., changes in water or food supplies or introduction of disease), indirect environmental effects (e.g., construction of roads that provide access for illegal cutting of forests), or cultural change in an indigenous population. Projects can also cause a range of effects that relate to land ownership and use, employment, revenue sharing, social services, and other such factors. Therefore, many nonenvironmental issues and impacts may need to be identified and addressed as part of comprehensive project evaluation, planning, and management. Anthropological, cultural, and other local experts may need to be consulted, especially where local indigenous people may be impacted.

Successful project implementation also depends on other considerations that affect the company's ability to operate effectively in the country and local community. The following are examples of such considerations.

1. Future regulatory requirements. Existing regulatory standards are likely to change over the life of the project. A company should develop its project-specific standards with a view to their long-term effectiveness in protecting the company's interests. It should also have a process to monitor effectively and, if possible, to participate in development of new regulatory requirements that may affect the project.

2. Permits and government approvals. Time lines, procedures, and substantive requirements for permits and other government approvals may not be set out clearly in the law or may be carried out much differently than the language of the law might suggest. A company needs to know as much as possible and as early as possible about how such processes actually work in practice.

3. Infrastructure limitations. Infrastructure essential to conducting some environmental activities may be limited or nonexistent in the area (e.g., sites for processing or disposal of solid or hazardous wastes or laboratories for analysis of waste discharges or emissions). The company itself may need to develop key facilities to achieve its desired level of environmental performance.

4. Local politics and NGO's. The project can be a catalyst for advocacy and other efforts by local political groups and/or by local or international environmental groups or other NGO's. Starting early in project evaluation and planning, a company needs to educate itself about local political, social, cultural, and other conditions, with particular attention given to local issues. Community and NGO outreach activities may be critically important so that issues that could adversely affect the project and the company's image can be avoided or addressed.

5. Cultural limitations. If the project is in an area of an indigenous population, consideration must be given to the impact on that culture and environment. Advance planning is required to research the area and develop a plan to meet the needs of the local population.

Therefore, project-specific requirements and implementation may need to take a wide variety of factors into account. Some will be external (e.g., legal, political, social, infrastructure) and others internal (e.g., company policies/standards, project schedule/cost). Project success will depend on the company's ability to manage effectively relevant legal, technical, political, social, economic, and cultural issues, both related to and separate from more traditional environmental concerns. The company should be prepared to allocate the time, resources, and expertise needed to achieve this objective.

## 9.9 Responsibilities for Government Institutional Development

International petroleum companies need to have a system that continuously monitors government efforts to develop new environmental policies and regulations that affect the E&P industry. The process must be capable of anticipating new requirements that may affect future projects. Lack of awareness of evolving environmental requirements could have substantial detrimental effects on the economics of a particular project.

In most situations, a well-articulated environmental regulatory structure is a benefit. In some countries, the lack of a clear environmental policy or regulatory framework that relates to oil and gas E&P activities can be more confusing and detrimental to project planning, timing, and finance than a seemingly stifling or excessively stringent regulatory structure.[24] (Ref. 6 presents some useful guidelines concerning what constitutes effective, efficient environmental laws and regulations.) A technically sound, well-articulated regulatory structure provides a framework for expectations; decision making; accountability; more certainty and predictability for project standards, scheduling, and financing; and a better basis for economic decisions. It delivers a more predictable and level playing field for all participants.

Unfortunately, even with a well-articulated regulatory structure, government institutions often lack the resources, technical knowledge and experience, and trained personnel required to implement an effective regulatory program. A government regulatory agency without knowledgeable, trained staff and clear guidelines within which to operate can present real challenges to a project. This situation often can leave a project open to costly closure measures because of increasing controls imposed during the course of the project. Therefore, companies should work proactively with governments to help develop a rational regulatory infrastructure and sound institutional capabilities to the extent that government laws and business ethics allow. It is important to educate regulators responsibly and honestly about the petroleum industry, where appropriate, especially in countries with little or no previous oil and gas E&P activity.

Company participation in development of government regulatory structures and institutions should be considered a form of "soft" technology transfer consisting of information on applicable, appropriate concepts and practices for industry regulation or self-regulation as practiced in other countries. This input should be provided as early in the legislative or regulatory development process as possible. Then, the government is more likely to perceive the offer of assistance as sincere rather than as just an attempt to fend off an unfavorable outcome for industry or a company once it has become apparent and likely. If this process is pursued properly and ethically, not only can it result in a more rational and predictable set of environmental requirements but also can foster the building of personal working relationships with regulators, establishing a trust and understanding of mutual expectations. This increases the probability that mutually acceptable requirements for environmental performance will result and also can serve as a source of early warning of governmental environmental concerns, allowing the company to respond to and address those concerns.

## 9.10 Case Studies

The two case studies provided here illustrate how some of the principles and approaches discussed earlier have been used by the E&P industry in practice. The first discusses regional developments in Africa that industry is pursuing collectively to address many of the environmental issues of concern in that area. The second focuses on a specific oil exploration and development project in the rain forest of South America.

**9.10.1 Legal and Regulatory Structures for Environmental Protection in Africa.** ***Introduction.*** The material in this case study describes some operational factors relevant to developing E&P environmental practices for Africa, with particular emphasis on oil-spill prevention and response. It does not lay out specific regulatory or performance requirements but instead recommends that companies set environmental management objectives taking into consideration the long-term effects of their operations on the environment and on the economic development of the country.

***Environmental, Cultural, and Operational Characteristics.*** Africa consists of approximately 54 independent countries. The world's second largest continent, it contains 22% of the world's land. Its extremely rich biodiversity is part of the world's natural resource heritage. Diverse physical and cultural conditions combined with equally variable geological and soil formations have resulted in bioclimatic zones ranging from hot desert to equatorial rain forest and from Mediterranean woodland to tropical savannah. Protection of the environment exposes E&P operations in Africa to virtually every ecological setting.

Countries range from those with well-developed general infrastructures and considerable environmental protection resources and knowledge, such as South Africa, to those without any environmental awareness or protection program. A strong trend, however, is reflected in new initiatives to develop such programs. One example of this is Angola, a country with enormous potential but little in the way of supporting infrastructure after years of civil war. Angola's emerging program on environmental awareness, protection, and regulation is proceeding with a sense of urgency. Its rapid post-war progress is the result, in large part, of strong industry/government cooperation.

The goal of most African governments is to develop technically sound environmental regulations that protect the environment while strongly encouraging rapid, but sustainable, economic development. Strategies to establish environmental planning, application, and reporting requirements for petroleum E&P projects are, for the most part, at early stages of development in much of Africa.

***Government Agencies/Ministries and Important Stakeholders/ Considerations.*** National laws for environmental protection vary throughout the African states. Most recently, governments have recognized the need for environmental legislation and have established a variety of governmental institutions to regulate environmental issues. Many governments have invited environmental and other NGO's, industry, outside governments, classification societies, or consultants to advise them about regulating activities that may impact the environment, in particular, those involved with petroleum operations and transport.

**TABLE 9.1—SOURCES OF INFORMATION ON ENVIRONMENTAL ISSUES IN AFRICA**

Organizations

Intl. Maritime Organization
4 Albert Embankment
London SE1 7SR U.K.
Phone: 44-171-735-7611
Fax: 44-171-587-3210

Intl. Petroleum Industry Environmental Conservation Assn.
87-93 Westbourne Grove
London W2 4UL U.K.
Phone: 44-171-221-2026
Fax: 44-171-229-4948
e-mail: ipieca@gn.apc.org.uk

Oil Spill Service Center
Lower William St.
Northam
Southampton
SO1 1QE U.K.
Phone: 44-703-331-551
Fax: 44-703-331-972

South Africa Oil Industry Environmental Committee
Phone: 027-214-198058
Fax: 027-214-198054

E&P Forum
25-28 Old Burlington St.
London W1X 1LB U.K.
Phone: 44-71-292-0600
Fax: 44-71-434 3721

Wood Mackenzie Consultants Ltd.
Kintore House
74-77 Queen St.
Edinburgh EH2 4NS Scotland
Phone: 01-31-225-8525
Fax: 01-31-243-4435

African Petroleum Producers Assn.
APPA Executive Secretary
P. O. Box 1097
Brazaville, Republic of the Congo
Phone: 011-242-836438
Fax: 011-242-836799
Telex 5552 KG

Oil Companies Intl. Marine Forum
96 Victoria St., 15th floor
London SW1E 5JW U.K.

Publications

IMO Emergency Procedures for Ships Carrying Dangerous Goods—Group Emergency Schedules

*Guidance Manual for the Inspection and Condition Assessment of Tanker Structures*, first edition, Intl. Chamber of Shipping (ICS)/OCIMF, London (1986)

*Inspection Guidelines for Bulk Oil Carriers*, second edition, OCIMF, London (1994)

ICS Safety in Oil Tankers (2nd Edition 1978)

"The Control of Pollution from Offshore Oil and Gas Activities," paper presented at the 1972 E&P Forum London Convention.

Intl. Tankers Owners Pollution Federation Ltd. (ITOPF)

*The International Convention for the Prevention of Pollution from Ships*, MARPOL (1973 and 1978)

Partnerships between government, industry, the IMO, and IPIECA have had positive effects on marine pollution control in Africa. (The list of organizations and publications in **Table 9.1** provides more specific information on sources of additional information and guidance concerning E&P environmental issues in Africa.) Principal among these are efforts to help countries establish national oil-spill response programs and regional cooperatives. IMO and IPIECA are valuable sources for information on maritime conventions and oil-pollution preparedness, response, and cooperation in Africa.

Oil Spill Response Ltd. (OSRL) owns and operates the world's largest oil-spill response facility with global responsibility. The operational base of OSRL is the Oil Spill Service Center (OSSC) in Southampton, England. The center is managed by OSRL for a consortium of international petroleum companies and is the principal provider of their African oil-spill response efforts and support strategies. It also acts as a response contractor for nonconsortium members and provides oil-spill training and related services on request. The OSSC is an excellent source for information and oil-spill expertise.

Because of its location, South Africa is constantly alert to the possibility of an environmental disaster. The South Africa Oil Industry Environmental Committee is responsible for coordinating South African oil industry efforts to protect the environment from all sources of land and water pollution related to the industry's business. The committee devotes substantial time and effort to planning and preparing to respond to oil pollution off the South African coast.

The Oil Industry Intl. E&P Forum is an association of international petroleum companies and industry organizations. Established in 1974, the organization represents its members' interests at the IMO and other specialist agencies of the U.N., governments, and other international bodies concerned with regulating gas and oil E&P.

The *Wood Mackenzie Reports* provides complete, detailed surveys and forecasts on petroleum industry developments in each African country. This information can be a useful reference for environmental planning.

The African Petroleum Producers Assn. (APPA), with 11 member states, meets annually to discuss environmental issues relating to E&P operations. APPA holds environmental conferences every 2 years (attended by industry representatives and government officials) that focus on specific problems, prospects, and technologies encountered in Africa. The conference is a good source of information to help industry understand evolving environmental issues and regulations throughout Africa. The petroleum ministry of a producing country is usually the representative to the APPA.

***Important Developments at the Regional Level in E&P.*** Oil spills are considered a severe threat to the marine environment and coastal zones of Africa. Governments have indicated their commitment to act on this threat by planning several initiatives that have yet to reach maturity in much of the continent. However, the U.N., petroleum companies, and other international organizations are directing considerable effort toward development of regional programs to combat pollution. These partnerships provide participating states with comprehensive experience and technology.

The Regional Seas Program is a global program implemented through regional components. For those regions of Africa where oil-spill resources and expertise are not readily available, IMO and industry are proposing a joint initiative, called the "IMO/Industry Global Oil Spill Planning Initiative." Its purpose is to strengthen pollution response capacity and establish viable and sustainable oil-spill response capabilities at national and regional levels. A key feature of this project is strengthening of industry and government cooperation. Countries are encouraged to develop and implement regional agreements between government organizations based on use of existing and future industry marine environmental protection capacity. To date, most oil-spill contingency planning and equipment organizations have been limited to the national level, sometimes as industry-only cooperatives and sometimes organized with governmental involvement. These cooperative agreements usually include provisions for equipment stockpiles, facilities, technical assistance, training and exercises, and financial concerns. The intent of the IMO/industry program is to provide assistance, starting at the development stage and concluding with a fully tested and exercised national contingency plan as well as a complete training schedule.

***Variations in Regional and National Jurisdiction for E&P.*** Except for the major producers and exporters, few African states have viable environmental management or national oil-pollution contingency plans. National governments must have adequate institutions, organization, manpower, and equipment to develop effective national environmental management plans. However, few African countries have developed these assets.

Industry can help to expand government and public awareness through industry-led campaigns to identify government focus points, establish roles and responsibilities, and facilitate discussion with government agencies on the benefits of national oil-pollution response. Considerable capacity exists within individual petroleum companies to manage oil-spill contingency and response preparation. Therefore, particular emphasis should be placed on government and industry cooperation to use these assets to develop national plans, stockpile pollution equipment, train response personnel, and perform emergency response exercises. Development of industry oil-spill equipment cooperatives is also an important tool in fighting pollution emergencies and furthering expansion of national programs.

***Existing Legislation, Regulation, Law, and Policy.*** Recently, certain oil-producing countries have developed a broad framework for environmental policy and are implementing legislation that provides effective and appropriate environmental protection on the marine environment. Nevertheless, few African states have environmental regulation for the purpose of enforcing internationally accepted standards and safety codes. Established standards require that E&P operators comply with IMO marine vessel inspection programs. However, while developed countries enforce strict maritime regulations, developing countries in Africa still can easily be served by substandard vessels unable to meet the strict international inspection program standards.

E&P operators should be diligent in managing these concerns to ensure that their facilities and, in particular, their vessels meet international standards for crew competence, safety standards, and adequate environmental protection during loading and unloading operations at terminals, ports, and in territorial waters. Industry should intensify its efforts to persuade governments to ratify the various IMO international conventions that provide access to emergency response funding and expertise.

***Contingency and Emergency Planning in Africa.*** When preparing contingency and response plans for a potential environmental crisis, companies should take into consideration the particular working conditions in which they operate. They need to devise procedures and have appropriate authorizations in place to implement the plans when required. African governments generally have large bureaucracies and less-developed infrastructures.

Customs and immigration restrictions, transportation availability, communications networks, construction equipment, and trained labor pools all present major challenges to companies' ability to accomplish emergency response actions. In principle, therefore, plans that require out-of-country personnel and equipment resources should include procedures for importing these assets under emergency circumstances and agreements with governmental agencies should be drawn up before a crisis. The objective of international support is to enhance operator capability to respond rapidly to oil spills in their region. If detailed importation and mobilization plans are not in place, authorities have been known to detain equipment, making it unavailable for use.

***Proposed and Planned Legislation, Regulation, Law, and Policy.*** Plans and funding guarantees for environmental decommissioning and site restoration are becoming an important requirement in the early phases of international projects and government partnerships. Regulation should address the role that local government and regulatory bodies must play in decommissioning and reclamation projects.

The IMO has established standards for removal of offshore installations. It also provides guidelines for a case-by-case evaluation of removal by industry. The Convention for the Protection of the Marine Environment of the North-East Atlantic[13] covers at-sea disposal of the parts of offshore installations that are removed. Elsewhere, particularly in the U.S. gulf coast, use of disposed of structures as artificial reefs has provided environmental benefits. African governments should be encouraged to examine these benefits.

***Law and Liability.*** Petroleum companies are responsible for direct and indirect consequences to the environment caused by their activities, actions, and omissions. Legislation that guides the development and implementation of environmental policy usually treats petroleum activities, for the most part, on the basis of fault liability. However, some African states are leaning toward development of environmental laws based on "strict liability." This means that E&P operators may be required to pay for or to repair damages caused by an accident that occurred for reasons outside their control, even after complying with all security procedures and applicable laws. Such legislation represents a major additional risk for petroleum companies over normal operating risks. Creation of strict liability in some African states may seem to ensure the preservation of the environment. In reality, however, it could discourage development of the country's resources and thus be detrimental to the country's ability to fund environmental control and improvement.

If strict liability is established in environmental laws, there are several ways to minimize its detrimental effects on petroleum operations. One is for the law to set a maximum limitation on monetary compensation that a claimant could be awarded under strict liability provisions. Another is to require entities that carry on activities that involve a high degree of risk to the environment and that are classified as such by regulation to have insurance (or be appropriately self-insured) for their civil liability. Another would be to exclude events caused by *force majeure* or third parties.

**9.10.2 Hydrocarbon Resource Development in the South American Rain Forest.** ***Introduction.*** Oil-rich resources are common in equatorial rain forests. The challenge of developing these resources responsibly is complicated by the need to preserve the delicate balance of the ecology and the cultures of those living in these regions. Rules for preserving these areas are undergoing development at a rapid pace but, unfortunately, are far from complete or adequate. The following case history[24] from Ecuador summarizes the principles and recommended approaches for an oil-development project (Ref. 24 provides a detailed description of the project).

***Formal Regulatory Infrastructure.*** Existing environmental controls in Ecuador are directed through the executive branch of the Republic of Ecuador. In June 1994, the president passed, by executive decree under Article 19(2) of the constitution, a set of environmental regulations for hydrocarbon activities. These regulations provide for implementation of a set of performance-based standards specific to different types of hydrocarbon-development projects. The programs are executed by the Undersecretary for the Environment of the Ministry of Energy and Mines through the Natl. Dept. of the Environment. Although the approximately 200 pages of rules are specific to the chemical, biological, and other constituents of discharges, the rules do not address all the concerns of a development project in a national park in the tropical rain forest. The needs not covered by rules were supplemented by a project-specific environmental management plan (EMP).

***EMP.*** The importance of an EMP to protect the sensitive environment of the rain forest was recognized as soon as it became known that there was a potential oil development project in Ecuador. The strategy for avoiding environmental and cultural mistakes was incorporated into the EMP, which served as the guidance document for the project. The purpose of the EMP was to outline the basic concepts of the project's environmental and human rights protection policies, to establish guidelines and standards to be followed in the exploration and construction phases of the project, and to describe how the oil-production operations would be conducted in an environmentally responsible manner.

The EMP was an integral part of the contract with the country for development of oil from Block 16 and the Tivacuno field. The EMP was a first in the country, and it set new standards for oil development for Ecuador and South America. The two key objectives of the EMP were to protect the environment and to minimize cultural and human-rights impacts of the project on the indigenous population.

In the area of human rights protection, the consortium agreed to respect the rights of the Waoroni (the primary indigenous tribe living in the block area) to decide for themselves the lifestyle they desired. At the request of the Waoroni and other native groups potentially affected by the development, the project consortium established an educational assistance program that provided newly constructed schools, educational materials (written in the native language), and teacher assistance (salary, transportation, and lodging) to those who wanted to participate. In addition, health screens were conducted on all workers entering the native groups' land. Workers and locals were vaccinated and medical assistance provided to the Waoroni during the development. All employees and contract workers were screened for controlled substance use, and controlled substances and alcohol were not allowed in the project area. These measures protected tribal members from outside diseases. The company also provided health training to the people.

The school and medical-assistance programs were implemented through local missionary organizations and previously underfunded government programs. The plan is now integrated with the existing systems, and the project has used the advice of the missionaries to enhance existing programs to optimize the assistance plan without creating a dependent "welfare state." The program has been designed to help those who wanted or needed help and those who would help themselves if provided the opportunity. It has not become a charity or "give-away" program, and it was not forced on the local residents.

The greatest threat to the Waoroni and the Yasuni Natl. Park continues to be colonization, not oil development. Before any oil development occurred in the Oriente, settlers or colonists had begun moving into the area. Cutting the trees, they left parts of the Oriente in disarray. The project agreed to pay the government and the Waoroni to set up a control system to exclude those who did not live in or have permission to work in the project area. In addition, it was agreed that a road into the area would be constructed by a route that crossed the 1.4-km-wide Napo River and that a ferry would be used rather than a bridge to limit access to the area. Those allowed to cross on the ferry were limited to workers with personal identification cards or tribal residents (verified by a census). This plan worked well with the country's desire to preserve the Yasuni Natl. Park area, while Ecuador had been able to allocate only limited funds to patrol the area to keep colonists out.

Finally, the EMP also specified measures to advance scientific research on the animals and plants in the area. Before construction, archeological investigations were conducted to help prevent the possibility of destruction of artifacts and historical remnants. The work on the various agreements began more than 3 years before the road or pipeline construction began in the block. This early start on implementing the EMP was essential to conduct scientific baseline studies before the area was disturbed by construction and oil development. Also, it was essential to develop trust with the tribal residents who were going to host the consortium for the next 20 years. An agreement was executed between the Waoroni, the President of Ecuador, and the consortium in the summer of 1992 to demonstrate that the consortium would keep its promises.

*Flora and Fauna.* Environmental planning was one of the first steps in the project. The consortium recognized that it must minimize deforestation by reducing cutting of trees during the road, pipeline, well-pad, and facilities construction phases of the project. In addition to the deforestation controls, provisions were made to reforest areas that were cut for construction purposes and could be replanted. The Missouri Botanical Garden and its Ecuadorian branch were contracted to provide the scientific assistance needed to inventory plants, to start nurseries with seeds of the cut trees to grow trees indigenous to the area, and to begin replanting work with local indigenous labor. This work continued throughout the construction phases of the work, with thousands of trees being planted each month. Several hundred new species of plants were identified during the course of this work, and the methods developed enhanced the reforestation efforts. This knowledge is being passed on through publication in technical journals, lectures, and professional discussions.

Animal life was studied through a 3-year research contract with a local consulting firm. This group of scientists, assisted by world-recognized environmental experts, continues to study stream, insect, reptile, and larger-animal populations. In addition to identification of completely new species of plants, much has been learned about the impact of development in the rain forest on animal and plant life and how to manage that impact. A scientific research station established at the Tivacuno River is now dedicated to Catholic U. of Quito for ongoing research in the Yasuni Natl. Park.

*Archeological.* Much of the area of the block development was remote and undisturbed by modern civilization. Penetration of the road and pipeline into virgin rain forest afforded a unique opportunity to uncover secrets of past civilizations that lived in the region. If the construction did not proceed carefully, these secrets could have been destroyed by the bulldozers and earth-moving equipment. To avoid such losses, an archeological team staffed by the Von Humbolt Foundation of Ecuador preceded the survey crews that were laying out facilities and road and pipeline routes. This team uncovered several areas of past civilizations, or "digs," and these areas were excavated by the archaeologists or marked and bypassed for future study. The more important discoveries were avoided by rerouting the road, pipeline, and facilities and saved as national historical treasures. It will take years to develop the findings from these archeological finds fully, but much already has been learned from this work about the past civilizations in this region.

*Waste Management.* Conventional elements of environmental management were incorporated into planning of the project. Although local environmental laws and regulations are much less developed than U.S. standards, the decision was made to meet U.S. controls and, specifically, to pattern the control measures after those in the State of Texas.

Perhaps the most costly environmental commitment of the project is reinjection of all produced water back into the underground reservoirs. This is particularly significant because it is projected that approximately 10 bbl of water will be produced for every 1 bbl of oil recovered, implying that injection rates could eventually reach 750,000 B/D. Previously, the national oil production company had not used underground injection. Almost all produced water was discharged into the area's rivers. However, injecting produced fluids from the project back into the underground reservoirs where they came from avoids damaging the environment and minimizes future liability to the consortium.

Air-quality-control measures include use of the best proven clean-burn technology on the large mainframe power-generating turbines, use of waste-heat recovery, use of closed process vessels that vent hydrocarbons to a flare system, and other state-of-the-art applications.

Solid and liquid wastes posed significant problems for the project during its construction and, ultimately, in the producing operation phases. During the construction phase, a peak of 3,000 to 4,000 workers had to be fed and housed. This waste-management problem was the equivalent of a small town that moved about 155 miles through the jungle over more than 2 years and spread out over several camps. In the initial phases of the project, a segregation and recycle program was adopted. Plastics, glass, metals, paper combustibles, and kitchen wastes were segregated in different containers. Plastics, glass, and metals were sent to Quito for recycling; paper and combustibles (not plastics) were incinerated at the camps; and ash and kitchen waste was land farmed, composted, or worm farmed.

Secondary aerobic biological sewage treatment plants were used to treat the camp waste waters, and the treated effluent was disinfected before discharge. The greatest waste-management difficulty was to educate the work force in acceptable trash collection and sanitary procedures for the block. Education and changing traditional habits has not been easy, but real progress is being made. Education is one of the real solutions to many of the environmental challenges faced.

Detailed spill prevention, control, and countermeasure plans were adopted that met American Petroleum Inst. and U.S. EPA standards and guidelines. Equipment was sent to the field, and training readied the crews for the first oil production. The decision was made to land farm oily wastes that could not be recycled. The tropical rain forest provides an excellent setting to decompose organics biologically, including oil. Land-farm cells lined with 2.4-in. high-density polyethylene were constructed at the northern production facility, and a biological culture was started. To limit the associated training and management challenges, only two land-farm locations were used.

In addition to oily wastes, drilling wastes were land farmed. These wastes were analyzed for toxic substances (e.g., heavy metals and naturally occurring radioactive materials). Because only water-based muds were used, they were acceptable for land farming. All toxic or hazardous materials were segregated and hauled to approved sites outside the block. Chemical drums were avoided wherever possible, and empty drums were returned to vendors or recycled.

Development of a sound EMP for such a project must focus on the local conditions and needs. For example, the block plan had to include both traditional E&P topics and local cultural and deforestation issues.

*Drilling Program.* Wells (6 to 12 per platform) were drilled directionally from a single platform. This significantly reduced the deforestation that would have been required to drill each well from its own local pad. The single-platform/multiple-well technique is very similar to offshore drilling methods; in this case, however, its purpose was to reduce the area that had to be cleared by 70% by limiting the number of well pads and roads.

Previously, only straight holes had been drilled in Ecuador, and virtually nothing was known about the conditions or cost of directional drilling in this region. The high-angle wells turned out to be very difficult to drill, and sticking was a continuing problem. Water-based muds were used in a closed-loop above-ground tank system with maximum recycle of the fluids. Drill cuttings were analyzed for toxic or hazardous substances. Because they were nontoxic, they were land farmed on the drillsite. Experience has both helped reduce the cost and proved the success of the directional-well drilling program.

*Road and Pipeline.* Road and pipeline design focused on cost-effectiveness, functionality, constructability, and environmental compatibility. To deny colonists easy access to the virgin rain forest, the chosen route and design required that the road and pipeline cross the wide Napo River by ferry or barge, not by bridge. Additionally, conventional wood-corduroy construction would be replaced by geotextile materials as the road base. Finally, the pipeline was buried to protect it from potential structural damage from falling trees or other accidental breakage and spills. The road and pipeline had to be constructed within a very confining 2-year schedule.

The road and pipeline system was designed with limiting deforestation as a key consideration. Road construction design restricted right-of-way width to approximately 79 ft rather than the current practice of 197 ft for a road and pipeline and also required hand-clearing techniques. The key element was use of geosynthetics, not the conventional wood-corduroy road base, for the road foundation. Finally, the pipeline was buried in the same right of way as the road to reduce the width of the clearing. The objective was to limit net deforestation to 988 acres for all facilities, well pads, the more than 93 miles of road, and 155 miles of pipeline.

With a Brazilian international road and pipeline contractor, construction of the road and pipeline began in the fall of 1992. The road design called for use of a locally available river-sand base (contained in a geotextile envelope) covered with a crushed-gravel surface. The gravel-wearing surface was approximately 15 to 20 ft wide, and the entire pipeline and road right of way was to be about 79 to 85 ft wide. Standard design is about three times as wide and uses cut trees as a base.

Most of the terrain crossed was rough and hilly, with poor soil conditions, and the narrow right of way dictated near-vertical slopes in some cut areas. This caused continuing slope failures, which will require constant maintenance until the slopes stabilize. The best previously reported rate for road construction in the rain forest was approximately 2.1 miles/D; the geotextile design allowed construction of an average of more than 7 miles/D. The road was completed in the required 2 years, about half the time it takes to build conventional wood-corduroy roads. The new design is proving to be much more durable, and the road and bridges have had no problem in supporting the 70-ton loads.

The 155 miles of 15.7- to 19.7-in. main pipeline and 3.9-in. fuel line were buried to provide added protection against line damage and spills. Previously, all oil lines in Ecuador had been laid on top of the ground. Construction began in the fall of 1992 at the Lago Agrio crude-oil tank farm north of the Napo River at the end of the pipeline. Securing right of way for the pipeline required difficult time-consuming negotiations with local landowners.

The pipeline right of way was reduced from the typical 197-ft width to less than 82 ft to minimize environmental damage. River crossings were drilled directionally under the river beds to help reduce the possibility of spills. Shutoff valves were installed at each major river crossing for use in the unlikely event of a break. There were two major river crossings. The Agua Rico River crossing was approximately 0.5 mile wide and was completed in about 1 month. The Napo crossing was almost 295 ft deep and 0.87 mile wide, the widest in South America. The Napo crossing pushed the envelope of technology and required more than 3 months to complete.

Development of the EMP benefited this project. It addressed all the conventional problems of air-, water-, and solid-waste management with national and international standards. It recognized the need to increase the level of environmental understanding and training of workers, residents, and everyone associated with the equipment and the operations. The EMP also specified that training be followed by numerous reminders and strict adherence to the internal standards of the project and not be limited to those of the country.

Many of the conventional technical problems of finding and producing oil in an international setting can be addressed by use of past experience, creativity, and proper application of the latest practices and technology. However, dealing with the people and the environment in a sensitive setting, such as the virgin rain forest of Ecuador, adds a new dimension. If properly addressed, the result can be that oil E&P is considered sustainable resource development.

## References


1. Smith, A.L.: "Sustaining Global Progress—In Partnership With the Environment," paper SPE 36041 presented at the 1996 SPE International Conference on Health, Safety and Environment in Oil and Gas Exploration and Production, New Orleans, 9–12 June.
2. "Making Development Sustainable, The World Bank Group and the Environment, Fiscal 1994," The World Bank, Washington, DC (1994).
3. Armstrong, K.E.: "Promoting Sustainable Development: Strategies for the Exploration and Production Industry," paper SPE 27113 presented at the 1994 SPE International Conference on Health, Safety and Environment in Oil and Gas Exploration and Production, Jakarta, 25–27 January.
4. Van Veldhuizen, H.D. and Anderson, R.B.: "Environmental Review: Policy and Guidelines," paper SPE 35857 presented at the 1996 SPE International Conference on Health, Safety and Environment in Oil and Gas Exploration and Production, New Orleans, 9–12 June.
5. Jones, F.V., Caudle, D.D., and Pappworth S.P.P.: "International Regulatory Compliance and EHS Planning," paper SPE 35930 presented at the 1996 SPE International Conference on Health, Safety and Environment in Oil and Gas Exploration and Production, New Orleans, 9–12 June.
6. Armstrong, K.E. and Frazier, W.K.: "Environmental Law for the E&P Industry: A Model for the Future," paper SPE 35858 presented at the 1996 SPE International Conference on Health, Safety and Environment in Oil and Gas Exploration and Production, New Orleans, 9–12 June.
7. *From Regulations to Industry Compliance: Building Institutional Capabilities*, Technical Report 11, UNEP Industry and Environment Program Activity Center, Paris (1992).
8. Todd, I. and Whewell, I.: "The Challenge of Developing and Enforcing a Goal-Setting Regulatory Regime," paper SPE 35756 presented at the 1996 SPE International Conference on Health, Safety and Environment in Oil and Gas Exploration and Production, New Orleans, 9–12 June.
9. Federal Clean Air Act Amendments, 42 USC 7401 *et seq.* (1990).
10. Browner, C.M.: "The Common Sense Initiative: A New Generation of Environmental Protection," presentation to the Center for Natl. Policy Newsmaker Luncheon, Washington, DC (10 July 1994).
11. Gunning, P. and Bergin, J.: "An Industry–Government Partnership Helping Companies Capture Revenues by Reducing Methane Emissions," *Pipeline & Gas J.* (December 1998) 42.
12. "Convention on the Prevention of Marine Pollution by Dumping of Wastes and Other Matter," Depositories: Mexico, USSR, U.K., U.S.; Places of Adoption, London, Mexico City, Moscow, Washington, DC (12 December 1972).
13. "The Convention for the Protection of the Marine Environment of the North-East Atlantic," Signatories: Belgium, Denmark, Finland, France, Germany, Iceland, Ireland, The Netherlands, Norway, Portugal, Spain, Sweden, and U.K. (21–22 September 1992).
14. "United Nations Framework Convention on Climate Change," U.N., New York City (20 June 1992–19 June 1993).
15. "The Basel Convention on the Control of Transboundary Movements of Hazardous Wastes and Their Disposal," Basel Convention Secretariat, Geneva, Switzerland (5 May 1992).
16. "The Convention on Biological Diversity," Intergovernmental Negotiating Committee, U.N., New York City (29 December 1993).
17. "Agenda 21, Rio Declaration," U.N., New York City (1992).
18. *Pollution Prevention and Abatement Handbook*, Environment Dept., World Bank, Washington, DC (1997).
19. "Environmental Benchmark for World Oil and Gas Exploration and Production," news release, UNEP and E&P Forum, London (17 June 1997).
20. "The Regional Seas Program," UNEP Water Branch, UNEP, Nairobi, Kenya (1974).
21. *Manual on Oil Pollution, Guidelines for Sampling and Identification of Oil Spills*, IMO, London (1995).
22. "The Business Charter for Sustainable Development," Publication 210/356A, Intl. Chamber of Commerce, Paris (1991).
23. Hanley, P.T., Campbell, G.R.B., and Wuestenfeld, K.S.: "Environmental Planning for Oil and Gas Projects in Alliance Organizations," paper SPE 36051 presented at the 1996 SPE International Conference on Health, Safety and Environment in Oil and Gas Exploration and Production, New Orleans, 9–12 June.
24. Chailer, J.D.: "Developing Countries: No Environmental Regulations Equals Project Stranglehold," paper SPE 30685 presented at the 1995 SPE Annual Technical Conference and Exhibition, Dallas, 22–25 October.
25. Hutton, W.C. and Skaggs, M.M. Jr.: "Renewable Resource Development in the Ecuadorian Rain Forest," paper SPE 30864 presented at the SPE 1995 Annual Technical Conference and Exhibit, Dallas, 22–25 October.


## SI Metric Conversion Factors

| | | |
|---|---|---|
| acre | × 4.046 856 | E − 01 = ha |
| bbl | × 1.589 873 | E − 01 = $m^3$ |
| ft | × 3.048* | E − 01 = m |
| mile | × 1.609 344* | E + 00 = km |

*Conversion factor is exact.

# Author Index

# Subject Index

## D



## E



## F



## G

## R



## S

# T



# U



# V



# W



# X



# Z